# Methods in Molecular Biology

*Series Editor*
**John M. Walker**
**School of Life and Medical Sciences**
**University of Hertfordshire**
**Hatfield, Hertfordshire, UK**

For further volumes:
http://www.springer.com/series/7651

For over 35 years, biological scientists have come to rely on the research protocols and methodologies in the critically acclaimed *Methods in Molecular Biology* series. The series was the first to introduce the step-by-step protocols approach that has become the standard in all biomedical protocol publishing. Each protocol is provided in readily-reproducible step-by-step fashion, opening with an introductory overview, a list of the materials and reagents needed to complete the experiment, and followed by a detailed procedure that is supported with a helpful notes section offering tips and tricks of the trade as well as troubleshooting advice. These hallmark features were introduced by series editor Dr. John Walker and constitute the key ingredient in each and every volume of the *Methods in Molecular Biology* series. Tested and trusted, comprehensive and reliable, all protocols from the series are indexed in PubMed.

# The Mitotic Spindle

## Methods and Protocols

Edited by

**Tomomi Kiyomitsu**

*Cell Division Dynamics Unit, Okinawa Institute of Science and Technology Graduate University, Okinawa, Japan*

Humana Press

*Editor*
Tomomi Kiyomitsu
Cell Division Dynamics Unit
Okinawa Institute of Science and Technology Graduate University
Okinawa, Japan

This work was supported by Okinawa Institute of Science and Technology Graduate University, UK Research and Innovation and University of Manchester.

ISSN 1064-3745 ISSN 1940-6029 (electronic)
Methods in Molecular Biology
ISBN 978-1-0716-4223-8 ISBN 978-1-0716-4224-5 (eBook)
https://doi.org/10.1007/978-1-0716-4224-5



Cover Illustration Caption: A mitotic spindle in a two-cell-stage medaka embryo showing chromosomes (RCC1-mCherry, magenta) and microtubules (EGFP-alpha-tubulin, green).

This Humana imprint is published by the registered company Springer Science+Business Media, LLC, part of Springer Nature.
The registered company address is: 1 New York Plaza, New York, NY 10004, U.S.A.

If disposing of this product, please recycle the paper.

# Preface

The mitotic spindle is a universal, supramolecular machine for chromosome segregation that captures sister chromosomes and segregates them in opposite directions to apportion genetic information equally to daughter cells. In addition, the mitotic spindle creates a vertical cleavage furrow at the midpoint of the elongated anaphase spindle, which not only partitions segregated chromosomes but also determines the size, location, and polarity of daughter cells. The latter factors are related to cell differentiation, tissue morphogenesis, and development in multi-cellular organisms.

The mitotic spindle consists of dynamic microtubule filaments in eukaryotes. Spatiotemporal regulation of spindle microtubules shapes the bipolar spindle and generates forces required for chromosomal alignment and segregation. In addition, astral microtubules emanating from centrosomes help to position the spindle and to form the cleavage furrow in animal cells.

Since the first observation of the mitotic spindle in the late nineteenth century, processes of spindle assembly, chromosome segregation, and cytokinesis have been extensively studied using diverse methods and organisms. As with other research fields, development of new technologies and methods has led to new discoveries, increasing our knowledge. In particular, live cell imaging technologies have greatly improved our understanding of the dynamic nature of microtubules during mitosis, which is essential to understand the mitotic spindle.

*The Mitotic Spindle*, edited by Paul Chang and Ryoma Ohi, was published in 2016. It covers key methods nicely and still contributes to mitotic spindle studies. In this fully new volume, I have asked leading researchers to present new technologies or methods that they have developed since the publication of the previous book. This volume is comprised of six parts: Advanced Microscopes (Chaps. 1, 2, 3, and 4), Spatiotemporal Manipulation of the Spindle and Target Molecules (Chaps. 5, 6, and 7), Quantitative Live Imaging (Chaps. 8, 9, 10 and 11), Screening (Chaps. 12 and 13), Centrosome Manipulation and Purification (Chaps. 14, 15, and 16), and Spindle Positioning in Synthetic or Complicated Contexts (Chaps. 17, 18, and 19). I greatly appreciate the efforts of all contributors and believe that these chapters will be very useful for the community. In particular, the "Notes" section is a hallmark of the Methods in Molecular Biology series. This section contains many useful tips and distinguishes this series from other methods publications. Please enjoy the Notes and utilize them to enhance the success of your experiments. We will all be gratified if this book contributes to your research.

I am writing this preface at Okinawa Institute of Science and Technology Graduate University (OIST), located in a southern island of Japan. The first president of OIST, Sydney Brenner, opined that progress in science depends on new techniques, new discoveries, and new ideas, probably in that order. Although his actual statement seems to have been paraphrased (Genome Biology, 2002, 3(9)), it is incontrovertible that progress in science depends on new techniques. For example, the advent of CRISPR technology has now made it possible to visualize both chromosomes and microtubules in live medaka fish embryos (shown on this book's cover), revealing unexpected features of mitotic spindle assembly in vertebrate embryos (Kiyomitsu et al., Nature Communications, 2024). In addition, development of chemical dyes, optogentic tools, and protein knockdown technologies, such as auxin-inducible degron, has enabled manipulation of endogenous target proteins and analysis of their functions in live mitotic cells. Considering the speed of recent technological developments in diverse fields, including super-resolution microscopy, mass spectrometry imaging, cryo-ET, protein structure prediction, and AI, future research will undoubtedly be exciting and will provide large data sets with high spatiotemporal precision and accuracy. Interdisciplinary studies combining these technologies will further accelerate our field, leading to better system-level understanding of the mitotic spindle.

***Okinawa, Japan*** ***Tomomi Kiyomitsu***

# Contents

## Contributors

AMLAN BARAI • *Equipe labellisée Fondation ARC, Aix Marseille University, CNRS, UMR 7288, IBDM, Turing Center for Living Systems, Marseille, France*

TAKUMI CHINEN • *Laboratory of Physiological Chemistry, Graduate School of Pharmaceutical Sciences, The University of Tokyo, Tokyo, Japan*

DELPHINE DELACOUR • *Equipe labellisée Fondation ARC, Aix Marseille University, CNRS, UMR 7288, IBDM, Turing Center for Living Systems, Marseille, France*

EMMANUEL DERIVERY • *MRC- MRC Laboratory of Molecular Biology, Cambridge, UK*

JULIEN DUMONT • *Université Paris Cité, CNRS, Institut Jacques Monod, Paris, France*

TATSUYA FUKUYAMA • *Department of Chemical Engineering, Kyoto University, Kyoto, Japan*

MAKOTO GODA • *Institute of Photonics Medicine, Hamamatsu University School of Medicine, Hamamatsu, Japan; Marine Biological Laboratory, Woods Hole, MA, USA*

SUSUMU GOYAMA • *Division of Molecular Oncology, Department of Computational Biology and Medical Sciences, Graduate School of Frontier Sciences, The University of Tokyo, Tokyo, Japan*

YU HATANO • *Faculty of Biology-Oriented Science and Technology, Kindai University, Kinokawa, Wakayama, Japan; Faculty of Life and Environmental Science, University of Yamanashi, Kofu, Yamanashi, Japan*

SHANE P. HERBERT • *Faculty of Biology, Medicine and Health, University of Manchester, Manchester, UK*

SHOTA HIRUMA • *Graduate School of Life Science, Hokkaido University, Sapporo, Japan*

TAKASHI HOTTA • *Department of Cell and Developmental Biology, University of Michigan, Ann Arbor, MI, USA*

KENJI IEMURA • *Department of Molecular Oncology, Institute of Development, Aging and Cancer (IDAC), Tohoku University, Sendai, Miyagi, Japan*

AKATSUKI KIMURA • *Marine Biological Laboratory, Woods Hole, MA, USA; National Institute of Genetics, Mishima, Japan; Genetics Program, Sokendai (Graduate University for Advanced Studies), Mishima, Japan*

KENJI KIMURA • *Division of Developmental Physiology, Institute for Genetic Medicine, Hokkaido University, Hokkaido, Japan*

DAIJU KITAGAWA • *Laboratory of Physiological Chemistry, Graduate School of Pharmaceutical Sciences, The University of Tokyo, Tokyo, Japan; Department of Physiological Chemistry, Graduate School of Pharmaceutical Sciences, The University of Tokyo, Tokyo, Japan*

ISABELLA KOPRIVEC • *Division of Molecular Biology, Ruđer Bošković Institute, Zagreb, Croatia*

LARA K. KRÜGER • *MRC- MRC Laboratory of Molecular Biology, Cambridge, UK*

HOLLY E. LOVEGROVE • *Faculty of Biology, Medicine and Health, University of Manchester, Manchester, UK*

YUSUKE T. MAEDA • *Department of Chemical Engineering, Kyoto University, Kyoto, Japan*

GILLIANE MATON • *Université Paris Cité, CNRS, Institut Jacques Monod, Paris, France*

AKIRA MATSURA • *Graduate School of Life Science, Hokkaido University, Sapporo, Japan*

FRANZ MEITINGER • *Okinawa Institute of Science and Technology Graduate University, Okinawa, Japan*

YUKO MIMORI-KIYOSUE • *Laboratory for Molecular and Cellular Dynamics, RIKEN Center for Biosystems Dynamics Research, Kobe, Japan; Department of Molecular Genetics, Institute of Biomedical Science, Kansai Medical University, Osaka, Japan*
NICOLAS MINC • *Université Paris Cité, CNRS, Institut Jacques Monod, Paris, France; Equipe Labellisée LIGUE Contre le Cancer, Paris, France*
MOMOKO MIYAZAWA • *Department of Physiological Chemistry, Graduate School of Pharmaceutical Sciences, The University of Tokyo, Tokyo, Japan; Division of Molecular Oncology, Department of Computational Biology and Medical Sciences, Graduate School of Frontier Sciences, The University of Tokyo, Tokyo, Japan*
MASAHIKO MORITA • *Image Processing Research Team, Center for Advanced Photonics, RIKEN, Saitama, Japan*
FUMIO MOTEGI • *Division of Developmental Physiology, Institute for Genetic Medicine, Hokkaido University, Hokkaido, Japan*
TAKASHI MURATA • *Kanagawa Institute for Technology, Atsugi, Japan*
TOMOMI NEMOTO • *Exploratory Research Center on Life and Living Systems, National Institute of Natural Sciences, Okazaki, Japan; National Institute for Physiological Sciences, Okazaki, Japan*
ESTHER FENG YING NG • *Okinawa Institute of Science and Technology Graduate University, Okinawa, Japan*
AUDE NOMMICK • *Université Paris Cité, CNRS, Institut Jacques Monod, Paris, France; Equipe Labellisée LIGUE Contre le Cancer, Paris, France*
RYOMA OHI • *Department of Cell and Developmental Biology, University of Michigan, Ann Arbor, MI, USA*
SHOTARO OKUDA • *Laboratory of Physiological Chemistry, Graduate School of Pharmaceutical Sciences, The University of Tokyo, Tokyo, Japan*
KOHEI OTOMO • *Exploratory Research Center on Life and Living Systems, National Institute of Natural Sciences, Okazaki, Japan; National Institute for Physiological Sciences, Okazaki, Japan; Juntendo University, Bunkyo-ku, Tokyo, Japan*
YUTA SHIMAMOTO • *Department of Chromosome Science, National Institute of Genetics, Shizuoka, Japan*
VALENTINA ŠTIMAC • *Division of Molecular Biology, Ruđer Bošković Institute, Zagreb, Croatia*
NOBUYUKI TAMAOKI • *Research Institute for Electronic Science, Hokkaido University, Sapporo, Japan*
AKIHIRO TANAKA • *Department of Chromosome Science, National Institute of Genetics, Shizuoka, Japan*
KOZO TANAKA • *Department of Molecular Oncology, Institute of Development, Aging and Cancer (IDAC), Tohoku University, Sendai, Miyagi, Japan*
MIKIKO TOKORO • *Faculty of Biology-Oriented Science and Technology, Kindai University, Kinokawa, Wakayama, Japan; Asada Institute for Reproductive Medicine, Asada Ladies Clinic, Aichi, Japan*
IVA M. TOLIĆ • *Division of Molecular Biology, Ruđer Bošković Institute, Zagreb, Croatia*
MOTOSUKE TSUTSUMI • *Exploratory Research Center on Life and Living Systems, National Institute of Natural Sciences, Okazaki, Japan; National Institute for Physiological Sciences, Okazaki, Japan*
RYOTA UEHARA • *Graduate School of Life Science, Hokkaido University, Sapporo, Japan; Faculty of Advanced Life Science, Hokkaido University, Sapporo, Japan*
JOSEPH L. WATSON • *MRC- MRC Laboratory of Molecular Biology, Cambridge, UK*

JING XIE • *Université Paris Cité, CNRS, Institut Jacques Monod, Paris, France; Equipe Labellisée LIGUE Contre le Cancer, Paris, France; Institut Curie, PSL Research University, CNRS UMR144, Laboratoire Biologie Cellulaire et Cancer, Paris, France; Institut Curie, PSL Research University, Sorbonne Universite, CNRS UMR3664, Laboratoire Dynamique du Noyau, Paris, France*
KAZUO YAMAGATA • *Faculty of Biology-Oriented Science and Technology, Kindai University, Kinokawa, Wakayama, Japan*
SHOHEI YAMAMOTO • *Department of Physiological Chemistry, Graduate School of Pharmaceutical Sciences, The University of Tokyo, Tokyo, Japan*
NORIO YAMASHITA • *Image Processing Research Team, Center for Advanced Photonics, RIKEN, Saitama, Japan*
TATSUMA YAO • *Faculty of Biology-Oriented Science and Technology, Kindai University, Kinokawa, Wakayama, Japan; Research and Development Center, Fuso Pharmaceutical Industries, Ltd., Osaka, Japan*
HIDEO YOKOTA • *Image Processing Research Team, Center for Advanced Photonics, RIKEN, Saitama, Japan*
NAO YONEZAWA • *Faculty of Biology-Oriented Science and Technology, Kindai University, Kinokawa, Wakayama, Japan*
KOYA YOSHIZAWA • *Graduate School of Life Science, Hokkaido University, Sapporo, Japan*

# Part I

## Advanced Microscopes

# Chapter 1

# Super-Resolution Imaging of Mitotic Spindle Microtubules Using STED Microscopy

Isabella Koprivec, Valentina Štimac, and Iva M. Tolić

## Abstract

Stimulated emission depletion (STED) microscopy is a powerful super-resolution imaging technique that only recently entered the field of mitosis, where it proved to be invaluable for studying various microtubule classes, kinetochore-microtubule attachments and chromosome segregation errors. Here, we describe immunofluorescence combined with STED microscopy as a method for analyzing microtubules and kinetochore-microtubule attachments in human mitotic spindles. We also describe live-cell STED microscopy as a method for single-plane short-term imaging of transient processes in crowded spindle areas. Finally, we outline image analysis approaches for the quantitative assessment of microtubule bundles within the spindle.

**Key words** STED microscopy, Cell division, Mitosis, Mitotic spindle, Microtubules, Kinetochores, Chromosomes, Nucleation, Attachments, Segregation errors

## 1 Introduction

Stimulated emission depletion (STED) microscopy is a super-resolution microscopy technique first introduced in 1994 [1, 2]. STED microscopy overcomes the diffraction limit of confocal microscopy by using a doughnut-shaped depletion laser to deplete the emitted fluorescence at specific positions, thus limiting emission only to the central "zero"-intensity laser spot [3]. Since STED can reach a resolution of up to 20 nm [4], it has been widely used to image structures and distribution of proteins within the cell. The power and versatility of STED microscopy are particularly evident in neuroscience, where visualization of cytoskeletal filaments and synaptic compartments has shed light on the architecture and motility of neurons, functions of synapses, dynamics of signal transmission, and neuron-glial interactions [5]. Moreover, STED microscopy

Authors Isabella Koprivec and Valentina Štimac have contributed equally to this chapter.

Tomomi Kiyomitsu (ed.), *The Mitotic Spindle: Methods and Protocols*, Methods in Molecular Biology, vol. 2872,
https://doi.org/10.1007/978-1-0716-4224-5_1,

proved invaluable for our understanding of complex organelles, as it elucidated the nanoscale distribution of mitochondrial proteins [6, 7] and cristae dynamics in mitochondria [8]. In mitosis, various super-resolution microscopy techniques have been used to resolve the centrosome [9] and the kinetochore [10] structure. Yet the visualization of fine or densely arranged microtubules within the spindle remained one of the biggest challenges. Luckily, the advent of STED microscopy in mitosis over the past few years enabled us to understand the complex landscape of spindle microtubules, their mutual interactions and the various attachments they form with kinetochores on chromosomes. The technique has since been used for studying mitosis in the malaria parasite (*Plasmodium berghei*), Indian Muntjac fibroblasts and various human cell lines, and even made its way to the clinic as part of the safety evaluation for limbal stem cells used in eye regeneration [11–19]. With its varied applications, STED has helped answer many of the long-standing questions about the shapes of microtubule bundles within the metaphase spindle [12], midzone microtubules during anaphase [13], early spindle formation [14], spindle microtubule growth [15], and previously indistinguishable microtubule classes and chromosome segregation errors [16].

Here, we compare STED microscopy to existing approaches (*see* **Note 1**) for studying spindle microtubules and discuss the advantages and disadvantages of STED microscopy (*see* **Note 2**). We also provide detailed protocols for immunostaining and live-cell STED super-resolution imaging of microtubules within human spindles, together with the accompanying image analysis approaches.

## 2 Materials

### 2.1 Cell Culture and Immunostaining

1. hTERT-RPE1 cells stably expressing either CENP-A-GPF or both CENP-A-GFP and Centrin1-GFP (*see* **Note 3**).
2. Cell culture medium: Dulbecco's Modified Eagle's Medium with 1 g/L D-glucose, pyruvate and L-glutamine (DMEM), supplemented with 10% (vol/vol) heat-inactivated Fetal Bovine Serum (FBS) and penicillin (100 IU/mL)/streptomycin (100 mg/mL) solution.
3. 35 mm uncoated glass coverslip dishes with 0.17 mm glass thickness (MatTek Corporation or Ibidi GmbH).
4. Cytoskeleton extraction buffer (CEB): 0.5% w/v Triton-X-100, 0.1 M PIPES, 1 mM EDTA and 1 mM $MgCl_2$ in milli-Q water. We use UltraPure 0.5 M EDTA (pH = 8.0, 15575020, Invitrogen). We recommend first preparing 5% w/v Triton-X-100, 1 M PIPES solution (pH = 7.0), and 1 M $MgCl_2$ solutions and diluting them in milli-Q water.

Store CEB for a maximum of three weeks at room temperature to achieve the best results. Before use, aliquot CEB into smaller volumes and pre-warm the aliquot to 37 °C to prevent mitotic spindle shrinkage at temperatures lower than 37 °C (*see* **Notes 4** and **5**).

5. Fixation solution: 3% paraformaldehyde and 0.1% glutaraldehyde solution in PBS. Add 3 mL of 4% paraformaldehyde, 8 μL of 50% glutaraldehyde, 0.4 mL 10× PBS and fill up with milli-Q water until 4 mL. The fixation solution should be made fresh each time and pre-warmed to 37 °C to prevent mitotic spindle shrinkage at temperatures lower than 37 °C.
6. Reduction buffer: 0.1% w/v sodium borohydride solution. Dissolve 10 mg of sodium borohydride in 10 mL of 1× PBS (hydrogen gas bubbles should appear upon dissolving). Make a fresh solution each time and handle sodium borohydride with care as it is reactive.
7. Quenching buffer: 100 mM glycine in 1× PBS. The solution can be stored for several months at 4 °C.
8. Blocking/permeabilization (B/P) buffer: 1% w/v NGS and 0.5 w/v Triton-X-100 in milli-Q water. The B/P buffer can be stored for several months at −20 °C.
9. Antibodies: Rat monoclonal tubulin (diluted 1:500, MA1-80017, Invitrogen), donkey anti-rat Alexa Fluor 568 or 594 (diluted 1:100, ab175475 and ab150156, Abcam); *see* **Note 6** and Table 1 for tips on how to choose the right antibody.

### *2.2 STED Imaging*

1. STED microscope: our STED images are acquired on the Expert Line easy3D STED microscope system (Abberior Instruments), equipped with a pulsed STED laser at 775 nm (*see* **Note 7**).

**Table 1**
**Choosing the right secondary antibody.**

| Alexa Fluor 594 | Alexa Fluor 647 | STAR RED |
|---|---|---|
| Very resistant to photobleaching | Very sensitive to photobleaching | Medium sensitivity to photobleaching |
| Enables imaging of the entire z-stack | Usually enables imaging of only 1–3 z-planes | Usually enables imaging of 3–10 z-planes |
| Each cell can be imaged multiple times | A cell can be imaged only once | A cell can be imaged only once |
| Requires high STED laser power to achieve an appropriate resolution | Requires less STED laser power to achieve the appropriate resolution | Requires very little STED laser power to achieve the appropriate resolution |

2. Sample with beads: we use the Abberior nanoparticle alignment slide (*see* **Note 8**).
3. 100×/1.4NA UPLSAPO100× oil objective (Olympus).
4. 60×/1.2NA UPLSAPO60× water objective (Olympus).
5. Immersion oil and lens cleaning tissues: Immersion oil type F30CC (stable at 23 °C) and type 37LDF (stable at 37 °C); Whatman lens cleaning tissue Grade 105 (*see* **Note 9**).
6. Tubulin and DNA dyes: SiR-tubulin kit (contains 50 nmol SiR-tubulin and 1 μmol verapamil, Spirochrome) for live-cell imaging; SPY-DNA dye series (Spirochrome) for live-cell imaging and DAPI for immunofluorescence (*see* **Note 10**).
7. Incubator chamber that controls the temperature at 37 °C and $CO_2$ at 5% for live cell imaging.
8. Software: Imspector software or Fiji/ImageJ.

## 3 Methods

### 3.1 Cell Culture and Immunostaining

For immunostaining, we optimized the first steps and chemicals of two previously published protocols for expansion microscopy [20, 21]. We used a fixation solution containing glutaraldehyde, which was demonstrated to be the best option for visualizing microtubules in the mitotic spindle [20]. For experiments described in this article, we used hTERT-RPE1 cells stably expressing either CENP-A-GPF or both CENP-A-GFP and Centrin1-GFP. All protocols are optimized for these cell lines.

1. Plate the cells 1–2 days before the fixation and keep them at 37 °C and 5% $CO_2$. The optimal confluency of cells on the day of fixation should be 80–90%, corresponding to the highest number of mitotic cells. We use 35 mm uncoated glass coverslip dishes with 0.17 mm glass thickness and seed the cells in 1 mL of the appropriate cell medium.
2. Remove the cell media from the dishes and add 500–1000 μL of pre-warmed CEB for 15 s to permeabilize cells and remove the cytoplasmic components that would otherwise result in unspecific binding of antibodies. CEB must be removed after 15 s because further exposure will cause the mitotic cells to detach, as CEB largely disrupts the membrane (*see* **Notes 5** and **6**).
3. Immediately after removing CEB, add 1 mL of pre-warmed fixation solution to the dish and incubate for 10 min at room temperature.
4. Aspirate the fixation solution and add 1 mL of the reduction solution for 7 min at room temperature. Sodium borohydride reduces the autofluorescence of glutaraldehyde from the fixation solution.

5. Aspirate the reduction solution and add 1 mL of the quenching solution for 10 min at room temperature. Glycine binds aldehyde groups from the fixation solution and reduces the unspecific binding of antibodies.
6. Remove the quenching solution and incubate the cells with 1 mL of B/P buffer for 2.5 h at 4 °C on the orbital shaker. This further permeabilizes the cells and prevents the unspecific binding of antibodies.
7. Incubate the sample with 300 μL of the primary antibody diluted in B/P buffer overnight at 4 °C on the orbital shaker. We achieved the best results using the rat monoclonal tubulin at 1:500 dilution.
8. The following day, wash the sample 3×, for 5 min each time, with 1 mL of 1× PBS at room temperature on the orbital shaker.
9. Incubate the sample with 300 μL of the secondary antibody diluted in B/P buffer for 1 h at room temperature on the orbital shaker. We use Alexa Fluor 568 or 594 antibodies at 1:1000 dilution (*see* **Note** 7).
10. Wash the sample 3×, for 5 min each time, with 1 mL of 1× PBS at room temperature on the orbital shaker.
11. Additionally, chromosomes can be stained using 1 mL of DAPI solution (1 μg/mL) for 10 min at room temperature on the orbital shaker.
12. Wash the DAPI solution 3×, for 5 min each time, with 1 mL of 1× PBS. Store the sample in 1 mL of 1× PBS at 4 °C for a maximum of three weeks. The fluorescence signal is the best when imaging is performed immediately after immunostaining.

### 3.2 Staining Tubulin with a Live-Cell Dye

1. Seed the cells 1–2 days before live-cell imaging on 35 mm uncoated glass coverslip dishes with 0.17 mm glass thickness in 1 mL of the appropriate medium and keep them at 37 °C and 5% $CO_2$. The optimal confluency of cells on the imaging day should be 80–90%, corresponding to the highest number of mitotic cells.
2. Dissolve the contents of SiR-tubulin kit vials according to the manufacturer's instructions. Add 50 μL of fresh anhydrous DMSO to the SiR-tubulin vial to make a stock concentration of 1 mM. Dissolve verapamil in 100 μL of fresh anhydrous DMSO to make a stock concentration of 10 mM (1000×).
3. Take 1 mL of cell medium from the dish with cells and add 0.1 μL of the SiR-tubulin dye to this 1 mL medium to make a final dye concentration of 100 nM. To avoid dye efflux, add 1 μL of the efflux pump inhibitor verapamil along with the dye

to 1 mL of cell medium. Resuspend the staining solution well and add it to the cells (if there is any remaining medium on the cells, remove it before adding the staining solution). Utilizing the old cell medium is essential because the fresh one can stop the cells from dividing.

4. Incubate for 1 h at 37 °C and 5% $CO_2$ before imaging.

### 3.3 STED Imaging of Fixed Samples

1. Before imaging, turn on the microscope system (use manufacturer's recommendations depending on the system) and put the dishes with fixed samples at room temperature for at least 30 min before the session. This ensures that the temperature of the sample is equilibrated to room temperature, preventing the sample drift during imaging.
2. Ensure the system is properly aligned using the sample with beads and follow the manufacturer's instructions for automatic and manual alignment.
3. For fixed sample imaging, choose the 100× objective and put a drop of immersion oil on the lenses. Put the dish with the sample on the stage and find focus using the brightfield or epifluorescence. Be aware that the epifluorescence can cause photobleaching before imaging, so use it only when necessary.
4. Find mitotic cells using eyepieces (for example, use the DAPI signal to discern mitotic phases based on the appearance of DNA).
5. Set up the protocol for STED imaging. Determine the size of the region of interest (ROI) to encompass the entire cell. Set the pixel size to 20 nm. The Expert Line easy3D STED microscope system can go down to 10 nm, but you should remember that lowering pixel size significantly increases the imaging time. Therefore, the pixel size should be adjusted to meet the requirements regarding the final image resolution and imaging duration.
6. Select either 2D or 3D STED based on whether you need better lateral (XY) or axial (Z) resolution, respectively. Since we usually require the best possible resolution between microtubules in the spindle midzone in the XY plane, we use 2D STED (*see* **Note 11**).
7. Determine the laser powers for each wavelength and the 775 nm STED line based on the photobleaching of your sample. We use the 488 nm laser to excite CENP-A-GFP (40% laser power) and the 561 nm laser to excite Alexa Fluor 594 (20% laser power). To achieve proper super-resolution of microtubules (*see* Subheading 3.6), the 775 nm STED laser for depleting the red line was set to 45%.

8. Additionally, adjusting dwell time (for how long the laser is applied to the sample) and line accumulation (the number of scans) is critical to achieve the best resolution. Keep in mind that increasing these two settings also increases the imaging time and photobleaching of the sample. We got the best results when we adjusted the dwell time to 10 μs and the line accumulation to 1. The pinhole size determines the amount of out-of-focus light reaching the detector. A better resolution is achieved when the pinhole size is smaller. For our purposes, we set the pinhole size to 1.0 AU. The described settings result in an average of 100–200 photons collected at the avalanche photodetector for each channel, corresponding to the optimum signal collection and the best resolution.
9. Before imaging, adjust the Z-stack's first and last planes and the distance between the planes. Alternatively, you can determine the middle plane of your image and set the total distance you want to acquire. *See* Table 1 for more information on fluorophore stability.
10. After the acquisition, shut down the system and clean the objective lenses with Whatman lens cleaning tissue.

### 3.4 Live-Cell STED Imaging

We imaged microtubules stained with 100 nM SiR-tubulin in an RPE1 cell line that stably expresses CENP-A-GFP and Centrin1-GFP and the following protocol was specifically adapted for this cell line. Imaging with SiR-tubulin dye at this concentration can be performed only for 45–60 min because the high concentration of the dye results in microtubule stabilization and the accompanying toxic effects after that point.

1. Heat the incubator chamber to 37 °C for live-cell imaging and adjust $CO_2$ to 5% after turning on the system. The cells must be kept in controlled conditions to ensure cell health.
2. Use either 100× oil objective or 60× water objective (*see* **Note 9** for tips on choosing the right objective). Use the appropriate immersion for the objective (if you decide to use the oil objective, we recommend an oil that is stable at 37 °C to avoid the drift of the sample). Focus the sample using the brightfield optics and find mitotic cells based on the CENP-A signal. Alternatively, SPY-DNA dyes at 20–40 nM concentration can be used for cell lines that do not express fluorescent proteins.
3. Follow **steps 5–9** from the protocol for *STED imaging of fixed samples*. ROI can include only a small part of the spindle (e.g., astral region or midzone) to reduce the photobleaching and expedite the imaging. We used the following laser powers for our sample: 488 nm to 15%, 640 nm to 30%, and the 775 nm STED laser to 10%. Pixel size was set to 25 nm, dwell time to 7 μs, line accumulation to 5, pinhole size to 1.0 AU and STED mode to 2D.

4. We used a single z-plane for the imaging of microtubules in the whole spindle because imaging is slow (~45 s per plane). In cases where you want to image a small ROI, the number of z-planes can be increased based on the dynamics of the process you want to image. If you wish to image cells during certain time periods, adjust the total time of imaging and the time interval between frames while also checking for signs of phototoxicity (e.g., spindle shrinkage).
5. After the acquisition, shut down the system and clean the objective lenses with Whatman lens cleaning tissue.

### 3.5 Using STED Microscopy for the Detection of Individual Microtubules in Crowded Environments

Compared to confocal microscopy, STED microscopy enables precise visual detection of various microtubule bundles within the spindle. In addition to providing stunning images of mitotic spindles, it also allows for visual detection and analysis of microtubules in crowded areas, including the astral region and the spindle midzone (Fig. 1a–d). Where confocal microscopy detects only a faint signal indistinguishable from that of the background, STED microscopy captures even the very thin structures (Fig. 1b, d). Additionally, where confocal microscopy shows a single microtubule bundle, super-resolution can distinguish between two separate entities within the bundle (Fig. 1b, d). Thus, using STED microscopy to study microtubules within the mitotic spindle significantly increases the accuracy and precision of analysis.

### 3.6 Determining the Resolution

To determine the resolution of our STED microscopy protocol compared to confocal microscopy, we image the same spindle using confocal and STED microscopy protocols on the same microscope (Fig. 2a). We use the Line tool within the Imspector software or Fiji/ImageJ to draw a line perpendicular to an isolated astral microtubule and create an intensity profile (Fig. 2a). We then estimate the resolution as the width of the tubulin intensity peak at its half-maximum value, measured from the background value obtained using the 25 × 25 pixel Square Tool (Fig. 2b). We consider the protocol appropriate for imaging if the measured width is, on average, less than 90–100 nm. In the example in Fig. 2, the resolution of the STED image was estimated to be 66 nm, compared to 234 nm in the confocal image. This implies that STED imaging improved the spatial resolution by a factor of 3.5.

### 3.7 Using STED Microscopy to Study Microtubule Bundle Composition and Nucleation

STED microscopy, combined with image analysis, can be a powerful tool for quantitative analysis of microtubules within the spindle. Using a protocol we previously developed to measure the tubulin intensity of a specific microtubule bundle [12], we can determine the number of microtubules within a particular bundle at any time point and any position within the spindle. Provided that the microtubule bundle is relatively isolated from its neighbors, we place a

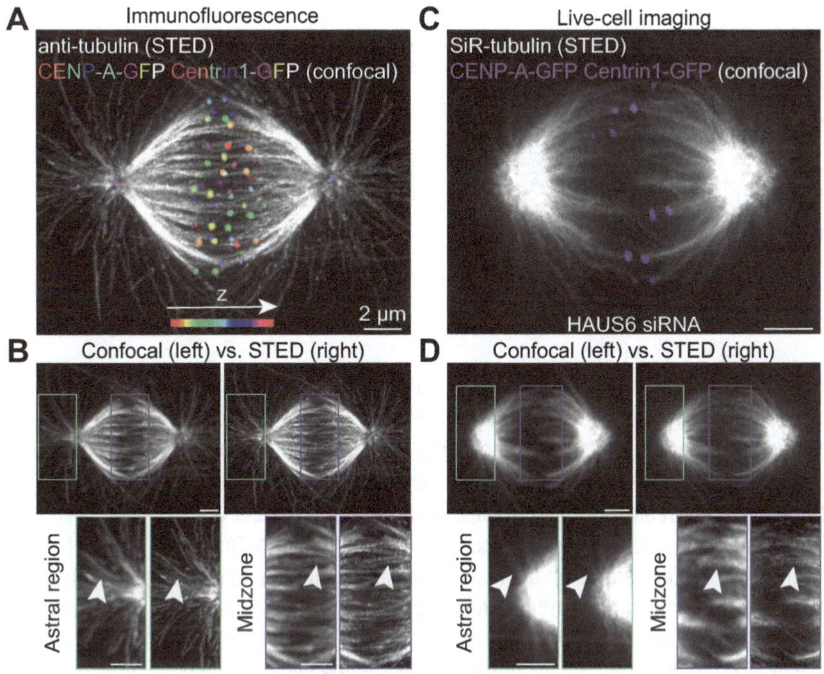


**Fig. 1** STED microscopy for discerning individual microtubules. (**a**) STED super-resolution image of microtubules immunostained for α-tubulin (gray) in RPE1 cells stably expressing CENP-A-GFP and Centrin1-GFP (rainbow, confocal). The image shows a maximum intensity projection of 8 central z-planes of the metaphase spindle. Kinetochores and centrosomes are color-coded for depth with the Spectrum LUT in ImageJ throughout the 8 z-planes, corresponding to 2 μm. (**b**) Comparison of tubulin signal obtained using either confocal or STED microscopy to image the spindle from (**a**). Insets represent close-ups of the astral region and the spindle midzone. (**c**) STED super-resolution image of microtubules dyed with 100 nM SiR-tubulin (gray) in HAUS6-depleted RPE1 cells stably expressing CENP-A-GFP and Centrin1-GFP (purple, confocal). The image shows one central z-plane of the metaphase spindle. (**d**) Comparison of tubulin signal obtained using either confocal or STED microscopy to image the spindle from (**c**). Insets represent close-ups of the astral region and the spindle midzone. Arrowheads point to structures that could only be resolved using STED microscopy. The brightness and contrast were adjusted so that astral microtubules are similarly visible in all spindles in STED microscopy or that all captured microtubules are visible in insets. Scale bars, 2 μm

25 × 25 pixel Square tool in Fiji/ImageJ in the middle of the microtubule bundle of interest and then place another 25 × 25 pixel square in the empty nearby area to measure the background, on a single-plane image of the spindle (Fig. 3a). The microtubule bundle's intensity equals the bundle's measured intensity (mean intensity within the square) minus that of the background. To obtain the number of microtubules within the bundle of interest, we compare it against astral microtubules, which consist of single microtubules. To measure the tubulin intensity of astral microtubules, we again place one 25 × 25 pixel square on the astral microtubule and another in the empty nearby area to measure the background (Fig. 3a). Subsequently, we subtract the two values to obtain the final intensity of the astral microtubule. This can be

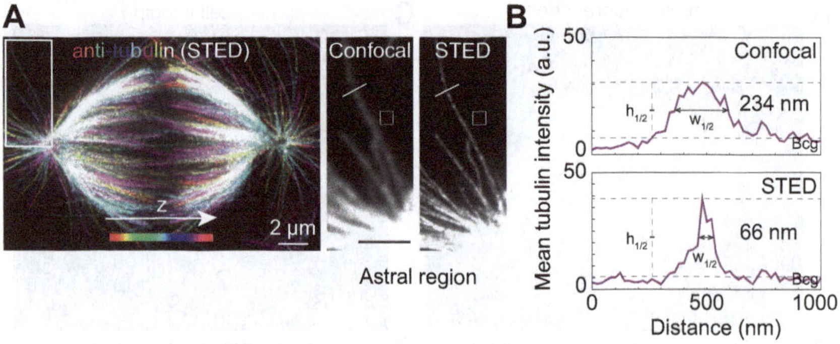


**Fig. 2** Determining the spatial resolution. (**a**) STED super-resolution image of microtubules immunostained for α-tubulin (rainbow) in RPE1 cells stably expressing CENP-A-GFP (not shown). The image shows a maximum intensity projection of 6 central z-planes of the metaphase spindle. Microtubules are color-coded for depth with the Spectrum LUT in ImageJ throughout the 6 z-planes, corresponding to 1.8 μm. Insets represent sum intensity projections of the astral region obtained using either confocal or STED microscopy to image the spindle from (**a**). Line Tool (length = 1 μm, thickness = 1) and Square Tool (25 × 25 pixel, corresponding to 0.5 × 0.5 μm) from ImageJ are drawn on the insets and represent tools to measure the intensity profile of the astral microtubule and the mean intensity of the nearby background, respectively. (**b**) Intensity profiles of the 1 μm line drawn perpendicularly to the astral microtubule in ImageJ for astral microtubules imaged using either confocal or STED microscopy. Resolution is defined as the width of the peak at its half-maximum, after subtracting the background, and is considered appropriate when this value amounts to <100 nm. The brightness and contrast were adjusted so that astral microtubules are similarly visible in all spindles in STED microscopy or that all captured microtubules are visible in insets. Scale bars, 2 μm

repeated many times, and the average can be used to make a comparison. Finally, to calculate the number of microtubules, we divide the intensity of the microtubule bundle of interest by the intensity of the astral microtubule.

STED microscopy can also be used to study specific nucleation processes that can hardly be visible when using confocal microscopy, particularly kinetochore-mediated nucleation. When using centrinone, an inhibitor of polo-like kinase 4 (PLK4) [22], to remove one centrosome, we were able to directly visualize sites where microtubule nucleation at kinetochores took place - including small microtubule stubs that arose from the kinetochores and clusters that started forming from them to create the future pole of the acentrosomal spindle side (Fig. 3b).

### 3.8 Using STED Microscopy to Study Kinetochore-Microtubule Attachments and Chromosome Segregation Errors

In addition to studying microtubule nucleation, STED microscopy is a powerful method to study kinetochore-microtubule attachments and chromosome segregation errors. STED microscopy allows direct visualization of any type of attachment, including mature, early end-on and lateral attachments within the two poles, but also more complex attachments that peripheral kinetochores form before they reach the area between the two spindle poles (Fig. 4a). This is particularly useful since direct visualization of kinetochore-microtubule attachments can be combined with cell

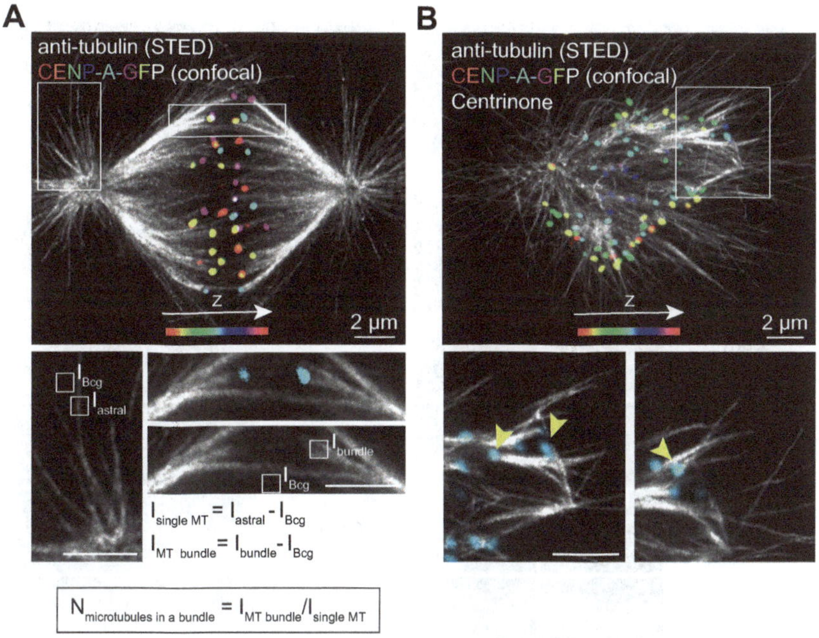


**Fig. 3** Measuring the number of microtubules within a bundle and visualizing microtubule nucleation at the kinetochore. (**a**) STED super-resolution image of microtubules immunostained for α-tubulin (gray) in RPE1 cells stably expressing CENP-A-GFP (rainbow, confocal). The image shows a maximum intensity projection of 6 central z-planes of the metaphase spindle. Kinetochores are color-coded for depth with the Spectrum LUT in ImageJ throughout the 6 z-planes, corresponding to 1.8 μm. Insets show the astral region and the microtubules associated with one kinetochore pair from the spindle in (**a**). Square Tool (25 × 25 pixel) from ImageJ is drawn on the insets and represents a tool to measure the intensity of the astral microtubule, microtubule bundle of interest and the associated backgrounds, respectively. A formula to calculate the number of microtubules within the bundle of interest is provided below. (**b**) STED super-resolution image of microtubules immunostained for α-tubulin (gray) in RPE1 cells stably expressing CENP-A-GFP and Centrin1-GFP (rainbow, confocal) and treated with 300 nM centrinone to remove one centrosome. The image shows a maximum intensity projection of the entire prometaphase spindle. Kinetochores and centrosomes are color-coded for depth with the Spectrum LUT in ImageJ throughout 6 z-planes, corresponding to 1.8 μm. Insets show single z-planes of kinetochore-mediated microtubule nucleation sites, marked with yellow arrowheads. The brightness and contrast were adjusted so that astral microtubules are similarly visible in all spindles in STED microscopy or that all captured microtubules are visible in insets. Scale bars, 2 μm

lines that enable simultaneous analysis of stably expressed proteins, such as Mad2, a spindle assembly checkpoint protein that binds to kinetochores lacking mature end-on attachments [23], or the kinetochore protein Mis12 [24]. This combined approach can provide extensive information about the nature of the visualized attachments.

The ability of STED microscopy to precisely detect attachments of kinetochores, even with single microtubules, is revolutionary when it comes to studying chromosome segregation errors, especially merotelic attachments in which an individual kinetochore is bound to microtubules extending from the opposite poles

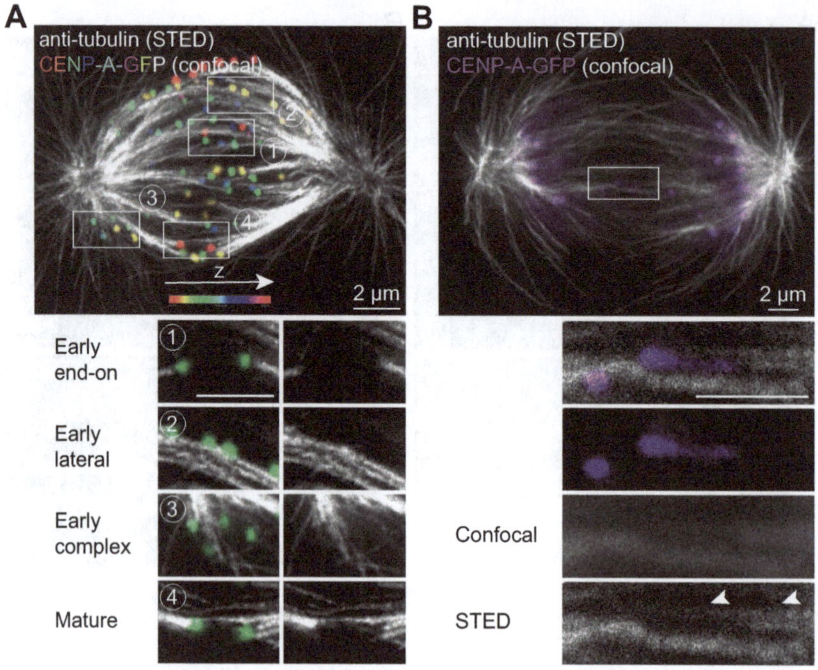


**Fig. 4** Classifying kinetochore-microtubule attachments and identifying errors. (**a**) STED super-resolution image of microtubules immunostained for α-tubulin (gray) in RPE1 cells stably expressing CENP-A-GFP (rainbow, confocal). The image shows a maximum intensity projection of the entire prometaphase spindle. Kinetochores are color-coded for depth with the Spectrum LUT in ImageJ throughout 7 z-planes, corresponding to 2.1 μm. Insets show one z-plane or a maximum intensity projection of two z-planes with various types of attachments from the spindle in (**a**). (**b**) STED super-resolution image of microtubules immunostained for α-tubulin (gray) in RPE1 cells stably expressing CENP-A-GFP (purple, confocal) and treated with 200 μM CK-666 inhibitor of the Arp2/3 complex for 3 h. The image shows a single central z-plane. Insets represent close-ups of the merotelic attachment from the spindle in (**b**). Arrows point to the additional microtubule from the opposite side imaged using STED microscopy, which is undetectable when using confocal microscopy. The brightness and contrast were adjusted so that astral microtubules are similarly visible in all spindles in STED microscopy or that all captured microtubules are visible in insets. Scale bars, 2 μm

[25]. In addition to the previously used indicators of merotelic attachments, including stretching of the kinetochore and its central location on the anaphase spindle, several microtubules that form the erroneous attachment can now be directly visualized using STED microscopy, whereas they are undistinguishable from the background when using confocal microscopy (Fig. 4b). Not only that but STED microscopy can also be used to detect merotelic attachments even before they result in a lagging chromosome in anaphase, while they are still located within a crowded metaphase plate [12].

## 4 Notes

1. Existing approaches for studying spindle microtubules and their attachments to chromosomes: Before STED, primary microscopy methods for detailed analysis of spindle microtubules were electron microscopy (EM) [26–30] or expansion microscopy (ExM) [20]. Even though EM is still a "gold standard" for studying spindles at a single-microtubule resolution, it is costly, time-consuming and unsuitable for live-cell imaging [31, 32]. Similarly, while much less expensive, ExM is also time-consuming and unsuitable for live-cell imaging. Thus, addressing numerous open scientific questions has relied on indirect approaches. For example, to study specific classes of microtubules within the spindle, several strategies have been developed: cold treatment was used to remove non-kinetochore microtubules and thus allowed the study of isolated kinetochore microtubules within the spindle [33]; a combination of hNuf2 and HSET RNAi was used to remove kinetochore microtubules and thus allowed the study of isolated non-kinetochore microtubules [34]; laser ablation was used to detect the connection between kinetochore microtubules and non-kinetochore microtubules that form a bridge between them, called bridging fibers [35]. Similarly, indirect approaches were employed to study kinetochore-microtubule attachments in healthy and error-prone cells: cold treatment was once again used to determine whether the kinetochore attaches to kinetochore microtubules or non-kinetochore ones [36, 37]; protein markers such as Mad1/2 or Bub1 were used to assess the stability of attachments [38, 39]; the interkinetochore distance, location in the central part of the spindle and stretching of the kinetochore were used as indicators of merotelic attachment [40, 41]. Yet, the arrival of STED microscopy to the cell division field allowed all of these phenomena to be visualized and studied more directly.
2. Advantages and disadvantages of STED microscopy: As with any method, there are several advantages and disadvantages to consider while determining whether STED microscopy is the right approach for a particular scientific question. Immunofluorescence combined with STED microscopy allows for a much shorter and simpler sample preparation protocol than electron microscopy and expansion microscopy—the entire sample preparation and imaging can be performed within three working days. Unlike the other two methods, STED also ensures that many cells are available for imaging in a dish or a slide, as the sample is minimally processed and rarely contains artifacts. In addition, STED can allow for user-friendly super-resolution

live-cell imaging with the SiR-tubulin dye [42], and it works well in combination with stably expressed proteins and tags [15]. Yet, the method is not without its limitations. Besides requiring a costly microscope system, it is important to carefully consider the choice of antibodies and dyes and to keep in mind that imaging the entire spindle can take up to several hours, which may be an issue if microscope availability is limited or large sample size is required. When using live-cell STED imaging, capturing the entire spindle is virtually impossible with the currently available systems, and one can only image a small region over a short period. With that in mind, STED microscopy remains a unique and powerful approach for studying spindle microtubules.

3. While we developed this protocol for RPE1 cells, it generally works as a good starting point for other cell lines in 2D and 3D cultures and for some organoids. However, further experimental optimization would likely be needed to achieve the appropriate resolution.

4. If the tubulin signal is weak or non-existent in the inner part of the spindle compared to the outer parts, in most cases, it means that the antibody did not penetrate the spindle. The problem might be that CEB is too old or some components have gone bad. We recommend preparing fresh ingredients and making an entirely new CEB.

5. If the spindles look shrunken or miss astral microtubules, the CEB and fixative were not properly pre-warmed. These two chemicals must be pre-warmed exactly to 37 °C.

6. To ensure you choose the appropriate secondary antibody, consult Table 1.

7. Be aware that the position of the system in the microscopy room is critical for obtaining the best super-resolution images. Avoid positioning the system close to the direct airflow from air conditioning or near any vibration. This will result in the drift of the sample or noise during imaging, respectively. (Note that vibrations can arise due to the music or mobile phone usage next to the system).

8. We recommend performing the system alignment before each session. Align all lasers in 2D and 3D and remember that the pinhole must be appropriately positioned to ensure the best results. Always check the beads after the automatic alignment procedure and do not simply rely on the precision of the automated protocol.

9. For the best possible super-resolution results, immersion and mounting media with the same refractive indexes should be used. While we found that the effect of this pairing is negligible

**Table 2**
**Things to consider when using Taxol-based dyes for live-cell imaging**

| Advantages | Disadvantages |
|---|---|
| Allow for live-cell imaging | Need to be used in high concentrations (100–200 nM) |
| Strong signal | Short imaging window (1 h) |
| Fast penetration into cells | Unwanted effects on microtubule stabilization |
| Easy to use | Affinity for stable microtubules |

when imaging spindles using the 2D STED mode, where PBS can be paired with an oil objective without a significant impact on the accuracy of the collected data, the effect can be much more significant when using 3D STED mode for imaging of very fine structures. As a general rule, we recommend using the 100×/1.4NA UPLSAPO100x oil objective for imaging mounted samples and the 60×/1.2NA UPLSAPO60x water objective for live-cell imaging.

10. For things to consider when using Taxol-based dyes for live-cell imaging, consult Table 2.
11. Choose 2D STED for primarily lateral (XY) super-resolution and 3D STED for axial (Z) super-resolution. Remember that switching from 2D to 3D STED increases the axial resolution but decreases lateral resolution and vice versa.

## Acknowledgments

The authors thank Alexey Khodjakov for the RPE1 cell lines; Marko Šprem and the Abberior team for help with the Expert Line easy3D STED microscope system setup and developing microscopy protocols; all members of the Tolić group for helpful discussions and advice. The authors acknowledge funding by the European Research Council (ERC Synergy Grant, GA Number 855158), the Croatian Science Foundation (HRZZ) through Swiss-Croatian Bilateral Projects (project IPCH-2022-10-9344) and Cooperation Programme with Croatian Scientists in the Diaspora "Research Cooperability" (project PZS-2019-02-7653), and projects co-financed by the Croatian Government and the European Union through the European Regional Development Fund—the Competitiveness and Cohesion Operational Programme: IPSted (Grant KK.01.1.1.04.0057) and QuantiXLie Center of Excellence (Grant KK.01.1.1.01.0004).

## References


1. Hell SW, Wichmann J (1994) Breaking the diffraction resolution limit by stimulated emission: stimulated emission-depletion fluorescence microscopy. Opt Lett 19:780–782. https://doi.org/10.1364/OL.19.000780
2. Klar TA, Hell SW (1999) Subdiffraction resolution in far-field fluorescence microscopy. Opt Lett 24:954–956. https://doi.org/10.1364/OL.24.000954
3. Vicidomini G, Bianchini P, Diaspro A (2018) STED super-resolved microscopy. Nat Methods 15:173–182. https://doi.org/10.1038/nmeth.4593
4. Göttfert F, Wurm CA, Mueller V et al (2013) Coaligned dual-channel STED nanoscopy and molecular diffusion analysis at 20 nm resolution. Biophys J 105:L01–L03. https://doi.org/10.1016/j.bpj.2013.05.029
5. Werner C, Sauer M, Geis C (2021) Super-resolving microscopy in neuroscience. Chem Rev 121(19):11971–12015. https://doi.org/10.1021/acs.chemrev.0c01174
6. Wurm CA, Neumann D, Lauterbach MA et al (2011) Nanoscale distribution of mitochondrial import receptor Tom20 is adjusted to cellular conditions and exhibits an inner-cellular gradient. PNAS 108(33):13546–13551. https://doi.org/10.1073/pnas.1107553108
7. Singh H, Lu R, Rodríguez PFG et al (2012) Visualization and quantification of cardiac mitochondrial protein clusters with STED microscopy. Mitochondrion 12(2):230–236. https://doi.org/10.1016/j.mito.2011.09.004
8. Stephan T, Roesch A, Riedel D et al (2019) Live-cell STED nanoscopy of mitochondrial cristae. Sci Rep 9:12419. https://doi.org/10.1038/s41598-019-48838-2
9. Mennella V, Keszthelyi B, McDonald KL et al (2012) Subdiffraction-resolution fluorescence microscopy reveals a domain of the centrosome critical for the pericentriolar material organization. Nat Cell Biol 14(11):1159–1168. https://doi.org/10.1038/ncb2597
10. Ribeiro SA, Vagnarelli P, Dong Y et al (2010) A super-resolution map of the vertebrate kinetochore. PNAS 107(23):10484–10489. https://doi.org/10.1073/pnas.1002325107
11. Pereira A, Sousa M, Almeida AC et al (2019) Coherent-hybrid STED: high contrast sub-diffraction imaging using a bi-vortex depletion beam. Opt Express 27(6):8092–8111. https://doi.org/10.1364/OE.27.008092
12. Novak M, Polak B, Simunić J et al (2018) The mitotic spindle is chiral due to torques within microtubule bundles. Nat Commun 9:3571. https://doi.org/10.1038/s41467-018-06005-7
13. Vukušić K, Ponjavić I, Buđa R et al (2021) Microtubule-sliding modules based on kinesins EG5 and PRC1-dependent KIF4A drive human spindle elongation. Dev Cell 56(9):1253–1267.e10. https://doi.org/10.1016/j.devcel.2021.04.005
14. Matković J, Ghosh S, Ćosić M et al (2022) Kinetochore- and chromosome-driven transition of microtubules into bundles promotes spindle assembly. Nat Commun 13(1):7307. https://doi.org/10.1038/s41467-022-34957-4
15. Almeida AC, Soares-de-Oliveira J, Drpic D et al (2022) Augmin-dependent microtubule self-organization drives kinetochore fiber maturation in mammals. Cell Rep 39(1):110610. https://doi.org/10.1016/j.celrep.2022.110610
16. Štimac V, Koprivec I, Manenica M et al (2022) Augmin prevents merotelic attachments by promoting proper arrangement of bridging and kinetochore fibers. eLife 11:e83287. https://doi.org/10.7554/eLife.83287
17. Almeida AC, Soares-de-Oliveira J, Maiato H (2023) Optimized protocol for live-cell analysis of kinetochore fiber maturation in Indian muntjac cells. STAR Protoc 4(1):102011. https://doi.org/10.1016/j.xpro.2022.102011
18. Zekušić M, Bujić Mihica M, Skoko M et al (2023) New characterization and safety evaluation of human limbal stem cells used in clinical application: fidelity of mitotic process and mitotic spindle morphologies. Stem Cell Res Ther 14(1):368. https://doi.org/10.1186/s13287-023-03586-z
19. Zeeshan M, Rea E, Abel S et al (2023) Plasmodium ARK2 and EB1 drive unconventional spindle dynamics, during chromosome segregation in sexual transmission stages. Nat Commun 14(1):5652. https://doi.org/10.1038/s41467-023-41395-3
20. Ponjavić I, Vukušić K, Tolić IM (2021) Expansion microscopy of the mitotic spindle. Methods Cell Biol 161:247–274. https://doi.org/10.1016/bs.mcb.2020.04.014
21. Zhang C, Kang JS, Asano SM et al (2020) Expansion microscopy for beginners: visualizing microtubules in expanded cultured HeLa

cells. Curr Protoc Neurosci 92:e96. https://doi.org/10.1002/cpns.96

22. Wong YL, Anzola JV, Davis RL et al (2015) Reversible centriole depletion with an inhibitor of Polo-like kinase 4. Science 348(6239): 1155–1160. https://doi.org/10.1126/science.aaa5111
23. Chen RH, Shevchenko A, Mann M et al (1998) Spindle checkpoint protein Xmad1 recruits Xmad2 to unattached kinetochores. J Cell Biol 143(2):283–295. https://doi.org/10.1083/jcb.143.2.283
24. Magidson V, Paul R, Yang N et al (2015) Adaptive changes in the kinetochore architecture facilitate proper spindle assembly. Nat Cell Biol 17(9):1134–1144. https://doi.org/10.1038/ncb3223
25. Gregan J, Polakova S, Zhang L et al (2011) Merotelic kinetochore attachment: causes and effects. Trends Cell Biol 21(6):374–381. https://doi.org/10.1016/j.tcb.2011.01.003
26. McDonald KL, O'Toole ET, Mastronarde DN et al (1992) Kinetochore microtubules in PTK cells. J Cell Biol 118:369–383. https://doi.org/10.1083/jcb.118.2.369
27. Mastronarde DN, McDonald KL, Ding R et al (1993) Interpolar spindle microtubules in PTK cells. J Cell Biol 123(6):1475–1489. https://doi.org/10.1083/jcb.123.6.1475
28. Sikirzhytski V, Renda F, Tikhonenko I et al (2018) Microtubules assemble near most kinetochores during early prometaphase in human cells. J Cell Biol 217(8):2647–2659. https://doi.org/10.1083/jcb.201710094
29. O'Toole E, Morphew M, McIntosh JR (2020) Electron tomography reveals aspects of spindle structure important for mechanical stability at metaphase. Mol Biol Cell 31(3):184–195. https://doi.org/10.1091/mbc.E19-07-0405
30. Kiewisz R, Fabig G, Conway W et al (2022) Three-dimensional structure of kinetochore-fibers in human mitotic spindles. elife 11: e75459. https://doi.org/10.7554/eLife.75459
31. Merdes A, Stelzer EHK, De Mey J (1991) The three-dimensional architecture of the mitotic spindle, analyzed by confocal fluorescence and electron microscopy. J Electron Microsc Tech 18:61–73. https://doi.org/10.1002/jemt.1060180110
32. McIntosh JR (2001) Electron microscopy of cells: a new beginning for a new century. J Cell Biol 153(6):F25–F32. https://doi.org/10.1083/jcb.153.6.f25
33. Brinkley BR, Cartwright J (1975) Cold labile and cold stable microtubules in the mitotic spindle of mammalian cells. Ann N Y Acad Sci 253:428–439. https://doi.org/10.1111/j.1749-6632.1975.tb19218.x
34. Cai S, O'Connell CB, Khodjakov A et al (2009) Chromosome congression in the absence of kinetochore fibres. Nat Cell Biol 11(7):832–838. https://doi.org/10.1038/ncb1890
35. Kajtez J, Solomatina A, Novak M et al (2016) Overlap microtubules link sister k-fibres and balance the forces on bi-oriented kinetochores. Nat Commun 7:10298. https://doi.org/10.1038/ncomms10298
36. Salmon ED, Cimini D, Cameron LA et al (2005) Merotelic kinetochores in mammalian tissue cells. Philos Trans R Soc Lond Ser B Biol Sci 360(1455):553–568. https://doi.org/10.1098/rstb.2004.1610
37. Etemad B, Kuijt T, Kops G (2015) Kinetochore–microtubule attachment is sufficient to satisfy the human spindle assembly checkpoint. Nat Commun 6:8987. https://doi.org/10.1038/ncomms9987
38. Kuhn J, Dumont S (2017) Spindle assembly checkpoint satisfaction occurs via end-on but not lateral attachments under tension. J Cell Biol 216(6):1533–1542. https://doi.org/10.1083/jcb.201611104
39. Etemad B, Vertesy A, Kuijt TEF et al (2019) Spindle checkpoint silencing at kinetochores with submaximal microtubule occupancy. J Cell Sci 132(12):jcs231589. https://doi.org/10.1242/jcs.231589
40. Cimini D, Moree B, Canman JC et al (2003) Merotelic kinetochore orientation occurs frequently during early mitosis in mammalian tissue cells and error correction is achieved by two different mechanisms. J Cell Sci 116:4213–4225. https://doi.org/10.1242/jcs.00716
41. Sen O, Harrison JU, Burroughs NJ et al (2021) Kinetochore life histories reveal an Aurora-B-dependent error correction mechanism in anaphase. Dev Cell 56(22): 3082–3099.e5. https://doi.org/10.1016/j.devcel.2021.10.007
42. Lukinavičius G, Reymond L, D'Este E et al (2014) Fluorogenic probes for live-cell imaging of the cytoskeleton. Nat Methods 11(7): 731–733. https://doi.org/10.1038/nmeth.2972

# Chapter 2

# Chromosome Counting at Meiosis and Mitosis of Mouse Oocytes and Embryos Using Super-Resolution Live-Cell Imaging and CRISPR/dCas9-Mediated Live-FISH

Yu Hatano, Nao Yonezawa, Mikiko Tokoro, Tatsuma Yao, and Kazuo Yamagata

## Abstract

Live cell imaging techniques are now essential for capturing chromosomal segregation in fertilized eggs. Although better spatiotemporal resolution of fluorescence observations could provide more information, higher phototoxicity may occur. Super-resolution microscopy is generally considered unsuitable for live cell imaging because of the considerable cell damage. Here, we developed a method for counting chromosomes in mouse living oocytes and early embryos using super-resolution microscopy based on disk confocal photon reassignment microscopy (OPRA). In this chapter, we describe the imaging conditions for minimally invasive, high-resolution observation of mouse oocytes and embryos and a method to count chromosomes via CRISPR/dCas-mediated live-cell fluorescence in situ hybridization.

**Key words** Mouse embryo, Super-resolution, Live-cell imaging, Live-FISH, Chromosome counting

## 1 Introduction

In preimplantation embryos, chromosome number directly influences pregnancy rates in assisted reproductive technology (ART) and animal breeding [1–5]. Furthermore, chromosomal segregation errors during meiosis and mitosis increase with maternal age [6–8]. In ART, the biopsy of a few cells in cleavage-stage embryos has been widely used to detect chromosomal aneuploidy before embryo transfer, followed by next-generation sequencing (NGS) of their DNA content. However, there is a concern that biopsy may damage the embryo, which may lead to a decrease in the pregnancy rate [9]. We previously developed a low-invasiveness, long-term live cell imaging of fertilized eggs using confocal microscopy and

Authors Yu Hatano and Nao Yonezawa have contributed equally to this chapter.

Tomomi Kiyomitsu (ed.), *The Mitotic Spindle: Methods and Protocols*, Methods in Molecular Biology, vol. 2872,
https://doi.org/10.1007/978-1-0716-4224-5_2,

demonstrated that early divisions of preimplantation embryos affect subsequent ontogeny [10, 11]. Moreover, using this methodology, we revealed that abnormal chromosome segregation during early division affects development to the blastocyst stage but does not affect the birth rate [12]. Similar results have been obtained by imaging bovine embryos [5, 13]. Although we detected obvious structural chromosomal abnormalities in all blastomeres during segregation, a resolution that allowed us to count the number of chromosomes in living embryos could not be achieved [12].

In recent years, super-resolution microscopy has been developed based on various principles. Stimulated emission depletion microscopy (STED): [14, 15]; Photoactivated localization microscopy (PALM): [16]; Fluorescence photoactivation localization microscopy (FPALM): [17]; Stochastic optical reconstruction microscopy (STORM): [18]; Structured illumination microscopy (SIM): [19, 20] are known and for example, in a previous study, mitosis of live U2OS cells was successfully observed and counted the number of chromosomes with Bessel SR-SIM [21]. In contrast, super-resolution microscopy is considered to be invasive. In particular, the observation of thick oocytes and preimplantation early-stage embryos requires repeated laser irradiation for 3D image acquisition and time-lapse observation, which causes damage to the embryos. Spinning disk confocal microscopy is suitable for the living-cell imaging of early embryos because it overcomes these limitations by scanning multiple points via disk rotation. Based on this feature, we succeeded in observing early embryos for a long time (more than 90 h) and obtained pups from imaged embryos via embryo transfer [10]. More recently, super-resolution has been achieved using a disk confocal photon reassignment microscope (OPRA), in which individual focal points projected onto a pinhole are optically reduced using microlenses (super-resolution via optical reassignment; SoRa) [22]. Using this super-resolution microscope, we successfully obtained pups from embryos and observed chromosome segregation in super-resolution, in which the grooves between sister chromatids of approximately 200 nm could be recognized [23]. In this chapter, we describe a minimally invasive, high-resolution chromosome observation method for mouse oocytes and embryos and a method for counting chromosomes via CRISPR/dCas-mediated live-cell fluorescence in situ hybridization targeting the sequence of the centromere region. This methodology includes the preparation of RNA-encoding fluorescent probes, their microinjection, super-resolution live-cell imaging, embryo transfer, and image analysis (Fig. 1). Here, we describe the methodologies used for chromosome counting during the first meiosis and first mitosis.

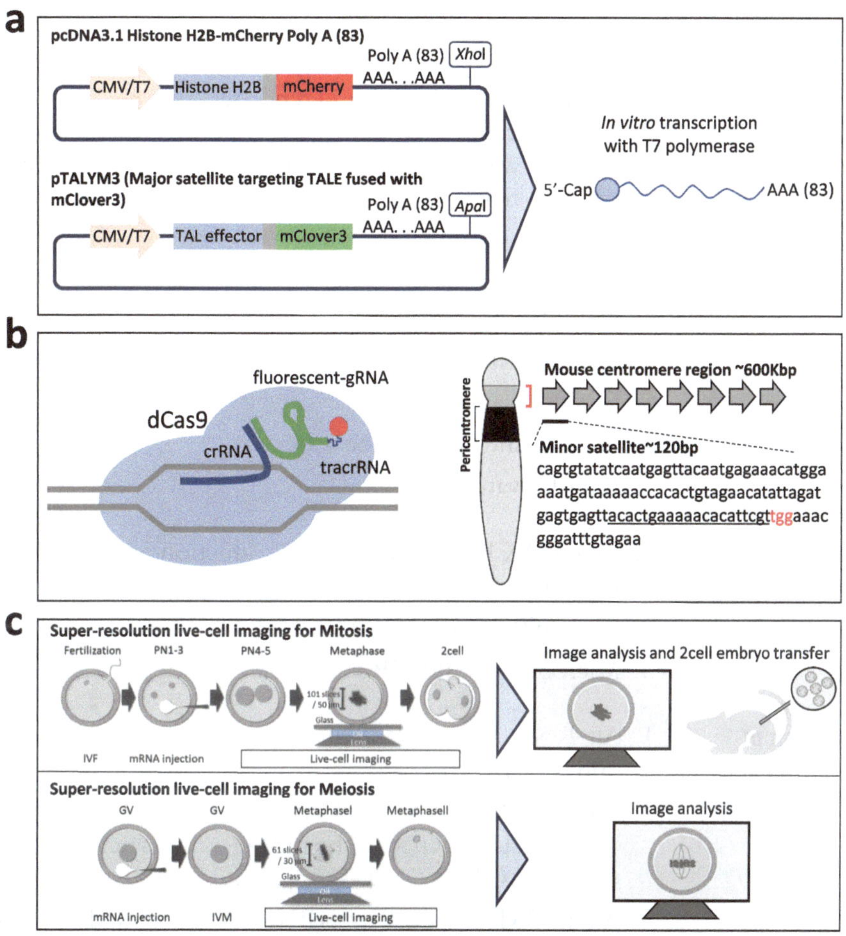


**Fig. 1** Scheme of an experiment for super-resolution live-cell imaging in mouse embryo and oocyte. (**a**) Synthesis of in vitro transcribed RNAs encoding Histone H2B fused with mCherry or major satellite targeting TALE fused with mClover3. The RNA is transcribed under the T7 promotor. The synthesized RNA has Cap structure and stretched adenine sequence at 5′ and 3′ ends, respectively, for the stabilization and efficient translation in oocytes and embryos. (**b**) Illustration of CRISPR-mediated live-FISH and the gRNA design. The ATTO550 or 647N fluorescent dye is conjugated to tracrRNA. The right panel shows the illustrated chromosome; the centromere contains a repeating sequence called minor satellites. The underline shows target sequences of crRNA against minor satellite. (**c**) Scheme of super-resolution live-cell imaging. The upper and lower panels show the experimental schemes in the first mitosis of fertilized mouse eggs and first meiosis. After the mRNA injection, the oocytes and embryos are incubated until fluorescence becomes visible. A thickness of 30 μm is observed around the equatorial plane of the oocyte. After the imaging, the embryos are transferred to the pseudo-pregnant mother to confirm the developmental capacities

## 2 Materials

### *2.1 Plasmids, Restriction Enzymes, and Other Chemicals*

1. pcDNA3.1-histone H2B-mCherry-poly (A). This plasmid encodes histone H2B fused to mCherry, which can be used to visualize the chromatin in living cells [24, 25]. This plasmid has a stretched polyadenine sequence (approximately 83 bp)

after the stop codon of mCherry to stabilize in vitro-transcribed RNA in the oocyte or embryo cytoplasm [26].

2. pTALYM3. This plasmid encodes the pericentromere (the major satellite), which targets TALE fused to mClover3 [27]. It was a gift from Maria-Elena Torres-Padilla and was available from Addgene (#47874). This plasmid also contains a stretched polyadenine sequence.
3. Restriction enzymes (*ApaI, XhoI*).
4. Phenol–chloroform.
5. Chloroform.
6. Ethanol.
7. DNase- and RNase-free water.
8. Milli-Q water.

### 2.2 In Vitro RNA Transcription

1. RiboMAX Large Scale RNA Production System T7 (Promega, P1320).
2. Ribo m7G Cap Analog (Promega, P1711).
3. MicroSpin S-200 HR Columns (Cytiva, 27512001).

### 2.3 Preparation of the dCas9, crRNA and Fluorescence-Labeled tracrRNA

1. Alt-R S.p. dCas9 Protein V3 (Integrated DNA Technologies, 1081066).
2. Alt-R CRISPR-Cas9 crRNA (Integrated DNA Technologies). The target sequence of the mouse minor satellite is 5′- ACACTGAAAAACACATTCGT-3′ [28].
3. Alt-R CRISPR-Cas9 tracrRNA, ATTO550 (Integrated DNA Technologies, 1075927).
4. Alt-R CRISPR-Cas9 tracrRNA, ATTO647N (Integrated DNA Technologies; *see* **Note 1**).
5. Nuclease-Free Duplex Buffer (Integrated DNA Technologies).

### 2.4 In Vitro Fertilization (IVF)

1. 10- to 18-week-old ICR female mice.
2. 12- to 24-week-old ICR male mice.
3. TYH medium [29].
4. Pregnant mare serum gonadotropin (PMSG, Asuka animal health, Serotropin).
5. Human chorionic gonadotropin (hCG, Asuka animal health, Gonatropin).
6. 35 mm/Non-Treated Dish.
7. Paraffin Liquid (Nacalai Tesque, 26117-45).

### 2.5 In Vitro Maturation (IVM)

1. 10- to 18-week-old ICR female mice
2. HEPES-IVG medium: 242.5 mL MEM α is mixed with 125.0 mg of Hepes sodium salt, 2.5 mL of GlutaMAX (Gibco, 35050061), 2.5 mL of Penicillin-Streptomycin-Amphotericin B Suspension (Merck, A5955), and 2.5 mL of 10% polyvinyl alcohol (average molecular weight 30,000–70,000, Merck) dissolved in saline. After mixing, the pH of the mixture is adjusted to 7.4 by adding 1N HCl and filtered with a 0.22 μm Filter Unit.
3. IVG medium: 4.75 mL of MEM α is mixed with 250 μL of Fetal Bovine Serum (FBS, Nichirei, 172012), 50 μL of GlutaMAX, 50 μL of Mercaptoethanol, 50 μL of ITS-A, 50 μL of Penicillin-Streptomycin-Amphotericin B Suspension, and 5 μL of Human follicle stimulating hormone (hFSH, MSD, Follistim). After mixing, the mixture is filtered using a 0.22 μm Filter Unit (*see* **Note 2**).
4. Polymer and serum-free pre-IVM medium: 1.9 mL of MEM α is mixed with 20 μL of Mercaptoethanol, 20 μL of Penicillin-Streptomycin-Amphotericin B Suspension, and 2 μL of hFSH. After mixing, the mixture is filtered using a 0.22 μm Filter Unit (*see* **Note 2**). After filtration, 2.5 μL of 10 IU Gonadotropin and 10 μL of Mouse Epidermal Growth Factor (mEGF, CST, 5331SC) are added per 1 mL of aliquots.
5. IVM medium containing twice the amount of protein: 1.9 mL of MEM α is mixed with 200 μL of FBS, 40 μL of GlutaMAX, 20 μL of Mercaptoethanol, 40 μL of ITS-A, 20 μL of Penicillin-Streptomycin-Amphotericin B Suspension, and 2 μL of hFSH. After mixing, the mixture is filtered using a 0.22 μm Filter Unit (*see* **Note 2**). After filtration, 2.5 μL of Gonadotropin and 10 μL of mEGF are added per 1 mL of aliquots.
6. Millinone (abcam, ab142322).
7. Chatot–Ziomek–Bavister (CZB) medium [30].
8. HEPES-CZB medium [30].
9. 35 mm/Non-Treated Dish.
10. Paraffin Liquid.
11. Needles for GV oocyte collection: 26G × 1/2 needle and 0.1 mL syringe.

### 2.6 Microinjection

1. HEPES-CZB medium [30].
2. HEPES-CZB contains 12% Polyvinylpyrrolidone K90 (PVP-HEPES-CZB, Nacalai Tesque, 28315-72).
3. $KSOM^{AA}$ medium [31].
4. Liquid paraffin.
5. Hyaluronidase (Merck; Type-IS, H3506).

6. Micromanipulator: An inverted microscope (e.g. Olympus, IX-73) is attached to a manipulator (Narishige, MN-4), three-axis Hanging Joystick Oil Hydraulic Micromanipulator (MMO-4), injector for Needle (Narishige, IM-21), injector for holding pipette (Eppendorf, CellTram Air 5176) and piezo-drive manipulator (Eppendorf, PiezoXpert®). A 4× (UPLFLN 4X numerical aperture (NA) 0.13) and 20× (LUCPLFLN 20X NA 0.45) objectives are used.
7. A narrow glass pipette (1 μm diameter) for the injection is made by pulling the glass capillary (e.g., Sutter, B100-75-10) using a puller (e.g., Sutter, P-1000IVF) and bending it by micro forge (e.g., Narishige, MF-900).

### 2.7 Super-Resolution Live-Cell Imaging

1. Super-resolution microscopy system based on a disk confocal photon reassignment (OPRA)-type microscope [22] is used (Fig. 2).
2. Optical power meter: TB200 (Yokogawa Electric).
3. Thermocouple thermometer (SATO, SK-1100).

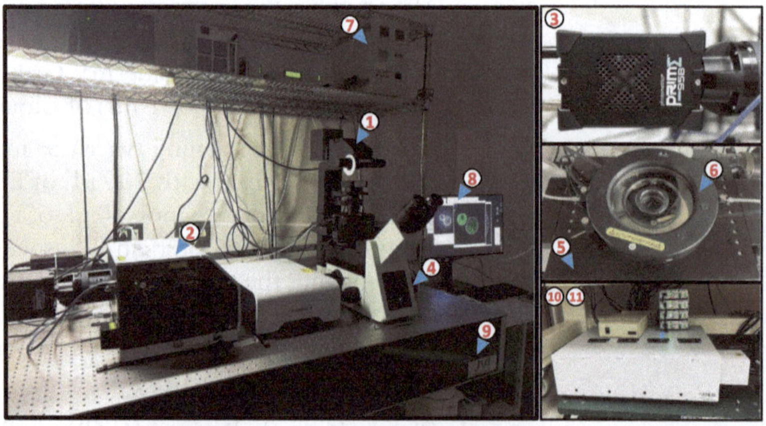


| No. | Devices | Manufacturer | Model |
|---|---|---|---|
| 1 | Inverted microscope | Olympus | IX-73 |
| 2 | Nipkow disk confocal unit | Yokogawa electric | CSU-W1 SoRa |
| 3 | sCMOS camera | Photometrics | Prime 95B |
| 4 | Z motor | Ludl | Mac5000 |
| 5 | XY auto stage | Sigmakoki | BIOS-206T |
| 6 | CO2 incubator | Tokai hit | MI-IBC |
| 7 | Gas mixer | Tokai hit | GM-8000 |
| 8 | Software | https://micro-manager.org | Micro-Manager |
| 9 | Vibration isolation table | Nippon boushin industry | SK-AS-1875T |
| 10 | Laser combiner | Sigmakoki | – |
| 11 | Lasers | Coherent | 405 nm |
| | | | 488 nm |
| | | | 561 nm |
| | | | 640 nm |

**Fig. 2** Outline of the super-resolution live-cell imaging systems. Photograph (upper) and table (lower) of the equipment for imaging

4. $KSOM^{AA}$ medium containing 0.00025% polyvinyl alcohol and 100 μM EDTA.
5. A 35 mm glass-bottomed dish (No. 1.5 coverslip, 14 mm glass diameter).

### 2.8 Embryo Transfer

1. Surrogate mothers (ICR strain) are purchased on day 1 of pseudo-pregnancy following mating with vasectomized ICR males.
2. Medetomidine (Meiji Animal Health).
3. Midazolam (Sandoz).
4. Butorphanol (Meiji Animal Health).
5. Atipamezole (Meiji Animal Health, Mepatia).

### 2.9 Image Analysis

1. ImageJ/Fiji image analysis platform (https://imagej.net/Fiji) [32].
2. FluoRender (https://www.sci.utah.edu/software) [33].

## 3 Methods

### 3.1 Preparation of In Vitro Transcribed RNA

1. Linearize plasmid DNA using restriction enzymes at a single digestion site located immediately downstream of the poly (A) sequence. In the cases of pcDNA3.1 Histone H2B-EGFP poly (A) and pTALYM3, *Xho*, I and *Apa* I are used, respectively (Fig. 1a).
2. Purify linearized plasmids by phenol-chloroform treatment and ethanol precipitation using RNase-free reagents. The purified linear plasmids are dissolved in fresh Milli-Q water.
3. Prepare an RNA transcription mixture with RiboMAXTM Large-Scale RNA Production System T7 and a linearized plasmid template. The Ribo m7G cap analog is added to an RNA production mixture containing rATP, rUTP, rCTP, and rGTP (*see* **Note 3**). Use 1.5–2 μg of plasmid template for a 20 μL volume reaction of RNA synthesis.
4. Treat the transcribed RNA with RQ-1 RNase-free DNase included in the RNA production kit.
5. Purify RNA by phenol-chloroform treatment and ethanol precipitation. Dissolve RNA with DNase-and RNase-free water.
6. Remove free rNTPs with MicroSpinTM S-200 HR columns. First, equilibrate the column with DNase and RNase-free water. Next, add the transcribed RNA solution to the column. Then, elute transcribed RNA multiple times using the same water as above.

7. Measure the concentration of the transcribed RNA in each fraction. Mix each fraction and adjust the RNA concentration to approximately 100 ng/μL. Confirm by electrophoresis that the appropriate size of RNA has been transcribed.

### 3.2 Preparation of the dCas9/gRNA Complex

1. Dissolve crRNA and fluorescence-labeled tracrRNA in a Nuclease-Free Duplex Buffer to a concentration of 1 μg/μL.
2. Mix crRNA and fluorescence-labeled tracrRNA at a ratio of 1: 2.
3. Hybridize using a thermal cycler (94 °C: 5 min, 60 °C: 5 min).
4. Dilute gRNA and dCas9 to the desired concentrations using DNase- and RNase-free water and then mix them (Fig. 1b). The final concentrations for mice embryos are 20 and 200 ng/ μL, respectively.
5. Incubate for 15–30 min at room temperature (*see* **Note 4**).

### 3.3 IVF

1. For live-cell imaging, any mouse strain can be used. In this section, we describe the case of ICR mice. To collect mouse eggs, use 10- to 18-week-old female ICR mice.
2. Superovulate by intraperitoneal injections of 10 IU pregnant mare serum gonadotropin (PMSG) and 10 IU human chorionic gonadotropin (hCG) at 48 h intervals.
3. Collect cumulus-oocyte complexes (COCs) 13–15 h after hCG injection and soak in 0.2 mL drops of TYH medium covered with liquid paraffin oil.
4. Collect sperm from the cauda epididymidis of ICR mouse (12–24 weeks) and soak in 0.2 mL drops of TYH medium.
5. Capacitate by incubation for 2 h at 37 °C under 5% $CO_2$.
6. Inseminate COCs with capacitated sperm (final concentration: 50–100 sperm/μL).
7. After 1.5 h incubation at 37 °C under 5% $CO_2$, disperse cumulus cells by brief treatment with hyaluronidase (150 U/mL) and collect anaphase/telophase II stage oocytes.

### 3.4 IVM

1. Collect oocytes at the germinal vesicle (GV) stage by puncturing the ovaries in HEPES-IVG using a needle.
2. Remove the cumulus cells by gentle pipetting. Sort fully grown oocytes and incubate them in IVG medium containing 2.5 μM Millinone at 6% $CO_2$, 5% $O_2$, and 37 °C until transferring to in vitro maturation (IVM) medium.
3. Microinject mRNA encoding fluorescent probe into GV oocytes in the HEPES-CZB medium.
4. For imaging during IVM, attach the oocytes first to the bottom of the glass-bottomed dish in each 2.5 μL droplet of protein

(FBS, GlutaMAX, ITS-A)-free pre-IVM medium, and then add another 2.5 μL of IVM medium containing twice the amount of protein (*see* **Note 5**).

### 3.5 Microinjection of mRNA or dCas9/gRNA Complex

1. Dilute the mRNA in Mili-Q water to a concentration of 10 ng/μL for histone H2B-mCherry and 100 ng/μL for TALE-mClover against the mouse major satellite sequence.
2. Place an aliquot (0.3 μL) onto a micromanipulation chamber.
3. Place 7 μL drops of HEPES-CZB medium and 7 μL drops of PVP-HEPES-CZB in the same injection chamber.
4. Transfer zygotes or GV oocytes to HEPES-CZB medium and inject mRNA solution or dCas9/gRNA complex using a piezo-drive microinjector with a narrow glass needle (approximately 1 μm diameter). Before the injection, aspirate PVP-HEPES-CZB once to wash and prevent stickiness of the inner wall of the glass needle.
5. Aspirate the mRNA solution into the needle once, and apply piezo pulses to the oocyte to break the zona pellucida and oolemma.
6. Introduce a few picolitres of solution into the oocyte and remove the needle quickly.
7. Incubate the cell until the fluorescence becomes visible. Normally, approximately 3 h or more is required for both embryos and oocytes, but depending on the purpose of observation, the embryos can be placed on the imaging device immediately after the injection.

### 3.6 Super-Resolution Live-Cell Imaging

1. Adjust the temperature setting of the air conditioner in the room to keep the room temperature stable at 30 °C at least one day prior to the start of imaging (*see* **Note 6**).
2. Adhere the zygotes to the bottom of the glass-bottomed dish in each droplet of protein-free $KSOM^{AA}$ medium containing 0.00025% polyvinyl alcohol and 100 μM EDTA. For oocyte imaging, soak the oocytes in pre-IVM medium for attachment to glass. Then, add the same volume of IVM medium containing twice the amount of protein (*see* **Note 5**).
3. Measure the power of the laser emitted from the tip of the objective lens using an optical power meter (TB200). The output power depends on the specimens and the probe; this chapter recommends 0.05 mW at 488 and 0.1 mW at 561 nm for the output power (*see* **Note 7**).
4. Transfer the glass-bottomed dish to the incubation chamber on the microscope stage, which is kept at 37 °C. Confirm the temperature in the dish with a thermocouple thermometer (*see* **Note 8**).

5. While taking multiple snapshots of the fluorescence image, find the location where the image appears brightest and clearest and adjust the correction ring of the objective lens (*see* **Note 9**).
6. Using μManager, set the positions (embryo or oocyte), number of planes in the Z-axis direction, the time interval between snapshots, total imaging time, camera sensitivity, data storage location, etc., and start imaging (for imaging conditions, *see* **Note 7**). Examples of time-lapse images of the first mitosis (Fig. 3a) and first meiosis (Fig. 3b) are shown.

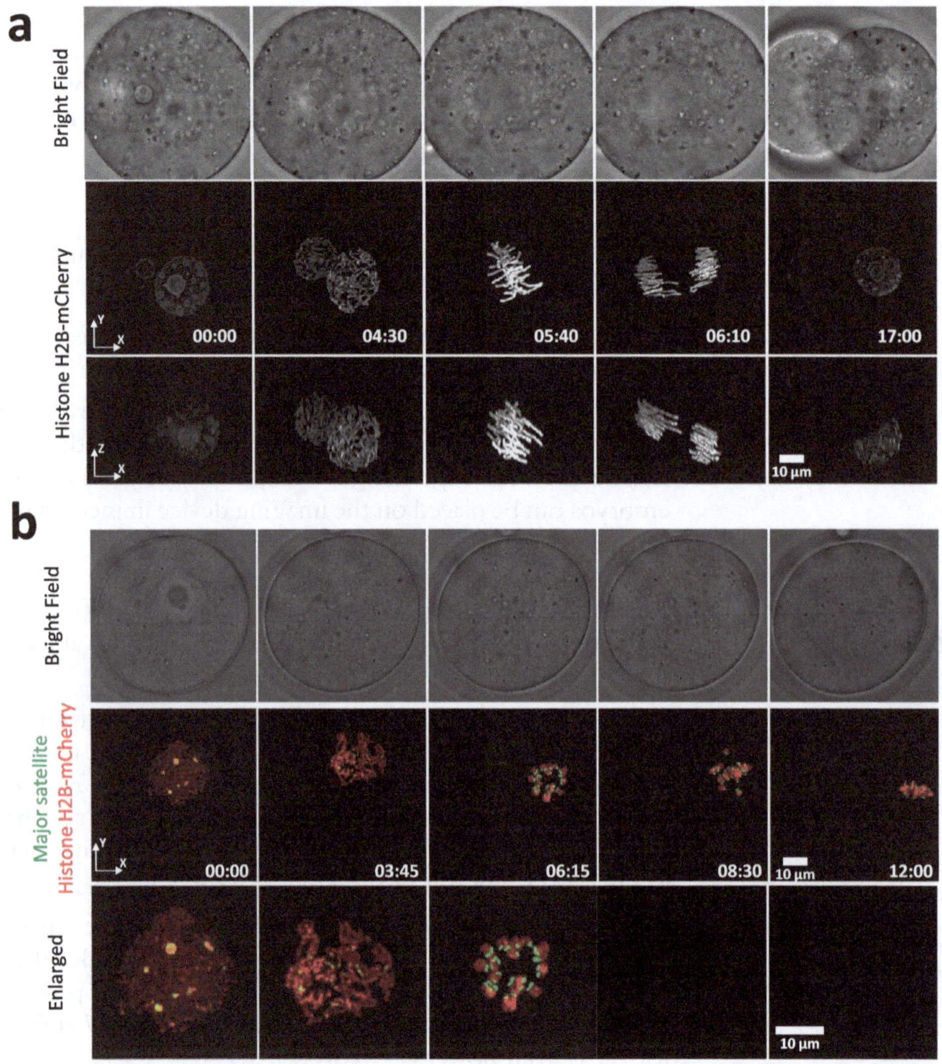


**Fig. 3** Images of super-resolution live-cell imaging of chromosomes in first mitosis (**a**) and first meiosis (**b**). (**a**) Time-lapse images of embryo injected with histone H2B-mCherry are shown. Bright-field and x-y and reconstructed x-z views of mCherry signals are indicated. This data was taken from Hatano et al. [22]. (**b**) Super-resolution images of chromosome (histone H2B-mCherry, red) and pericentromere (major satellite targeting TAL effector, green) in meiosis of mouse oocytes. Numbers in the lower right of each panel indicate hours and minutes from the start of imaging

### 3.7 Embryo Transfer

1. Anesthetize the pseudo-pregnant ICR mice by injecting a mixture of medetomidine, midazolam, and butorphanol.
2. Prick the oviduct of the anesthetized mice using micro-forceps.
3. Transfer two-cell embryos into the oviducts of day 0.5 pseudo-pregnant mice using a glass needle.
4. Awaken the mice by administering atipamezole after embryo transfer.
5. Perform a cesarean section on the 18th day of pregnancy to determine the number of born pups.

### 3.8 Chromosome Counting

We use the ImageJ/Fiji software for chromosome counting based on histone H2B-mCherry signals and FluoRender software for the same purpose in oocytes.

1. Generate a point spread function (PSF) image using the Diffraction PSF 3D plug-in (http://fiji.sc/Diffraction_PSF_3D) to mitigate the effects of distortion and determine the expected appearance of a given image.
2. Perform deconvolution using the Tikhonov regularized inverse filter algorithm and the PSF image in the DeconvolutionLab2 plug-in [34] to improve the contrast and sharpness of the original images.
3. Remove noise in the deconvolved images using the top-hat algorithm within the morphological filters (3D) function in the MorphoLibJ plug-in [35].
4. Binarize using Otsu's algorithm.
5. Separate the proximate objects in the image using the distance transform watershed algorithm with the City-Block (1, 2, 3) option in the MorphoLibJ plug-in.
6. Count the number of chromosomes in the post-processing images using Ext.Manager3D_Count function. The example of images of rendering and counting chromosomes are shown in (Fig. 4).
7. Quantify the volume and fluorescent intensity (minimum, average, maximum, sum) of each chromosome in the images using Ext.Manager3D_Quantif3D function.

### 3.9 Centromere Counting

We use the FluoRender software to count dot signals at the centromere.

1. Create z-stack image files for each time point using ImageJ/Fiji and load them into FluoRender.
2. Render using the "Paint Brush" function.
3. Check [Edge Detect] in the Selection Setting and adjust the threshold parameter according to the intensity.

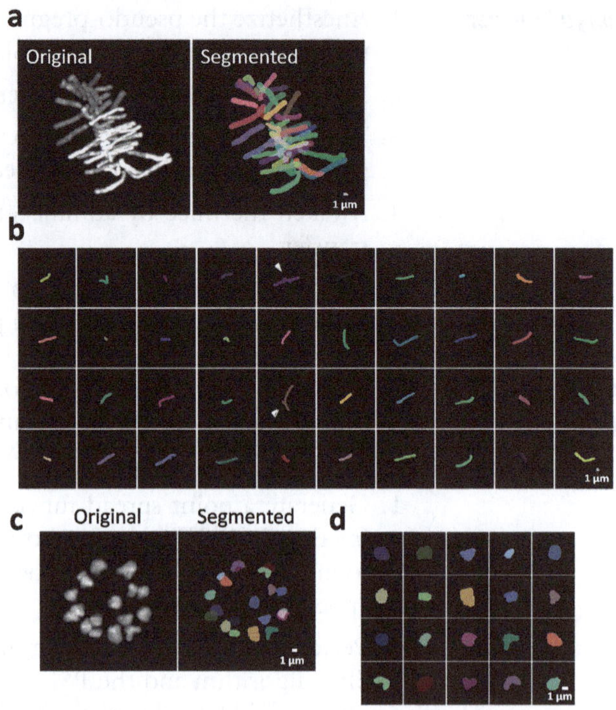


**Fig. 4** Chromosome counting of an embryo (**a**, **b**) and an oocyte (**c**, **d**) using histone H2B-mCherry mRNA. (**a**) Chromosome counting of embryos in the first mitosis. The left panel shows the original image. The right image shows the segmented chromosomes in the left panel. The embryo images were processed using ImageJ/Fiji. (**b**) Recognized object by the segmentation is shown in each panel. The white arrowheads indicate a processed chromosome with a risk of miscount. Panels of A and B were data from Hatano et al. [22]. (**c**) Chromosome counting of oocytes meiosis I. The left panel shows the original image. The right image shows the processed image. The oocyte images were processed using FluoRender. (**d**) Each panel shows the recognized object of segmented chromosome. In this image of the oocyte, we succeeded in counting exactly 20 chromatids

4. Manually render the dot signals of the centromere and count them one by one. The example images illustrating chromosomal centromere rendering and centromere counting are shown in Fig. 5.

## 4 Notes

1. This RNA-fluorochrome complex is build-to-order.
2. During filtration, the first approximately 900 μL of eluate is discarded to avoid potential contamination by toxic substances present in the filter, and only the subsequent eluate is used for the experiment.

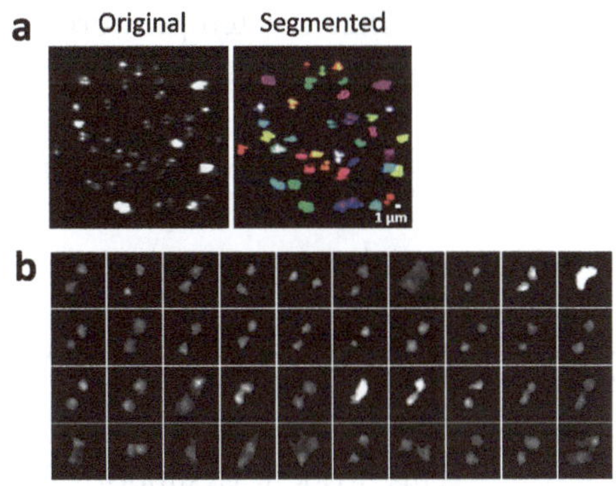


**Fig. 5** CRISPR-mediated live cell imaging of the centromere region and dot signal counting in mouse first mitosis embryos. (**a**) The left panel shows the original image. The right image shows the processed image. (**b**) Each panel shows the recognized object. In this image of the embryo, 40 chromosomes could be counted. These data were from Hatano et al. [22]

3. The volumes of 100 mM rATP, 100 mM rUTP, 100 mM rCTP, 100 mM rGTP, and Ribo m7G Cap Analog in 20 μL reaction are 1.5 μL, 1.5 μL, 1.5 μL, 0.12 μL. and 1.5 μL, respectively.
4. The resulting gRNA concentration is approximately 333 ng/μL. dCas9/gRNA complex, which should be stored at −80 °C and used within 1 month.
5. When embryos or oocytes are placed in a medium containing proteins or high polymer concentrations, they move and disappear during imaging. Therefore, when using this method, it is necessary to ensure that the sample adheres to the glass surface.
6. To stabilize the temperature of the equipment, avoid setting the temperature below 30 °C to prevent equipment temperature instability and potential overload of the incubator chamber on the stage, which can lead to malfunctions. Conversely, if the temperature is set above 30 °C, the heat may cause breakdown of the machine or computer equipment.
7. We evaluated phototoxicity using embryo development rate as an indicator. For example, for the 561 nm laser, 46.7% (7/15) of the pups were obtained at 0.1 mW and 10-min time interval conditions. Under 0.1 mW, 5-min time interval condition, only 7.7% (1/13) of pups were obtained [23]. The 561 nm laser causes less phototoxicity to the embryo, whereas the 488 nm laser did not ensure blastocyst development, even at very weak power.
8. A wire-type thermocouple thermometer, such as a Thermocouple thermometer (SATO, SK-1100), can be used to measure the temperature of the culture medium in the dish. Ensure

that the temperature has stabilized at 37 °C before starting imaging.

9. The glass in the glass-bottom dish may expand or contract with changes in temperature. Therefore, to accurately align the correction rings, it is necessary to wait 30–120 min until the glass temperature stabilizes.

## Acknowledgments

This work was supported by JSPS KAKENHI Grant Numbers JP25712035, JP25116005, JP18H05528, and JP18H02357 to KY. The funders had no role in the study design, data collection, interpretation, or decision to submit the work for publication. We would like to thank Editage for English language editing.

### References

1. Magli MC, Jones GM, Gras L et al (2000) Chromosome mosaicism in day 3 aneuploid embryos that develop to morphologically normal blastocysts in vitro. Hum Reprod 15: 1781–1786
2. Sandalinas M, Sadowy S, Alikani M et al (2001) Developmental ability of chromosomally abnormal human embryos to develop to the blastocyst stage. Hum Reprod 16:1954–1958
3. Rubio C, Rodrigo L, Mercader A et al (2007) Impact of chromosomal abnormalities on preimplantation embryo development. Prenat Diagn 27:748–756
4. Mantikou E, Wong KM, Repping S, Mastenbroek S et al (2012) Molecular origin of mitotic aneuploidies in preimplantation embryos. Biochim Biophys Acta 1822:1921–1930
5. Yao T, Suzuki R, Furuta et al (2018) Live-cell imaging of nuclear–chromosomal dynamics in bovine in vitro fertilised embryos. Sci Rep 8(1): 7460
6. Hassold T, Hunt P (2001) To err (meiotically) is human: the genesis of human aneuploidy. Nat Rev Genet 2(4):280–291
7. Handyside AH (2012) Molecular origin of female meiotic aneuploidies. Biochim Biophys Acta (BBA)-Mol Basis Dis 1822(12)
8. Sakakibara Y, Hashimoto S, Nakaoka Y et al (2015) Bivalent separation into univalents precedes age-related meiosis I errors in oocytes. Nat Commun 6(1):7550
9. Mastenbroek S, Twisk M, van Echten-Arends J et al (2007) In vitro fertilization with preimplantation genetic screening. N Engl J Med 357(1):9–17
10. Yamagata K, Suetsugu R, Wakayama T (2009) Long-term, six-dimensional live-cell imaging for the mouse preimplantation embryo that does not affect full-term development. J Reprod Dev 55(3):343–350
11. Yamagata K, Suetsugu R, Wakayama T (2009) Assessment of chromosomal integrity using a novel live-cell imaging technique in mouse embryos produced by intracytoplasmic sperm injection. Hum Reprod 24(10):2490–2499
12. Mashiko D, Ikeda Z, Yao et al (2020) Chromosome segregation error during early cleavage in mouse preimplantation embryo does not necessarily cause developmental failure after blastocyst stage. Sci Rep 10:854
13. Yao T, Ueda A, Khurchabilig A et al (2022) Micronucleus formation during early cleavage division is a potential hallmark of preimplantation embryonic loss in cattle. Biochem Biophys Res Commun 617:25–32
14. Hell SW, Wichmann J (1994) Breaking the diffraction resolution limit by stimulated emission: stimulated-emissiondepletion fluorescence microscopy. Opt Lett 19:780–782
15. Hell SW, Kroug M (1995) Ground-state-depletion fluorescence microscopy: a concept for breaking the diffraction resolution limit. Appl Phys B 60:495–497
16. Betzig E, Patterson GH, Sougrat R (2006) Imaging intracellular fluorescent proteins at nanometer resolution. Science 313:1642–1645

17. Hess Samuel T, Thanu PK, Girirajan, Michael D, Mason (2006) Ultra-high resolution imaging by fluorescence photoactivation localization microscopy. Biophys J 91(11): 4258–4272
18. Rust MJ, Bates M, Zhuang X (2006) Sub-diffraction-limit imaging by stochastic optical reconstruction microscopy (STORM). Nat Methods 3(10):793–796
19. Gustafsson MG (2000) Surpassing the lateral resolution limit by a factor of two using structured illumination microscopy. J Microsc 198:82–87
20. Huff J (2015) The Airyscan detector from ZEISS: confocal imaging with improved signal-to-noise ratio and super-resolution. Nat Methods 12:i–ii
21. Gao L, Shao L, Higgins CD (2012) Noninvasive imaging beyond the diffraction limit of 3D dynamics in thickly fluorescent specimens. Cell 151(6):1370–1385
22. Azuma T, Kei T (2015) Super-resolution spinning-disk confocal microscopy using optical photon reassignment. Opt Express 23(11): 15003–15011
23. Hatano Y, Mashiko D, Tokoro M et al (2022) Chromosome counting in the mouse zygote using low-invasive super-resolution live-cell imaging. Genes Cells 27(3):214–228
24. Kanda T, Sullivan KF, Wahl GM (1998) Histone-GFP fusion protein enables sensitive analysis of chromosome dynamics in living mammalian cells. Curr Biol 8(7):377–385
25. Ueda J, Maehara K, Mashiko D et al (2014) Heterochromatin dynamics during the differentiation process revealed by the DNA methylation reporter mouse, MethylRO. Stem Cell Rep 2(6):910–924
26. Yamagata K, Yamazaki T, Yamashita M (2005) Noninvasive visualization of molecular events in the mammalian zygote. Genesis 43(2): 71–79
27. Miyanari Y, Ziegler-Birling C, Torres-Padilla ME (2013) Live visualization of chromatin dynamics with fluorescent TALEs. Nat Struct Mol Biol 20(11):1321–1324
28. Anton T, Bultmann S, Leonhardt H (2014) Visualization of specific DNA sequences in living mouse embryonic stem cells with a programmable fluorescent CRISPR/Cas system. Nucleus 5(2):163–172
29. Toyoda Y, Yokoyama M, Hosi T (1971) Studies on the fertilization of mouse eggs in vitro. I. in vitro fertilization of eggs by fresh epididymal sperm. Jpn J Anim Reprod 16(4):147–151
30. Chatot CL, Ziomek CA, Bavister BD (1989) An improved culture medium supports development of random-bred 1-cell mouse embryos in vitro. J Reprod Fertil 86:679–688
31. Biggers JD (2002) Thoughts on embryo culture conditions. Reprod Biomed Online 4:30–38
32. Schindelin J, Arganda-Carreras I, Frise E et al (2012) Fiji: an open-source platform for biological-image analysis. Nat Methods 9: 676–682
33. Wan Y, Otsuna H, Chien C B et al (2012) FluoRender: an application of 2D image space methods for 3D abbbbnd 4D confocal microscopy data visualization in neurobiology research. In: 2012 IEEE Pacific visualization symposium. IEEE, pp 201–208
34. Sage D, Donati L, Soulez F et al (2017) DeconvolutionLab2: an open-source software fordeconvolution microscopy. Methods 115: 28–41
35. Legland D, Arganda-Carreras I, Andrey P (2016) MorphoLibJ: integrated library and plug-ins for mathematical morphology with ImageJ. Bioinformatics 32:3532–3534

# Chapter 3

# Cell Division Imaging of Tobacco BY-2 Cells in a 3D Volume by Two-Photon Spinning-Disk Confocal Microscopy

**Takashi Murata, Motosuke Tsutsumi, Kohei Otomo, and Tomomi Nemoto**

## Abstract

Plant cells have a unique organization of microtubules during mitosis and cytokinesis. However, it is difficult to observe cell division in plant cells because of the distortion of images caused by thick samples. We addressed this issue by two-photon excitation coupled with spinning-disk confocal microscopy. Herein, we describe the method for observing microtubules and nuclei during cell division in tobacco BY-2 cells by two-photon spinning-disk confocal microscopy. The protocol includes the transformation of BY-2 cells with a plasmid for the expression of tubulin and histone markers, preparation of cells for microscopy, and operation of the two-photon spinning-disk microscope.

**Key words** Cell division, Microtubule, Two-photon excitation, Spinning disk confocal microscopy

## 1 Introduction

### 1.1 Background

Vegetative cells of land plants have lost centrosomes during evolution, and microtubule organization during cell cycle progression is unique [1] (Fig. 1). In interphase, parallel arrays of microtubules run along the plasma membrane. These arrays regulate the orientation of cellulose microfibrils in the cell wall, which controls the direction of cell expansion. Upon approaching mitosis, microtubules along the plasma membrane gather to form the preprophase band (PPB). The PPB recruits proteins that control the orientation of the cell plate, although microtubules of the band depolymerize during spindle formation, i.e., before the initiation of cell plate formation. During spindle formation, microtubules polymerize in the absence of centrosomes. At anaphase, chromosomes are separated in the absence of focused spindle poles. Finally, cytokinetic microtubule arrays, the phragmoplast, deposit the cell plate that partitions two daughter cells.

The organization of microtubules during cell division has been investigated in root cells of the model plant *Arabidopsis thaliana*

Tomomi Kiyomitsu (ed.), *The Mitotic Spindle: Methods and Protocols*, Methods in Molecular Biology, vol. 2872, https://doi.org/10.1007/978-1-0716-4224-5_3,

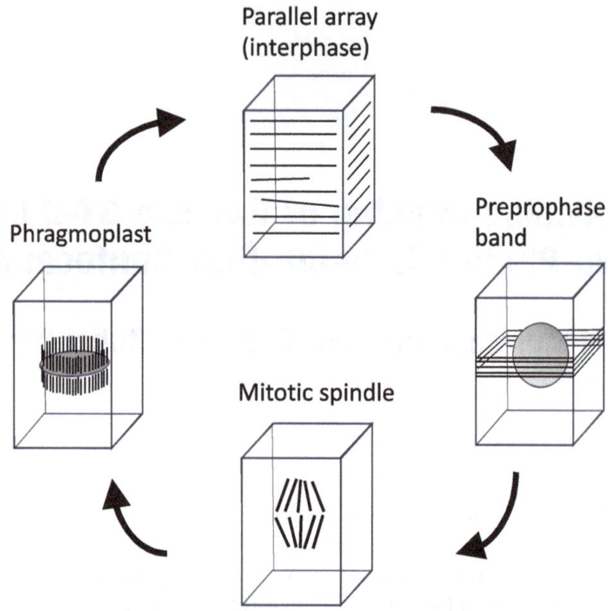


**Fig. 1** Schematic drawings of microtubule organization in plant cells during cell cycle progression

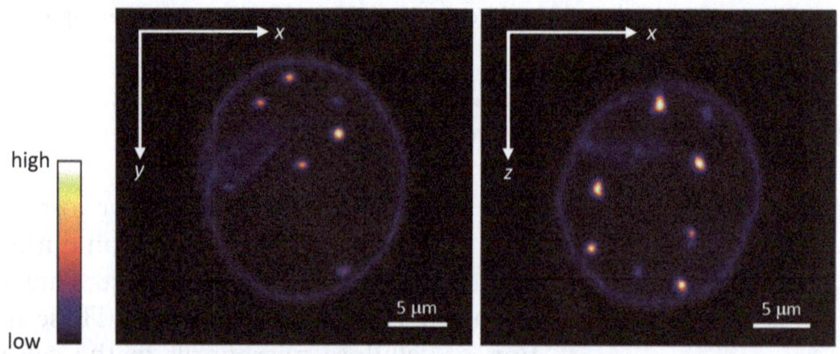


**Fig. 2** Two-photon spinning disk microscope images of centromeres in the tobacco BY-2 nucleus. mCitrine-CenH3 and SUN-mCitrine are expressed for centromere and nuclear envelope labeling, respectively. Slices of the xy plane and xz plane are shown in the left and right, respectively

[2] and suspension cultured cells of the tobacco plant [3]. Protonemal cells of the moss *Physcomitrium patens* are also a good material [4]. In all cases, it is necessary to observe through thick materials; the dividing cells of *A. thaliana* in a root tip are covered by epidermal or root cap cells. Tobacco suspension culture cells and moss protonemal cells have a width of approximately 30–40 and 20 μm, respectively. Therefore, for visualizing the microtubule arrays of dividing plant cells, techniques different from glass-adhered animal cultured cells are necessary.

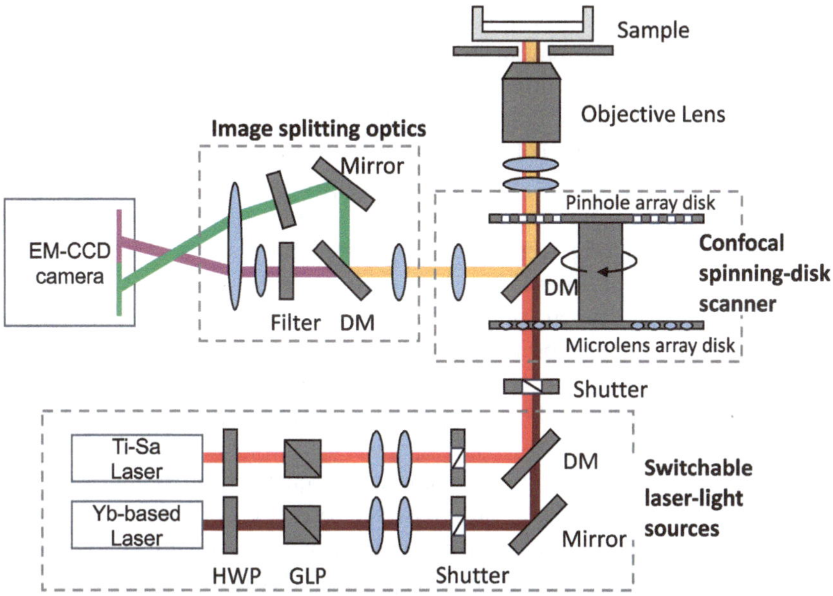


**Fig. 3** Optical settings of the two-photon spinning-disk microscope with switching between two laser light sources and the use of image-splitting optics. *HWP* half-wave plate, *GLP* glan-laser prism, *DM* dichroic mirror

Two-photon excitation microscopy has been used to observe thick specimens such as the mouse brain [5]. Nevertheless, its resolution is lower than that of conventional single-photon microscopy because the resolution of optical microscopes depends on the wavelength of light [6]. The resolution of two-photon microscopy can be improved by introducing a pinhole in the light path, similar to that for confocal microscopy. Shimozawa et al. [7] combined a specially designed spinning-disk pinhole array and two-photon excitation, demonstrating improved 3D images of thick specimens. We improved the area of the field of view of the two-photon spinning-disk confocal microscope [8] and demonstrated low phototoxicity using dividing plant and animal cultured cells [9].

As demonstrated by Shimozawa et al. [7], image distortion along the z-axis, often observed by conventional confocal microscopy, is minimized by two-photon spinning-disk microscopy. We confirmed the isometric images of centromeres in the nucleus of tobacco BY-2 cells (Fig. 2). This feature is suitable for observing mitosis and cytokinesis in plant cells in 3-dimensional views.

Figure 3 depicts an overview of the two-photon spinning-disk microscope constructed at the National Institute for Physiological Sciences (NIPS) [9]. The microscope has two excitation laser light sources, i.e., a wavelength-tunable Ti-Sa laser light source and a fixed wavelength Yb laser light source. A Ti-Sa laser light source is used at approximately 800–980 nm for the excitation of green or yellow fluorescent dyes, and a Yb laser light source is used at approximately 1040 nm for the excitation of red fluorescent dyes.

The laser light sources are switched by motorized shutters, enabling the sequential acquisition of dual-wavelength excitation images. The specimen is excited by the laser beams through a specially designed confocal spinning-disk scanner consisting of Nipkow disks: microlens array and pinhole array. The emitted fluorescence is scanned by the spinning-disk scanner, separated by image-splitting optics, and finally acquired by an EM-CCD camera. The objective lens is essential for observation. The system has a 60× water immersion lens with high NA (NA 1.27), enabling higher resolution imaging and focusing in thick living cells. Because the microscope is custom-constructed, the laser light wavelength and power are controlled independently from the image acquisition software. The equipment details are described in Subheading 2.12.

### 1.2 Special Consideration for Two-Photon Excitation

For two-photon microscopy, selection of the fluorescent protein is important. The cross-section (i.e., efficiency) of two-photon absorption is not always proportional to single-photon absorption. For instance, the two-photon absorption of mCherry, one of the most frequently used red fluorescent proteins, is approximately 2.5 times lower than that of tdTomato but 4.3 times lower in single-photon absorption [10]. In addition to the cross-section, the absorption spectrum differs from the deduced value in which the excitation wavelength is multiplied twice that of single-photon excitation [11]. Because the wavelength and laser beam power available for two-photon microscopes are limited, the researcher must carefully select the fluorescent proteins based on a two-photon cross-section before preparing transformed cells, plants and animals.

After the start of observation, precise control of laser beam power is vital. Although nonlinear excitation efficiency results in specific excitation of the focused plane of the objective lens, excitation with excessive laser beam power causes photobleaching of samples (Fig. 4) and destruction of materials in the worst case. Therefore, preliminary observations for determining the optimal laser beam power and wavelength are essential when new fluorescent proteins are introduced.

For dual-color observation, we generally use excitation with a single wavelength and image separation by the image-splitting optics (Fig. 3). For successful observation, it is important to combine two fluorescent proteins and optimize the excitation wavelength. We recommend a combination of mCherry and mCitrine (YFP) with 1040-nm excitation by a Yb laser light or 960- to 980-nm excitation by a Ti-Sa laser light. A combination of mNeonGreen and mCherry with 960-nm excitation is also good. Although the combination of EGFP and mCherry is frequently used for conventional dual-color observation, it has a difficulty of two-photon excitation efficiency. EGFP is efficiently excited at approximately 950 nm, and the excitation efficiency of mCherry

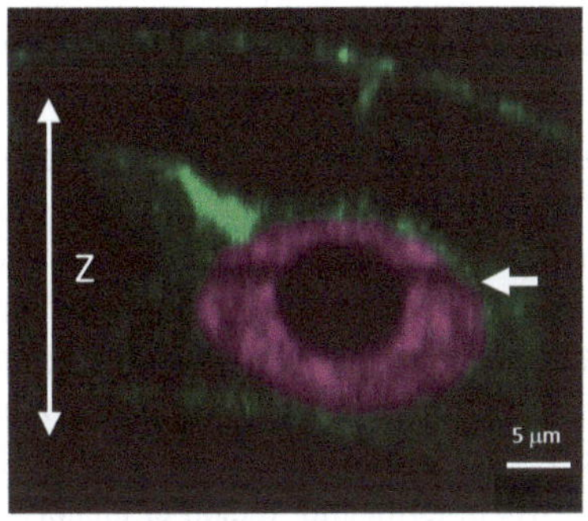


**Fig. 4** Example of photobleaching by two-photon excitation. A side view of a mCitrine-β-tubulin (shown in green) and H2B-mCherry (shown in magenta) dual-labeled cell. Excitation with an over-dose laser beam in the plane of the nucleus caused photobleaching of the excited plane (indicated by an arrow)

by the wavelength is not so high. Herein, we describe the method for the transformation of BY-2 cells with a plasmid harboring mCitrine–tubulin and histone H2B–mCherry and observation by two-photon spinning-disk microscopy.

### 1.3 How to Access the Two-Photon Spinning Disk Microscope

The two-photon spinning-disk microscope was originally constructed at Riken Institute, Japan [7]. The microscope is currently not available, and the scanner unit for two-photon excitation is commercially available from Yokogawa Co. as a custom order. To our knowledge, the only microscope that can be utilized by general users is constructed at NIPS, Okazaki, Japan. Domestic users in Japan can apply for Advanced Bioimaging Support (ABiS), and the Nemoto laboratory will support observation. Overseas users have to contact Prof. Nemoto directly for collaboration.

## 2 Materials

### 2.1 Culture of Tobacco BY-2 Cells

Tobacco (*Nicotiana tabacum*) BY-2 cells are distributed by RIKEN Bioresource Center, Japan (http://www.brc.riken.jp/lab/epd/Eng/catalog/pcellc.shtml). The cells are weekly subcultured in modified LS medium [12], i.e., Murashige and Skoog basal salt mixture supplemented with 3% sucrose, 0.2 g/L $KH_2PO_4$, 0.1 g/L myo-inositol, 1 mg/L thiamine-HCl, and 0.2 mg/L 2,4-dichloroacetic acid. The pH is adjusted to 5.6 with KOH. The medium is sterilized at 120 °C for 15 min (*see* **Note 1**). See detailed methods described in [12].

### 2.2 Plasmid for mCitrine-tubulin and H2B-mCherry Expression

The cells expressing mCitrine-TUB8 (β-tubulin8 of *Nicotiana tabacum*) and histone H2B-mCherry are described in [13]. The plasmid pCAMBIA1300-H2BmCherry-mCitrineTUB8 for dual expression of the two fusion proteins should be requested to Prof.

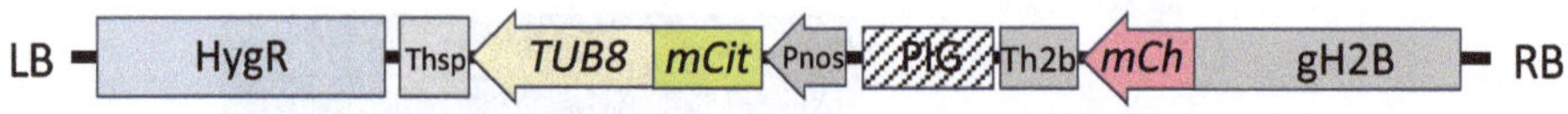


**Fig. 5** Schematic drawing of the map of the plasmid pCAMBIA1300-H2BmCherry-mCitrineTUB8 for the expression of mCitrine-β-tubulin and H2B-mCherry. Because the region between the left border sequence (LB) and the right border sequence (RB) of the plasmid is integrated into the plant genome, the map shows the region between LB and RB. The plasmid has a hygromycin-resistant gene driven by a 35S promoter (HygR), TUB8 cDNA of *Nicotiana tabacum* (*TUB8*) fused with mCitrine coding sequence (*mCit*) driven by a Nos promotor (Pnos) and a histone H2B genomic sequence containing a promotor region (gH2B) fused with mCherry coding sequence (*mCh*). Thsp: Heat shock protein terminator of *Arabidopsis thaliana*. PIG: intergenic region of *Physcomitrium patens*. Th2b: terminator region of tobacco histone H2B gene

Takashi Murata (Kanagawa Institute of Technology: murata@bio.kanagawa-it.ac.jp). Figure 5 shows a schematic of the plasmid map.

### 2.3 Agrobacterium Competent Cells

The strain LBA4404 of *Agrobacterium tumefaciens* is generally used for the transformation of tobacco BY-2 cells [14]. Competent cells of the strain are commercially available from Takara Co.

### 2.4 LB Agar Plate

A conventional LB medium containing 1.5% agar is used for the transformation of *Agrobacterium tumefaciens* LBA4404. For the selection of transformed cells, add kanamycin at 50 mg/L (final concentration).

### 2.5 YEB Medium

The YEB medium is used for the preculture of *Agrobacterium* cells. For making YEB medium, dissolve 0.5% beef extract, 0.1% yeast extract, 0.5% polypeptone, 0.5% sucrose, and 0.05% $MgSO_4$ with distilled water. After autoclaving, add kanamycin at 50 mg/L (final concentration).

### 2.6 3% Sucrose

3% sucrose in distilled water should be autoclaved at 120 °C for 15 min.

### 2.7 Agar Medium for the Selection of Transformed BY-2 Cells

After infecting BY-2 cells with *Agrobacterium*, the cells are cultured on agar plates of BY-2 culture medium containing 1% agar (*see* **Note 2**), 500 mg/L carbenicillin for killing *Agrobacterium* and 15 mg/L hygromycin for the selection of transformed cells. To prepare agar plates, drop stocks of hygromycin and carbenicillin into the petri dish and pour 30–35 mL of melted medium with agar. The plate should be stored at 4 °C, and used within a month.

### 2.8 Antibiotics-Containing BY-2 Medium

For the second selection of the transformed BY-2 cells, a liquid medium is used. The medium contains 100 mg/L carbenicillin and 15 mg/L hygromycin. Add stocks of hygromycin and carbenicillin to the liquid medium. The medium should be made just before use.

### 2.9 Glass-Based Dishes

35-mm diameter glass-based dishes with No.1 thickness (0.15 mm) or No. 1S thickness (0.17 mm) should be used. A larger area of glass is better for attaching cells onto the glass surface.

### 2.10 Plasma-Discharge Device

For cleaning and hydrophilic treatment of the glass-based dishes, a plasma discharge device (e. g., PIB-10, Vacuum Device, Japan; or PDC-001, Harrick Plasma, Ithaca, New York) is used. The glass-based dish is put into the device and discharged.

### 2.11 Poly-L-Lysine Solution

Poly-L-lysine hydrobromide with a large molecular weight (150,000–300,000) is dissolved in distilled water at 1 mg/L. The solution can be stored at −25 °C.

### 2.12 Microscope Equipment

For two-photon imaging, we use a custom-built microscope system [8, 9] based on an inverted microscope (Nikon Ti-E, Nikon) equipped with a Ti-Sa laser light source (MaiTai eHP DeepSee; Spectra-Physics; tunable wavelength: 690–1040 nm, repetition-rate: 80 MHz) and a Yb-based laser light source (FemtoTrain; Spectra-Physics; wavelength: 1040-nm, repetition rate: 10 MHz), an objective lens (CFI Plan Apo IR 60 × WI/1.27 NA; Nikon), a spinning disk scanner with 100 μm-diameter pinholes aligned on a Nipkow disk (CSU-MPφ100; Yokogawa Electric), an image-splitting optical unit (W-VIEW GEMINI; Hamamatsu Photonics), and an EM-CCD camera (iXon Ultra 897; Andor Technology). For splitting the fluorescence of mCitrine and mCherry, a 560 nm dichroic mirror (DM; FF560-FDi01-25 3 36; Semrock), a 500–550 nm bandpass filter (BrightLine 525/50; Semrock) for mCitrine and a 600–660 nm bandpass filter (D630/60M; Chroma Technology) for mCherry are used. The microscope has a motorized XY-stage and Z-drive. Each laser source can be controlled by a motorized shutter, and this microscope system allows two-wavelength excitation and simultaneous two-wavelength observation.

### 2.13 Software

The oscillation of laser light is controlled using the manufacture's dedicated software: MaiTai control software (MaiTai Customer Ver. 2.0.23.0; Spectra-Physics) for MaiTai, and Sprint software (Sprint.vi—1.0.0.27—CV; Spectra-Physics) for FemtoTrain. The wavelength tuning of MaiTai laser light is also performed using the MaiTai software. The motorized shutters, Z-drive of the microscope, and camera are controlled by NIS-Elements software (NIS-Elements AR 5.20.02 64-bit; Nikon). Furthermore, image analyses of acquired images are performed on the NIS-Elements software.

## 3 Methods

### 3.1 Routine Subculture of Tobacco BY-2 Cells

Subculture 1.5–2 mL of 7-day grown tobacco BY-2 cells to 80 mL of fresh medium at 1-week intervals. 4-day grown cells are used for transformation (*see* **Note 3**). Because the timing of co-cultivation with *Agrobacterium* is very important, the subculture day in the week should be scheduled.

### 3.2 Transformation

1. Transform agrobacterium LBA4404 competent cells with the plasmid pCAMBIA1300-H2BmCherry-mCitrineTUB8 by electroporation and spread on LB agar plate containing kanamycin. Incubate for 3 days at 28 °C. Confirm colony formation. Store the plate in a refrigerator until use (*see* **Note 4**).
2. (at day 0) Add 1.5 mL of 7-day old (i.e., stationary phase) BY-2 cells to 80 mL fresh medium, as usual subculture.
3. (at day 3) Inoculate *Agrobacterium* to 1 mL of YEB medium containing kanamycin. Culture *Agrobacterium* under shaking at 28 °C for 18–24 h. Store the *Agrobacterium* plate in a refrigerator until ensuring the successful transformation of BY-2 cells.
4. (at day 4) Transfer 4 mL of the 4-day grown BY-2 suspension to an empty petri dish. Add 100 mL of the *Agrobacterium* culture. Make sure bacterial growth before adding the culture. Mix well and spread on the surface of the empty dish (Fig. 6a). Seal with surgical tape. Incubate for 2 days at 26 °C (*see* **Note 5**).
5. (at day 6) After a 2-day culture, cells proliferate in the petri dishes (Fig. 6b). Harvest the culture with 3% sucrose into a 15-mL plastic tube. Brake down cell clusters by gentle pipetting. Wash 3 times with 3% sucrose by repeated centrifugation (160× *g*, 3 min), followed by resuspension with 3% sucrose.
6. For a 4 mL culture of BY-2 cells, a cell pellet of 1.5 mL should be obtained. Add half of the volume (i.e., 0.75 mL) of 3% sucrose.
7. Take 0.5 mL of the diluted pellet and spread by gentle shaking on the BY-2 agar plate with antibiotics (Fig. 6c). Seal the plate with surgical tape.
8. (at day 7: optional) Observe the plate with a dissecting fluorescence microscope with an RFP filter. If transformation is successful, several red spots of the transformed nucleus should be observed (*see* **Note 6**).
9. (at days 17–20) After 2 weeks of culture on the antibiotic selection medium, several colonies should appear (Fig. 6d, *see* **Note 7**). Transfer the colonies on the agar plate to liquid culture. Take one-third of the colonies from the plate with a

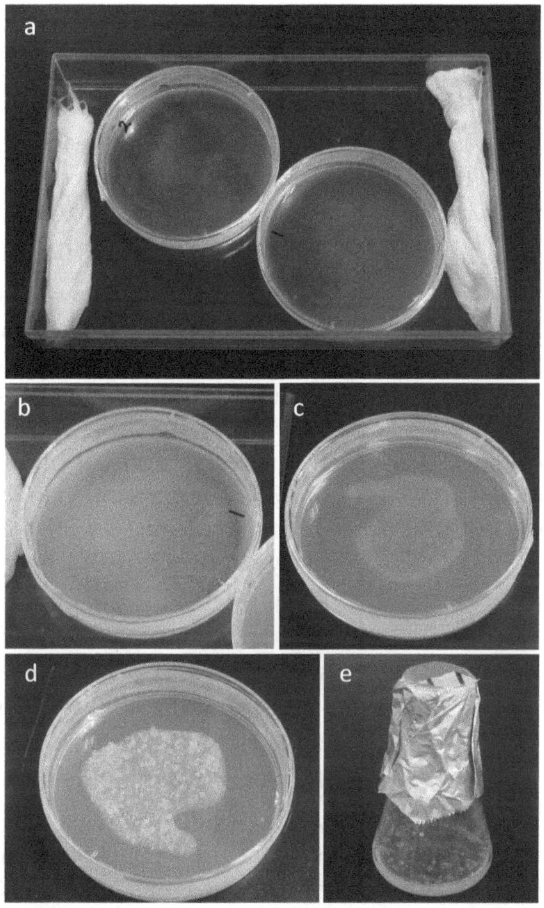


**Fig. 6** Transformation of tobacco BY-2 cells. (**a**, **b**) Petri dishes at the start (**a**) and the end (**b**) of co-culture of BY-2 cells and *Agrobacterium*. (**c**, **d**) Selection of agar plates just (**c**) and 2 weeks (**d**) after spreading. (**e**) Colonies resuspended into 15 mL of medium with carbenicillin and hygromycin

sterilized spatula (*see* **Note 8**) and put them into the 100 mL sterilized flask filled with 15 mL of antibiotics-containing BY-2 medium (Fig. 6e).

10. (approximately 1 month after the start) After 5–7 days of **step 9**, the suspension becomes uniform and dense. Transfer 2 mL of the suspension to 40 mL of the antibiotics-free medium and continue subculture.
11. After 5–7 days of **step 10**, the cells can be subcultured as untransformed cells and are ready to use.

### 3.3 Subculture for Observation of Mitosis

1. Tobacco BY-2 cells expressing fluorescent fusion proteins are subcultured at 1-week intervals by adding 2 mL of the culture to 80 mL of fresh medium. After subculturing to a fresh medium, the cells start dividing. The cells continuously divide

for 3–4 days (log phase), after which the division activity gradually ceases, and elongation becomes dominant (stationary phase). In the log phase, cells tend to grow as a cluster, and conditions for microscopy are suboptimal. We use cells one day after subculture; there is a high frequency of division and good conditions for microscopy.

2. Transfer 1.5 mL of a 7-day-old suspension to 15 mL of fresh culture medium in a 100 mL flask and culture for 18–20 h.
3. As an option, the 7-day-old culture can be stored in a refrigerator for 1 or 2 days. The stored suspension is similarly used for the subculture of observation. Longer storage, up to 3 days, is possible, but the timing of cell division tends to be delayed.

### 3.4 Attachment of Cells onto a Poly-lysine-Treated Glass-Bottom Dish

1. It is necessary to perform hydrophilic treatment of the surface of a 35-mm-diameter glass-bottom dish. Put the dish in a glow discharge plasma for 30–60 s. Exposure to the plasma effectively removes hydrocarbons and other contaminants from the glass surface.
2. Add approximately 2 μL of poly-lysine solution. If the glass surface is very clean, the solution will spread by gravity (Fig. 7a). If the solution is not spread well, spread the solution with the tip of the micropipette.
3. Wait until the solution is air-dried.
4. Wash the glass surface with distilled water flows more than 5 times. After the washes, air-dry. The glass is ready for attachment. Use within a day (*see* **Note 9**).
5. Mix 0.5 mL of the cell suspension with 5 mL of 3% sucrose in water in a 50-mL Falcon tube. Plastic micropipette tips are used for handling cells, and approximately 5 mm has been cut off from the end.
6. Take 0.5 mL of the mixture and pipette vigorously onto the surface of the poly-lysine-coated culture dish. Repeat 3–4 times

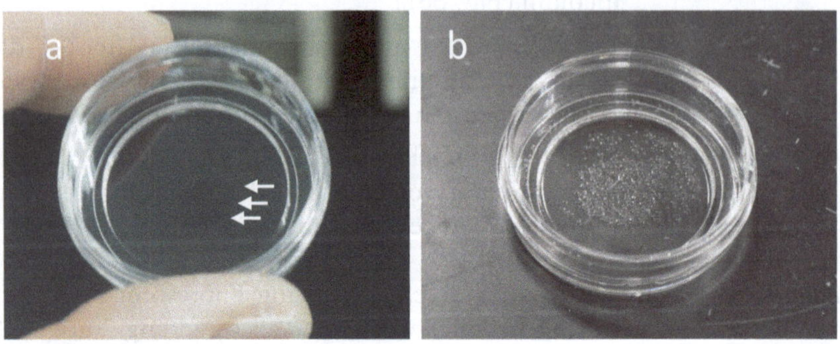


**Fig. 7** A glass-based dish during poly-L-lysine spreading (**a**) and cells attached on a glass-based dish (**b**). Arrows in (**a**) indicate poly-L-lysine solution spread on the glass surface

to attach sufficient cells onto the glass surface. Make sure the cells are attached to the surface of the glass (Fig. 7b).

7. Remove the excess solution and add 0.7 mL of fresh BY-2 medium. The cells are ready to observe.

### 3.5 Optical Setting of the Microscope

1. Before the start of the observation, set a dichroic mirror and fluorescence filters of the W-VIEW.
2. Adjust a correction ring of the objective lens to the position of 0.15 or 0.17 for No.1 or No.1S glass, respectively (*see* **Note 10**).
3. Drop distill water onto the objective lens and set a glass-based dish with cells.
4. Search a "test" cell for preliminary observations to optimize the optics and laser light conditions under wide-field fluorescent excitation and ocular observation.
5. Select a laser light wavelength of 960 nm or 1040 nm (*see* **Note 11**). For 960 nm, set the MaiTai wavelength at 960 nm by the MaiTai software and open the motorized shutter for MaiTai by the NIS-Elements software. For 1040 nm, the motorized shutter for FemtoTrain is also opened.
6. Set the laser beam power by a rotation of half-wave plates (HWP). Recommended starting values are 66 and 10% for 960 and 1040 nm, respectively.
7. As an option, insert the microscope's magnification optics (1.5×). The Nikon inverted microscope has internal magnification optics. The use of the optics increases the resolution of images by narrowing a relative pinhole size, not merely increasing magnification. Bright samples are required to use the option.
8. Open the shutter and register the *xy*-position of two-channel (mCitrine and mCherry) images by adjustment screws on the W-VIEW optics.
9. Take a few *z*-stack and check photobleaching. If fluorescent signals quickly decline, decrease the laser beam power and try again. If the image intensity is too low, use a binning mode of the EM-CCD camera (binning x2 means that 4 neighboring pixel signals are integrated into one pixel together) and increase exposure time.

### 3.6 Acquiring Timelapse Images

1. Search a dividing cell for observation under fluorescent excitation. For good image quality, choose cells close to the surface of the glass-based dish.
2. Set a Z-stack condition by "ND acquisitions" function on the NIS-Elements software. Set Z-interval at 0.5 μm. This value is satisfied the Nyquist sampling rate in this optical condition.

3. Set the top and the bottom ends of *z*-optical sectioning. Usually, the top is above the cell top and the bottom is below the cell bottom. If the acquisition time for a single *z*-series (Z-stack) is over 1 min, reduce the Z-range.
4. Set also the time intervals and number of repeats by "ND acquisitions" function. Then, start the image acquisition.
5. After the completion of image acquisition, save the image to the data storage. Images are not saved automatically in the software's default settings (automatic saving can be set as an option). The image data contain only partial imaging parameters (objective lens, camera setting, and *z*-step size). It is recommended that the pixel size, excitation wavelength information of dichroic mirrors, fluorescence filters, etc., be recorded as a part of the filename: e.g., "100nmpix_960nm_DM560_BP525_50."

### 3.7 Viewing 3D Stacks

1. After the completion of image acquisition, images are displayed on the NIS-Elements software.
2. Check the volume view using "Show Volume View" function of the software. An example of a volume view is illustrated in Fig. 8.

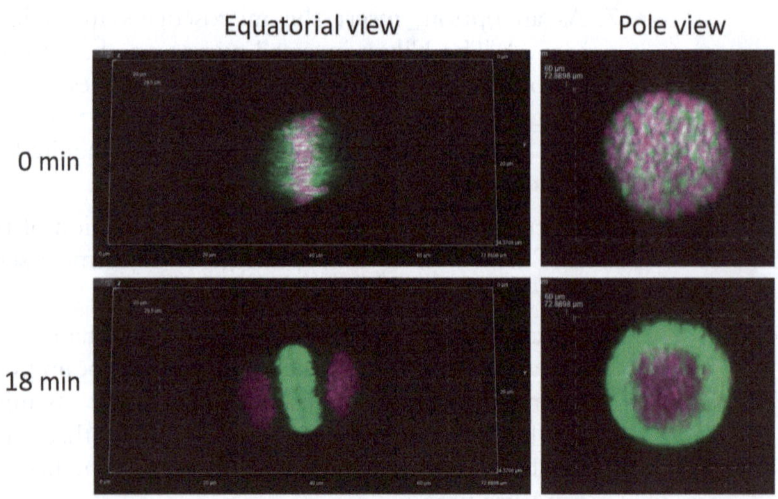


**Fig. 8** Images of dividing tobacco BY-2 cells taken using the two-photon spinning-disk microscope. 3D images from stacks were reconstructed using the NIS-Elements software. Equatorial view: reconstructed image from the side of the cylindrical cell. Pole view: reconstructed image from the top of the cylindrical cell. 0 min: just before the onset of anaphase. 18 min: late stage of cytokinesis depicting the cytokinetic apparatus phragmoplast

## 4 Notes

1. Autoclaving for a longer duration of 20 min often causes a change in the medium color to brown.
2. The survival of BY-2 cells on agar plates depends on the quality of agar. Cells on the agar plate may die at low density culture if the quality of agar is poor. We use Agarose ME (purified agarose for electrophoresis).
3. The transformation rate depends on the duration of the pre-culture before mixing with *Agrobacterium*. Cells cultured for 3 days or shorter time are not competent for *Agrobacterium* transformation as shown by An (1985) [14].
4. The *Agrobacterium* plate can be stored for a month.
5. To prevent drying of the dish, the plate should be kept in a container with a moistened paper towel.
6. For the DNA construct used in this protocol, bright spots of nuclei should be observed. However, no detectable fluorescence may be observed for the constructs of genes driven by native promotors. For such cases, continue culturing and check the fluorescence after preparing the suspension culture (**step 10**).
7. If no colonies appear on the plate, check the growth conditions of BY-2 cells. If the cells are contaminated or stressed for some unknown reason, they must not be transformed. Alternatively, mutation of plasmids/*Agrobacterium* is one possibility for the failure of transformation. For such cases, culture the different colonies of *Agrobacterium* on the plate and try transformation again.
8. We do not isolate a single colony of the BY-2 cells. Analysis of clones may be dangerous because of unanticipated mutations in the clone. This problem can be overcome by analyzing the cell population containing several clones.
9. The poly-lysine treated dish can be stored before washing with distilled water. The poly-lysine-treated air-dried dish should be stored in a freezer before the washing step.
10. For focusing water-immersion lenses, glass-thickness correction is essential. Mis-setting of the correction ring increases spherical aberration and causes a decrease of fluorescent signals.
11. The 960-nm wavelength is more effective in exciting mCitrine than exciting mCherry, and vice versa. To obtain a high S/N image of mCitrine, we recommend 960 nm. To obtain a high S/N image of mCherry, 1040 nm is better.

## Acknowledgments

We thank Dr. Kentaro Kobayashi (RIES, Hokkaido University) for technical support and helpful discussion. The plasmid used for the protocol was built by Mss. Yoko Matsuzaki and Mika Ikeda in Hasebe laboratory, National Institute of Basic Biology, Japan. This work is supported by Collaborate Research Laboratory (CORE-Lab), Network Joint Research Center for Materials and Devices (Grant number 20166001), and in part by Planned collaborative project, National Institute for Physiological Science (Grant numbers 22NIPS241, 23NIPS232), General collaborative project, National Institute for Physiological Science (Grant number 20-122), General Joint Research, Exploratory Research Center on Life and Living Systems, National Institute of Natural Sciences (Grant number 21-410), Japan Science and Technology Agency (JST) CREST (Grant number JPMJCR20E4) and JSPS KAKENHI (Grant numbers JP18H02471, JP19K15406, JP20H05669, JP22H02756, JP22K14578, JP23K05835).

### References

1. Wasteneys GO (2002) Microtubule organization in the green kingdom: chaos or self-order? J Cell Sci 115:1345–1354
2. Ho C-MK, Lee Y-RJ, Kiyama LD et al (2012) Arabidopsis microtubule-associated protein MAP65-3 cross-links antiparallel microtubules toward their plus ends in the phragmoplast via its distinct C-terminal microtubule binding domain. Plant Cell 24:2071–2085
3. Murata T, Sano T, Sasabe M et al (2013) Mechanism of microtubule array expansion in the cytokinetic phragmoplast. Nat Commun 4: 1967
4. Hiwatashi Y, Obara M, Sato Y et al (2008) Kinesins are indispensable for interdigitation of phragmoplast microtubules in the moss *Physcomitrella patens*. Plant Cell 20:3094–3106
5. Svoboda K, Yasuda R (2006) Principles of two-photon excitation microscopy and its applications to neuroscience. Neuron 50:823–839
6. Koho SV, Slenders E, Tortarolo G et al (2020) Two-photon image-scanning microscopy with SPAD array and blind image reconstruction. Biomed Opt Express 11:2905–2924
7. Shimozawa T, Yamagata K, Kondo T et al (2013) Improving spinning disk confocal microscopy by preventing pinhole cross-talk for intravital imaging. Proc Natl Acad Sci USA 110:3399–3404
8. Otomo K, Hibi T, Murata T et al (2015) Multipoint scanning two-photon excitation microscopy by utilizing a high-peak-power 1042-nm laser. Anal Sci 31:307–313
9. Kamada T, Otomo K, Murata T et al (2022) Low-invasive 5D visualization of mitotic progression by two-photon excitation spinning-disk confocal microscopy. Sci Rep 12:809
10. Drobizhev M, Tillo S, Makarov NS et al (2009) Absolute two-photon absorption spectra and two-photon brightness of orange and red fluorescent proteins. J Phys Chem B 113:855–859
11. Drobizhev M, Makarov NS, Tillo SE et al (2011) Two-photon absorption properties of fluorescent proteins. Nat Methods 8:393–399
12. Nagata T, Nemoto Y, Hasezawa S (1992) Tobacco BY-2 cell-line as the Hela-cell in the cell biology of higher-plants. Int Rev Cytol 132:1–30
13. Kimata Y, Yamada M, Murata T et al (2023) Novel inhibitors of microtubule organization and phragmoplast formation in diverse plant species. Life Sci Alliance 6:e202201657
14. An GH (1985) High-efficiency transformation of cultured tobacco cells. Plant Physiol 79: 568–570

# Chapter 4

# Spatial Statistics of Three-Dimensional Growth Dynamics of Spindle Microtubules

**Norio Yamashita, Masahiko Morita, Hideo Yokota, and Yuko Mimori-Kiyosue**

## Abstract

The latest high-resolution 3D live-cell imaging technology, lattice light-sheet microscopy (LLSM), has successfully tracked the dynamics of microtubule growth throughout the entire mitotic spindle with unparalleled precision. By using green fluorescent protein-labeled end-binding protein 1 (EB1-GFP) as a marker for growing microtubule ends, LLSM has generated an extensive collection of multidimensional datasets mapping the positions and trajectories of these growing microtubule ends. Processing this data requires statistical analysis in three-dimensional space. This chapter describes the spatial statistical methods developed for this purpose, illustrated with practical examples. Finally, we discuss future prospects for analyzing complex, large-scale image data.

**Key words** Lattice light-sheet microscopy, LLSM, Mitosis, Spindle, Microtubule dynamics, EB1, Spatial statistics

## 1 Introduction

Microtubules are dynamic structures that continuously undergo cycles of growth and shortening [1, 2]. This dynamic behavior allows for the rapid formation of patterns with different orientations along the polarization axis of the cell in response to environmental changes [3], the formation of structures such as mitotic spindles [4, 5], and the generation of the pulling force needed to separate chromosomes into two daughter cells during cell division [6, 7]. However, until very recently, the microtubule dynamics could not be tracked in three dimensions because conventional imaging techniques lacked sufficient spatiotemporal resolution [8].

Lattice light-sheet microscopy (LLSM) [9] has dramatically improved spatiotemporal resolution, enabling the three-dimensional (3D) tracking of dynamic cellular structures such as

Tomomi Kiyomitsu (ed.), *The Mitotic Spindle: Methods and Protocols*, Methods in Molecular Biology, vol. 2872, https://doi.org/10.1007/978-1-0716-4224-5_4, © The Author(s) 2025

microtubules and organelles [10]. LLSM leverages the advantages of light-sheet microscopy, including high-speed acquisition and non-invasiveness [11]. Using Bessel beam technology [12], it generates an ultra-thin light sheet (thickness < 1 μm) by arranging many such beams in a two-dimensional lattice [9, 13]. This enables 3D scanning of cells with high-spatiotemporal resolution, particularly enhancing the resolution along the optical axis and obtaining highly accurate positional information in all spatial directions [10, 14]. This makes LLSM a state-of-the-art technology that can capture the 3D location and dynamics of intracellular structures with unparalleled precision.

Using LLSM, we visualized 3D microtubule growth dynamics at the whole-cell level by imaging end-binding protein 1 (EB1), a microtubule plus-end tracking protein, fused to green fluorescent protein (EB1-GFP) [15]. EB1-family proteins fused to GFP have been shown to bind selectively to the growing ends of microtubules, demonstrating the utility of LLSM as an analytical tool for the study of microtubule dynamics in cells and animals [16–20]. At growing microtubule ends, EB1-GFP forms a comet-shaped pattern with a short-axis diameter of approximately 25 nm and an elongated tail of up to 1000 nm parallel to the long axis [15]. Time-lapse imaging and tracking of EB1-GFP comets allow the generation of microtubule growth trajectories and measurement of the growth speed [21], but such methods were limited to two dimensions until the advent of LLSM.

LLSM enables 3D tracking of EB1-GFP dynamics. However, the performance of LLSM has generated an overwhelming amount of multidimensional data, making the development of computational techniques for data analysis an urgent challenge. Therefore, we developed algorithms to spatially analyze the dynamics of individual spindle microtubules [8]. Using this technique, we have succeeded in characterizing microtubule dynamics and arrangement patterns that have been difficult to analyze by human observation and manipulation. For example, we determined the 3D distribution of the initial microtubule elongation and the direction of microtubule growth inside the spindle, where the density of microtubules is so high that it is difficult to distinguish individual microtubules. If the microtubule growth axis does not project toward the centrosome, it is assumed to be a non-centrosomal microtubule. LLSM enables quantitative analysis of the ratio of centrosomal and non-centrosomal microtubules, which has long been a matter of debate because of the difficulty of direct observation in cells [22]. Such examples demonstrate the high potential of spatial statistics to analyze digital data extracted from LLSM images with highly accurate information.

In the following, we describe step-by-step the methods we have developed, outline their applications, and provide some updates.

## 2 Materials

### 2.1 Cells

1. HeLa(A1) cells expressing EB1-GFP [15] and histone H2B protein fused to red fluorescent protein (H2B-tagRFP), which can be prepared as described previously [9] (*see* **Note 1**).
2. HeLa(A1)-AURKA cells expressing EB1-GFP and H2B-TagRFP, with myc-tagged aurora kinase A (myc-AURKA) inserted into the pLVSIN-EF1α-IRES-bsr plasmid [23]. AURKA cDNA can be obtained from Addgene, MA, USA (pWZL Neo Myr Flag AURKA, Addgene plasmid #20427; http://n2t.net/addgene:20427; RRID: Addgene_20427 [24]).
3. HeLa(A1) cells can be prepared by introducing H2B-tagRFP into HeLa cells expressing EB1-GFP (clone 2F10), with exogenous EB1-GFP expression levels at approximately 40% of endogenous EB1 [23]. In HeLa(A1)-AURKA cells, myc-AURKA expression level is about four times higher than endogenous AURKA.

### 2.2 LLSM Imaging

LLSM imaging was performed in the Betzig Laboratory at the Janelia Research Campus of the Howard Hughes Medical Institute, USA, as described elsewhere [9].

### 2.3 Computation

1. IMARIS software (Bitplane/Oxford Instruments, Abingdon, UK) for measurements and viewing multidimensional images.
2. VCAT5 software for image preprocessing [25]. This is developed by the RIKEN Image Processing Research Team and is available from http://logistics.riken.jp/vcat/vcat/en.
3. MATLAB software (The Mathworks Inc., Natick, Massachusetts) for code generation and image processing.

## 3 Methods

### 3.1 LLSM Imaging

1. Set the coverslip with seeded cells into the sample holder and maintain it at 37 °C during observation.
2. Collect data in dual-color and single-color images in a high-speed dithered mode at intervals of 1.510 s and 0.755 s, respectively.
3. Convert 3D datasets acquired by sample scanning in $x$, $y$, $s$ (skew) coordinates into conventional $x$, $y$, $z$ coordinates ("deskewed") and process using deconvolution operations with experimentally measured point spread functions for each emission wavelength before visualization (*see* **Note 2**). The voxel size depends on the device state, and here it is calculated to be $0.100 \times 0.100 \times 0.217\ \mu m^3$ based on actual measurements.

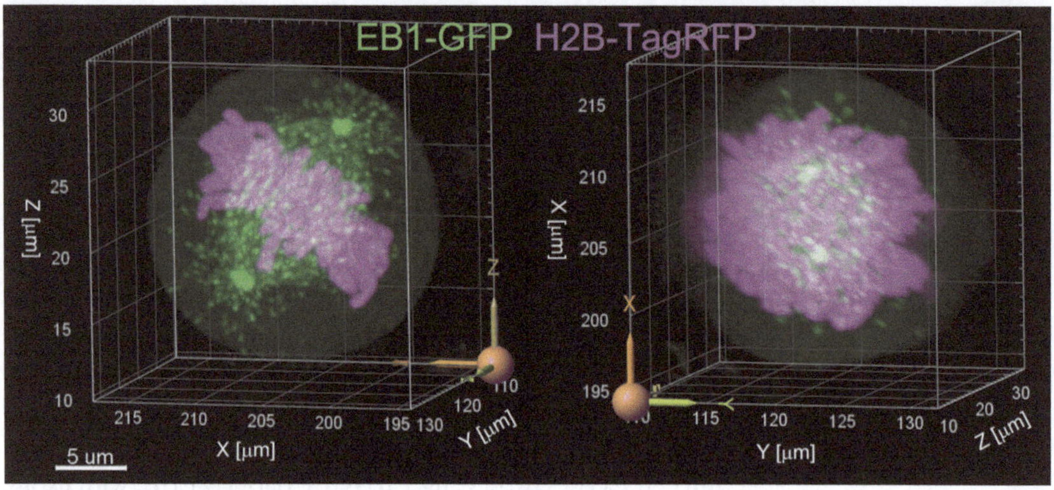


**Fig. 1** Example of a LLSM image. The image was imported into the IMARIS software and displayed in three-dimensional (3D) coordinates. The left image was rotated as in the direction of the arrows representing the *x*, *y*, and *z* axes

4. Identify phases in mitosis by the shape of the chromosomes.

### 3.2 Image Visualization

1. Import LLSM time-lapse imaging data (stored as individual multi-stack images in tagged image file format [TIFF] with sequential numbers added for each time point) into multidimensional image processing software for visualization. We use IMARIS software for this purpose.
2. Edit the voxel size and time intervals using the Edit/Image Properties function.
3. Visualize 3D images from any direction and through any slice, and make various measurements. Scale bars and time stamps can be displayed in any size and location (Fig. 1).
4. The original image or the measurement results can be exported after editing as a snapshot, a time-series movie, or comma-separated values (CSV) files of numerical data (*see* **Note 3**).

### 3.3 Image Preprocessing

Depending on the purpose, preprocessing such as cropping, bleach correction, and object motion correction is applied to facilitate highly accurate data analysis (*see* **Note 4**).

1. *Drift correction of spindle position*. It is necessary to correct for changes in spindle position and angle to compare all datasets in a common frame of reference because the spindle moves within the cell over time. Track the positions of the two centrosomes automatically and correct their displacements for the whole dataset using the Correct Drift function (Fig. 2). Export the drift-corrected data as sequentially numbered TIFF files for further processing.

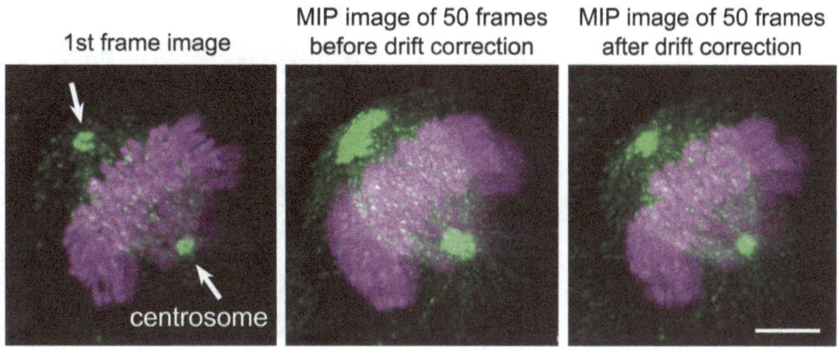


**Fig. 2** Effects of drift correction. Time-lapse LLSM images (1.510 s interval, 50 frames) of a HeLa cell (clone A1) were imported into IMARIS and displayed in 3D coordinates. The first frame is shown in the left panel. The center and right panels show the maximum intensity projection (MIP) of all 50 frames before and after drift correction, respectively. The spindle apparatus can often rotate in any direction in cells, as shown in the center panel. This rotational drift was corrected for by using the positions of the two centrosomes as a reference (right). Scale bar: 5 μm. (Reprinted with permission from Ref. [8])

2. *Intensity equalization.* Using VCAT5 software, correct fluctuations in fluorescence intensity caused by photobleaching of fluorescent proteins and unevenness due to various factors at the time of image acquisition (*see* **Note 5**). Normalize the intensity of a 3D image at each time point using the average voxel intensity in the whole image volume as the normalization factor.
3. *Signal enhancement.* Enhance EB1-GFP comet signals by employing a top-hat transformation [26] (Fig. 3; *see* **Note 6**).

### 3.4 Automated Tracking of EB1-GFP Comet Movement

1. Import the preprocessed images into IMARIS and edit the pixel size and time parameters.
2. Automatically detect EB1-GFP comets using the local contrast mode and subsequently execute automated tracking using the Autoregressive Motion algorithm, which predicts whether the spot will again move the same distance and in the same direction (Fig. 4).
3. To eliminate possible tracking errors, apply several filter algorithms, including Max speed, track duration, and track straightness, on the assumption that microtubule growth is essentially straight, with a variable but limited elongation rate. Adjust the parameters in different mitotic phases. When there are few track points, such as only two, it is difficult to find continuity if the tracks are crowded, which can easily result in connection errors. Therefore, trajectories with three or fewer consecutive points were excluded. Correct false-positive detection manually by comparing the original image with the tracking result using

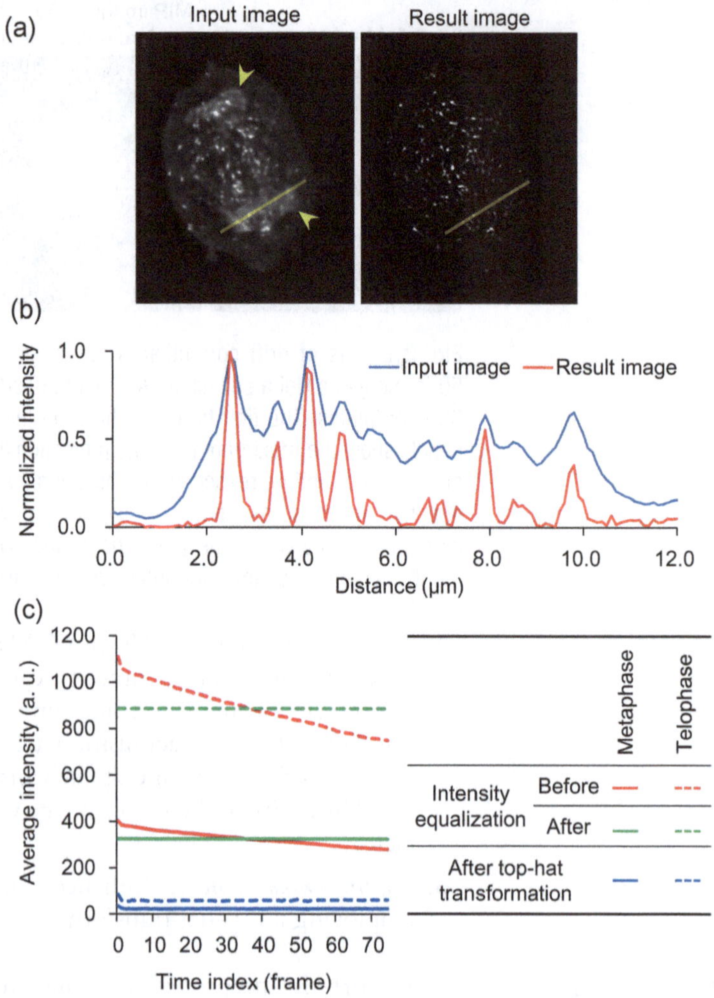


**Fig. 3** Effects of image preprocessing. After intensity equalization, the images were enhanced using a top-hat transformation. (**a**) Example of the resulting image. Arrowheads indicate the bleed-through from the red channel (red-fluorescent protein-tagged histone H2B). (**b**) Line profiles (yellow lines in panel a) of signal intensity (three-pixel linewidth). (**c**) Effect of intensity equalization and top-hat transformation on the 3D volume intensity. Average 3D volume intensities at each time point before and after preprocessing are plotted for a metaphase cell and a telophase cell. For top-hat transformation, because the area occupied by a single EB1-GFP comet ranges from $3 \times 3$ pixels$^2$ for a small comet to $7 \times 10$ pixels$^2$ for the largest comets, use an opening radius of $R = 2.0$ pixels in the *xy* plane and $R = 1.0$ pixels in the *z* direction, taking into account the 3D voxel pitch ($0.100 \times 0.100 \times 0.217$ μm$^3$). (Reprinted with permission from Ref. [8])

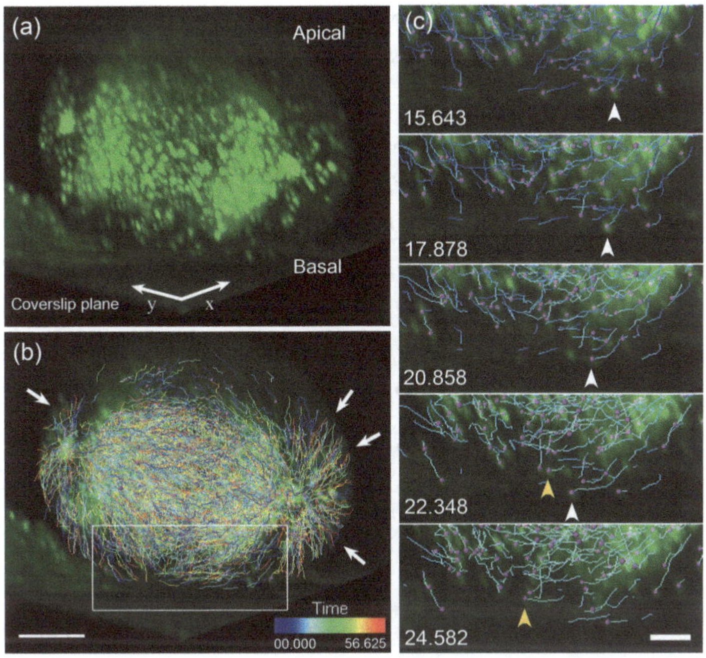


**Fig. 4** Visualization of interactions between EB1-GFP comets and the cell cortex. (**a**) Side view of an anaphase cell. The *xy* plane of the coverslip is indicated. (**b**) Superimposition of EB1-GFP trajectories generated by automated 3D comet tracking. Arrows indicate trajectories that disappeared shortly after reaching the cortex. Color bar: time (s). Scale bar: 5 μm. (**c**) Enlarged time-lapse images of the region enclosed by the white rectangle in panel b. Some trajectories run along the basal cell cortex (white and yellow arrowheads). Time (s) is shown. Scale bar: 2 μm. (Reprinted with permission from Ref. [8])

the manual editing options (deletion, creation, connection, and disconnection) (*see* **Note 7**).

4. Exclude the first three and last three frames from further statistical analysis because short tracks lasting less than three frames are removed during the filtering process. Excluding these frames also avoids misidentification of growing microtubules as new microtubules in the first frame.
5. Export parameters, such as the coordinates of individual spots, trajectory direction, length, and velocity, in CSV format for further computational analysis.

### 3.5 Spatial Statistics Analysis of EB1-GFP Comet Movement

The dynamic and spatial properties of EB1-GFP trajectories can be further analyzed using computational techniques, and the calculated results can be geometrically represented in 3D space by developing a pipeline implemented in MATLAB software.

1. Notation

   $N$: Number of EB1-GFP comet

   $t$: Time index (frame number)

   $\Delta T$: Time interval

   $P_i$: EB1-GFP comet ($i = 1, 2, 3 \ldots, N$)

   $Q_j$: Centrosome ($j = 0, 1$)

   $\mathbf{p}_i(t)$: Position of EB1-GFP comet $P_i$ at time $t$ ($i = 1, 2, 3 \ldots, N$)

   $\mathbf{q}_j(t)$: Position of centrosome $Q_j$ at time $t$ ($j = 0, 1$)

   $\mathbf{v}_i(t)$: Velocity of EB1-GFP comet $P_i$ at time $t$ ($i = 1, 2, 3 \ldots, N$)

   $\mathbf{P}_i$: Trajectory of EB1-GFP comet $P_i$

   $t_i^{\text{start}}$: Start time index of trajectory $\mathbf{P}_i$

   $t_i^{\text{end}}$: End time index of trajectory $\mathbf{P}_i$

   $t_i^{\text{far}}$: Time index at which the displacement of trajectory $\mathbf{P}_i$ from the centrosome is maximal.

2. Classification of trajectories according to mean speed

   Calculate the velocity $\mathbf{v}_i(t)$ of each EB1-GFP comet $P_i$:

$$\mathbf{v}_i(t) = \left[\mathbf{p}_i(t+1) - \mathbf{p}_i(t)\right]/\Delta T \qquad i = 1, 2, 3 \ldots, N; \quad t = t_i^{\text{start}}, t_i^{\text{start}} + 1, \ldots, t_i^{\text{end}} - 1.$$

Classify the trajectory $\mathbf{P}_i$ of each EB1-GFP comet $P_i$ according to its mean speed, defined as:

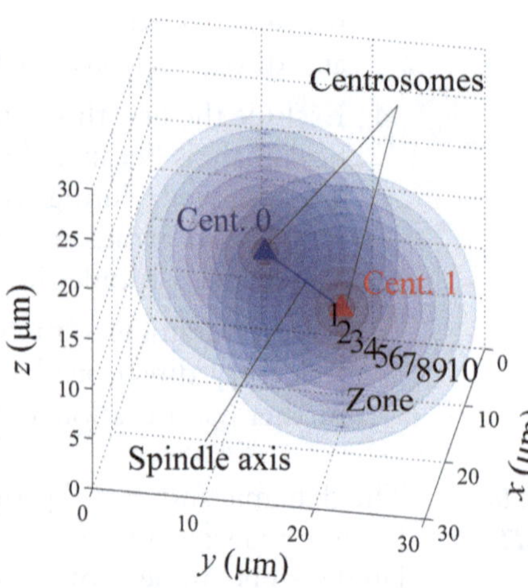


**Fig. 5** Illustration of the method used to analyze the distribution of EB1-GFP trajectories in 3D space. Spheres centered at centrosomes (red and blue triangles) and with radii equal to the intercentrosomal distance were generated and then divided into 10 equally sized spherical zones

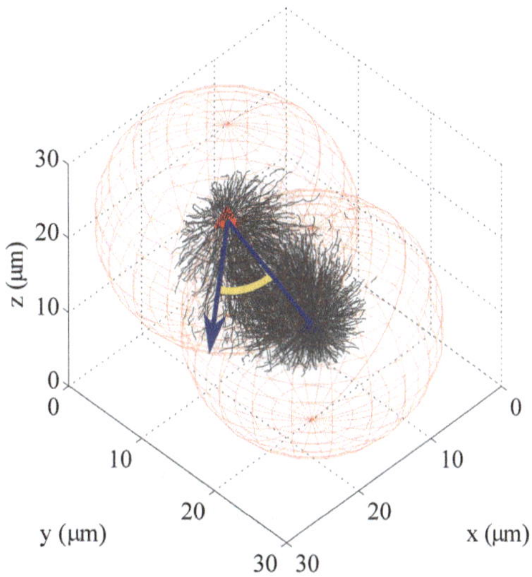


**Fig. 6** Definition of the centrosomal axis. The centrosomal axis is indicated with a blue line. The angle (yellow arc) between each EB1-GFP comet trajectory (blue arrow) and the centrosomal axis was calculated. (Reprinted with permission from Ref. [8])

$$v_{i\_\text{mean}} = \sum_{t=t_i^{\text{start}}}^{t_i^{\text{end}}-1} \frac{|\mathbf{v}_i(t)|}{t_i^{\text{end}} - t_i^{\text{start}}}.$$

3. Start and reach positions of EB1-GFP comet trajectories

Analyze the start position $\mathbf{p}_i(t_i^{\text{start}})$ and the reach position $\mathbf{p}_i(t_i^{\text{far}})$ of each EB1-GFP trajectory $\mathbf{P}_i$. The reach position is defined as the farthest point of a trajectory from the centrosome $Q_j$, closest at $t_i^{\text{start}}$ ($|\mathbf{p}_i(t_i) - \mathbf{q}_j(t_i)|$ is a maximum at $t_i^{\text{far}}$ for trajectory $\mathbf{P}_i$). Group the results into spherical zones on the basis of the start position or the reach position (Fig. 5), and view them using a combined line graph and histogram that is normalized by the total trajectory number.

4. Direction of instantaneous EB1-GFP trajectory

(a) *Definition of the intercentrosomal axis*

The straight line between the two centrosomes is defined as the intercentrosomal axis (Fig. 6). Derive the intercentrosomal axis vector as follows:

$$\mathbf{v}_{\text{cent}}(t) = \mathbf{q}_1(t) - \mathbf{q}_0(t)$$

(b) *Definition of the direction of movement*

Determine the angle between the direction of instantaneous EB1-GFP movement and the centrosomal axis, $\theta_i(t)$ ($0° \leq \theta_i(t) \leq 180°$). This can be obtained from the centrosomal axis vector, $\mathbf{v}_{\text{cent}}(t)$, and the velocity vector of

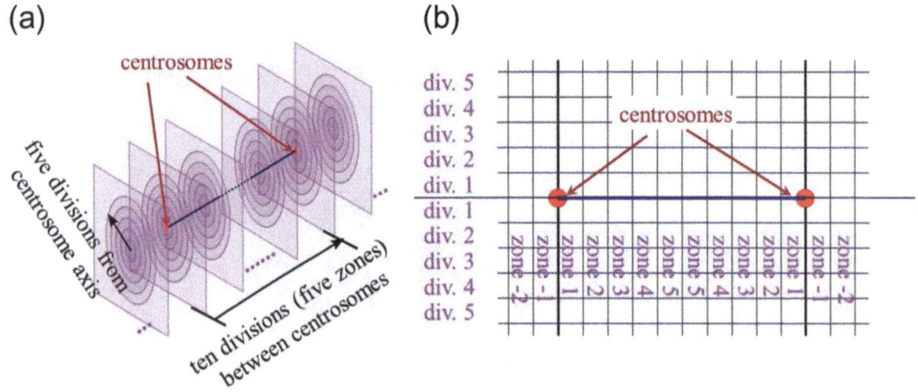


**Fig. 7** Illustration of the method used to analyze the distribution of EB1-GFP trajectories in 3D space. (**a**) The 3D space was divided vertically and cylindrically about the centrosomal axis into 70 bins. (**b**) The vertical zones and the cylindrical divisions were numbered as shown. (Reprinted with permission from Ref. [8])

the EB1-GFP comet, $\mathbf{v}_i\,(t)$, while taking into consideration the symmetry of the centrosome position:

$$\theta_i(t) = \arccos\left(\frac{\alpha \mathbf{v}_{\text{cent}}(t).\mathbf{v}_i(t)}{|\mathbf{v}_{\text{cent}}(t)||\mathbf{v}_i(t)|}\right) \quad \alpha = 1, if\ \left|\mathbf{p}_i(t_i^{\text{start}}) - \mathbf{q}_0(t_i^{\text{start}})\right| \leq \left|\mathbf{p}_i(t_i^{\text{start}}) - \mathbf{q}_1(t_i^{\text{start}})\right|$$

$$\alpha = -1, if\ \left|\mathbf{p}_i(t_i^{\text{start}}) - \mathbf{q}_0(t_i^{\text{start}})\right| > \left|\mathbf{p}_i(t_i^{\text{start}}) - \mathbf{q}_1(t_i^{\text{start}})\right|$$

(c) *Direction of instantaneous EB1-GFP movement in different regions*

Obtain the angle between the direction of EB1-GFP movement and the centrosomal axis, $\theta_i\,(t)$ $(i = 1, 2, 3\ldots, N)$, for all positions $\mathbf{p}_i\,(t)$ of EB1-GFP comets in 70 spatial bins (Fig. 7). First, vertically divide the intercentrosomal region into 10 equal zones and create an additional four zones outside the centrosomes (i.e., two zones outside each centrosome). Number each zone symmetrically around the center point (i.e., zones $5, 4, 3, 2, 1, -1, -2$). Next, divide the space around the centrosomal axis into five cylindrical zones with radial intervals equal in size to the intervals between the vertical zones. Number each division in order of increasing distance from the axis (i.e., divisions $1, 2, 3, 4, 5$).

### 3.6 Merging Multiple Datasets

The above describes the analysis of individual cell data, but for quantitative statistical analysis, multiple data must be merged, taking into account experimental error and biological variation. In this section, we describe how to merge individual datasets using the spindle structure as a reference frame.

1. Conversion of coordinates

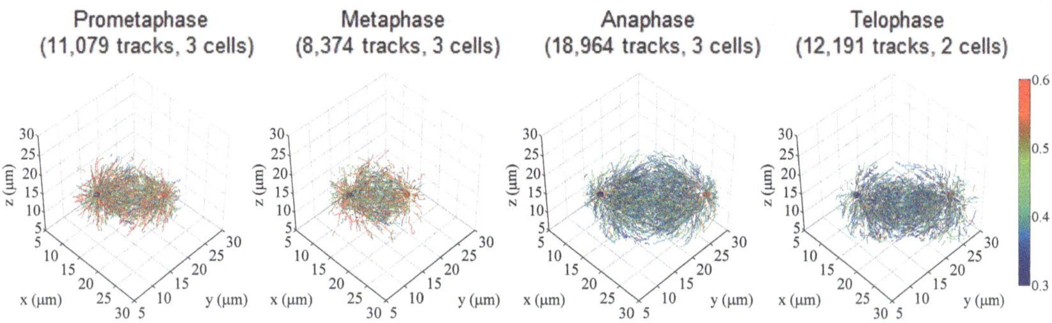


**Fig. 8** Merged trajectories in various mitotic phases. Blue and red triangles indicate centrosomes. Orange and cyan dots indicate the start and end positions of trajectories, respectively. The colored bar indicates the mean speed of EB1-GFP comet trajectories (0.3–0.6 μm/s). Trajectories after randomly pruning 13% (three leftmost panels) or 20% (rightmost panel) of total trajectories from the merged datasets

Convert the positions of EB1-GFP comets and centrosomes so that the position of the first centrosome (cent. 0) in the first frame comes to a designated position $\mathbf{O}_{\text{desig}} = (12.5, 12.5, 15.0)$ μm. Then, rotate the second centrosome (cent. 1) so that the spindle axis in the first frame aligns with the designated vector $\mathbf{e}_{\text{desig}} = \langle 1, 1, 0 \rangle / \sqrt{2}$. Repeat the same processing for all remaining frames. Converted positions of the comets ($\mathbf{p}_{i\_\text{conv}}(t)$) and centrosomes ($\mathbf{q}_{j\_\text{new}}(t)$) are represented as follows:

$$\mathbf{p}_{i\_\text{conv}}(t) = \mathbf{M}_{\text{rot}}(t)(\mathbf{p}_i(t) - \mathbf{q}_0(t)) + \mathbf{O}_{\text{desig}}$$

$$\mathbf{q}_{j\_\text{conv}}(t) = \mathbf{M}_{\text{rot}}(t)\left(\mathbf{q}_j(t) - \mathbf{q}_0(t)\right) + \mathbf{O}_{\text{desig}}$$

Here, $\mathbf{M}_{\text{rot}}(t)$ is a rotation matrix to align the direction of intercentrosomal axis vector $\mathbf{v}_{\text{cent}}\ (t)$ to the designated direction $\mathbf{e}_{\text{desig}}$. After the transformation, the centrosomal axis becomes

$$\mathbf{v}_{\text{cent\_desig}}(t) = \mid \mathbf{v}_{\text{cent}}(t) \mid \mathbf{e}_{\text{desig}}$$

2. Next, spatial statistics analysis described in Subheading 3.4 is conducted for the individual cell data in the converted coordinates.
3. Merging datasets for cells with various spindle sizes

   For the merging of $n$ cells, the length between centrosomes is normalized. Here, the transformed centrosomal axis vector of $k$th cell ($1 \leq k \leq n$) is denoted $\mathbf{v}^k_{\text{cent\_desig}}(t)$. For all comets in the cells to be merged, multiply the scale factor $\mid \mathbf{v}^k_{\text{cent}}(t) \mid / \mid \mathbf{v}^1_{\text{cent}}(t) \mid$ in the direction of the centrosomal axis vector $\mathbf{v}^k_{\text{cent\_desig}}(t)$ so that the length of the centrosomal axis is normalized. Because the normalization process changes the trajectory shapes and the distances between objects, features

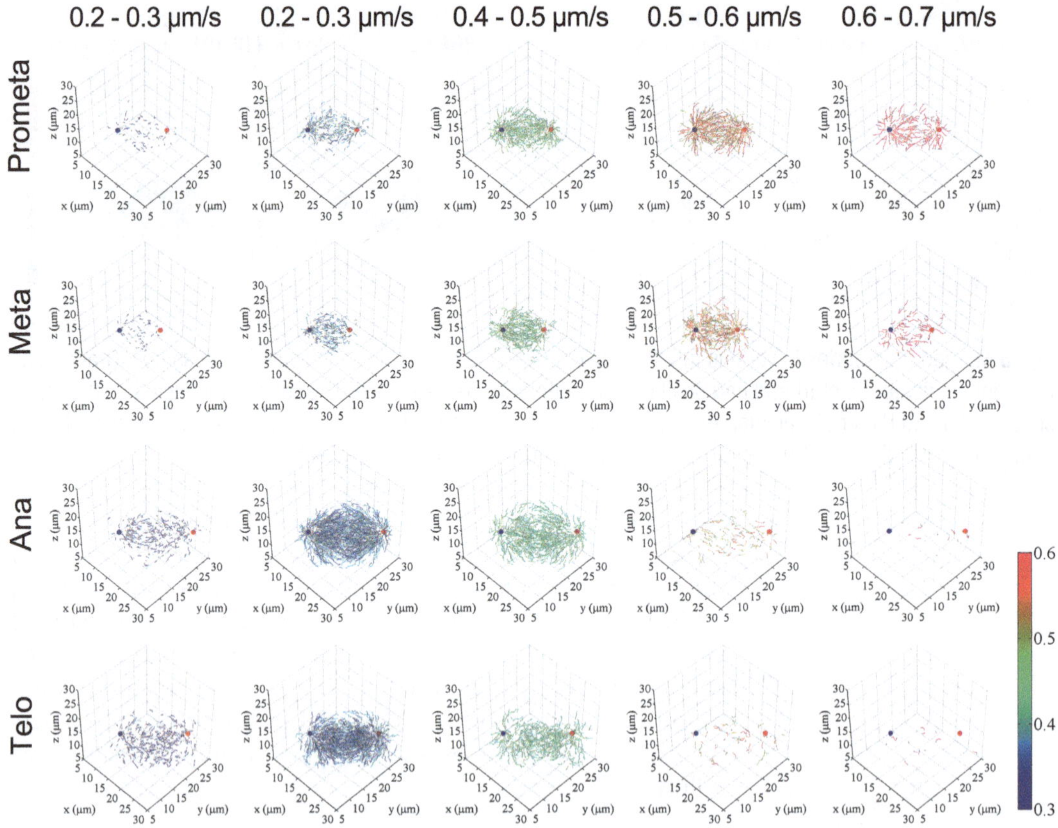


**Fig. 9** Merged trajectories were divided into 10 classes spanning the entire speed range (0–1 μm/s, 0.1 μm/s steps). Data included in the 0.2–0.7 μm/s range are shown after randomly pruning 40% of the trajectories from each cell. Blue and red triangles indicate centrosomes. Orange and cyan dots indicate the start and end positions of trajectories, respectively. The colored bar indicates the range of the mean speeds of EB1-GFP comet trajectories (0.3–0.6 μm/s)

are not recalculated; therefore, the comet speed, the mean speed of each trajectory, the growth angle of each comet relative to the spindle axis, and initial and reach zones of trajectories remain unchanged. Obtain the total number of comets in each zone by combining data from all cells.

4. Geometrical display of trajectories

   Display the merged datasets in 3D coordinates (Fig. 8). Since the amount of data will be very large, perform random data pruning, if necessary, to improve visibility.

5. Data classification and result representation to interpret the merged datasets

   (a) *Visualization of the speed distribution of EB1-GFP comet trajectories.* Divide merged trajectories into 10 classes spanning the entire speed range (0–1 μm/s, 0.1 μm/s steps). Prune data in some plots to improve visibility by reducing the number of trajectories shown (Fig. 9).

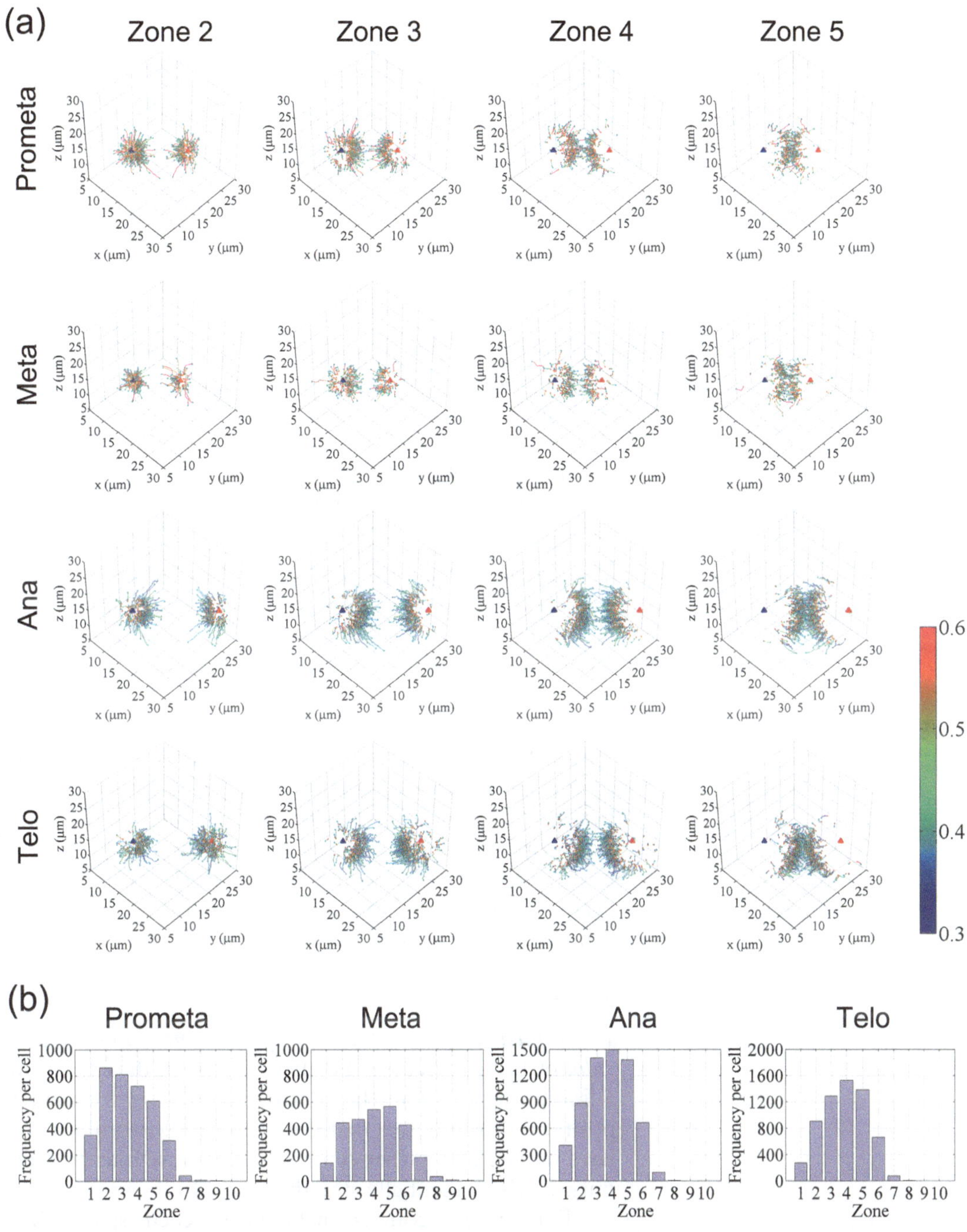


**Fig. 10** (**a**) Trajectories of EB1-GFP comets, grouped according to the position of the first track point and displayed separately for each indicated spherical zone. Orange and cyan dots indicate start and end positions, respectively. The colored bar indicates the mean speed (μm/s). (**b**) Histogram of the numbers of start points in each zone

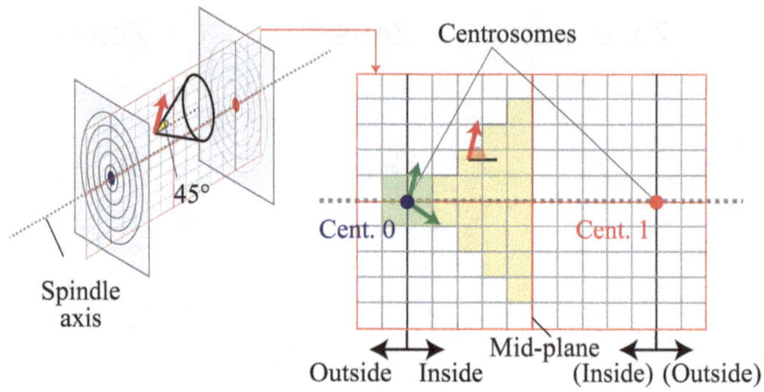


**Fig. 11** Method for detecting off-axis microtubule growth. Merged trajectories are divided vertically and cylindrically about the centrosomal axis into 70 bins, as shown in Fig. 7. If the angle between trajectory and the centrosomal axis is larger than 45° (red arrow) in the yellow-colored region, then the corresponding microtubule is unlikely to originate from the centrosomes, based on the assumption that microtubules grow in a straight line

(b) *Visualization of the distribution of EB1-GFP comet trajectories in different regions.* Set the spherical classification bin (zone) as shown in Fig. 5. For each zone, extract trajectories according to their start position and plot them in separate 3D coordinate systems (Fig. 10) (*see* **Note 8**).

*Classification of trajectories according to the angle of travel.* Attribute to each trajectory the centrosome closest to its start position (cent. 0 or cent. 1). Calculate the angle between the trajectory vector and the intercentrosomal axis. For "cent. 0" comets, define the direction of intercentrosomal vector as being from cent. 0 to cent. 1. For "cent. 1" comets, define the direction of the intercentrosomal vector as being from cent. 1 to cent. 0. Extract the comets in each spatial bin (Fig. 7), and analyze their travel angles.

(c) *Examination of the percentage of non-centrosomal microtubules.* If the direction of comet movement deviates from a line projected from the centrosome toward the comet position, this indicates that the microtubule is unlikely to have initiated from that centrosome because microtubule outgrowth is essentially straight (Fig. 11).

Extract the comets included in each spatial bin (Fig. 7) and display their travel angles in a histogram at 15° intervals (Fig. 12). These results indicate the subpopulation of microtubules that grow at an off-axis orientation and can be used to identify non-centrosomal microtubule nucleation (*see* **Note 9**).

(d) *Difference extraction of the EB1-GFP comet number.* Some images may be crowded with a large number of EB1-GFP

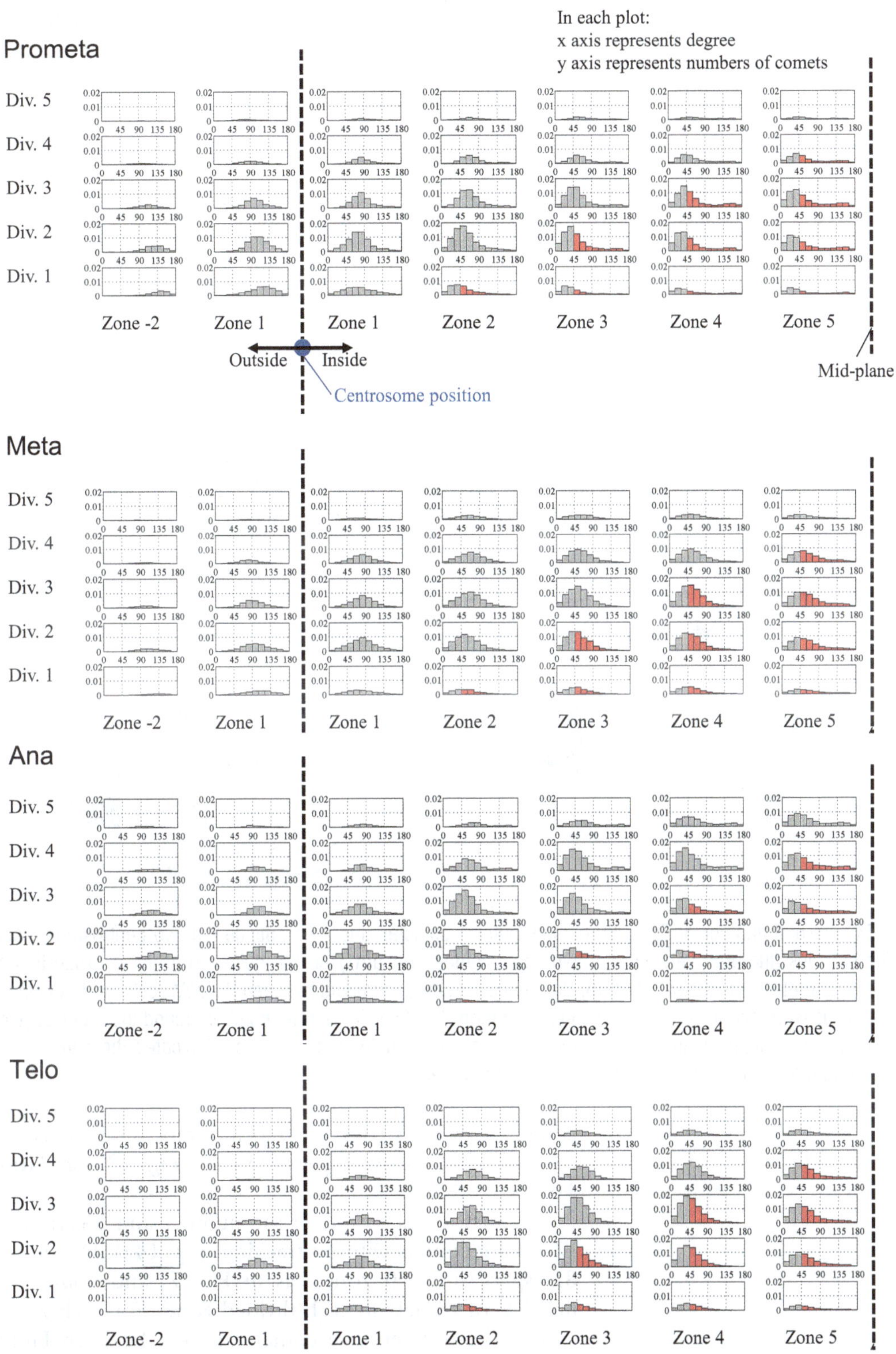


**Fig. 12** Histograms showing the distributions of trajectory angles in each spatial bin. The fractions of trajectories judged to be off-axis (≥45°) are displayed in red

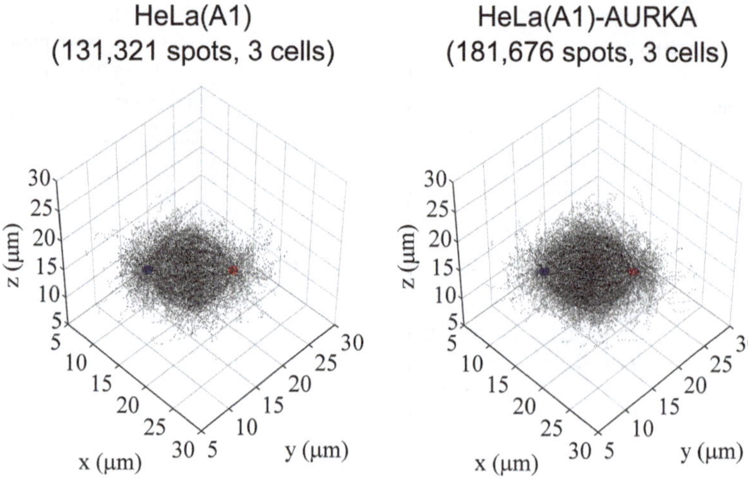


**Fig. 13** EB1-GFP comet distribution after merging data for control HeLa(A1) cells and HeLa(A1) cells expressing aurora kinase A (HeLa(A1)-AURKA). All comet trajectories captured in frames 1–10 from three cells are shown

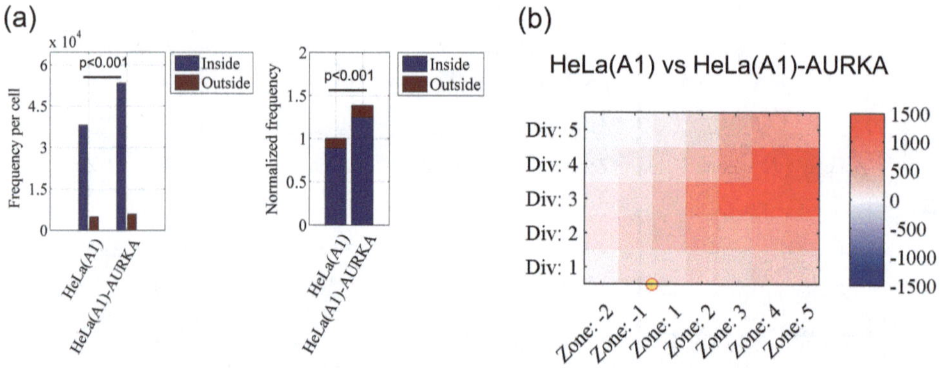


**Fig. 14** (**a**) Total number of EB1-GFP comets outside spindles (the area outside the two centrosomes) and inside spindles (the area between the two centrosomes). HeLa(A1)-AURKA cells had a higher proportion of EB1-GFP comets inside the spindles. (**b**) Differences in comet numbers between HeLa(A1) cells and HeLa(A1)-AURKA cells were extracted and plotted in a heat map. The EB1-GFP comet number around the outer edge of the spindle was higher than that near the center of the spindle. The yellow dot indicates the centrosome position, and the colored bar indicates the comet number

comets, making EB1-GFP tracking impossible, as is the case for AURKA-overexpressing cells (Fig. 13) (*see* **Note 10**).

In such cases, compare the number and location of EB1-GFP comets without tracking (Fig. 14a). Divide the 3D space into 70 zones, as in Fig. 7, and count the number of comets in each divided zone. For each zone, extract the difference in comet numbers between HeLa

(A1) cells and HeLa(A1)-AURKA, and plot the results in a heat map (Fig. 14b) (*see* **Note 11**).

This chapter described a method for spatial statistical analysis of the distribution and dynamics of objects in LLSM multidimensional images. The current methods require programming tailored to each specific objective, posing significant challenges for non-specialists. Simplifying these methods with AI and machine learning may be possible, but applying these technologies to complex, large-scale image data is difficult. A hybrid approach, where AI and machine learning are applied to quantified object information (e.g., coordinates and trajectories of EB1-GFP comets, etc), might be an effective solution (*see* **Note 12**).

## 4 Notes

1. Although we used early-generation GFP in this study, the image quality, spatiotemporal resolution, and tracking accuracy can be improved using a fluorescent protein with higher brightness and stability.
2. In addition to deskew and deconvolution processing, some preprocessing of LLSM images can be performed using the graphical user interface of LLSpy, a Python library for LLSM data processing (Lambert, T., and L. Shao. 2019. tlambert03/LLSpy: v0.4.8. Zenodo. https://doi.org/10.5281/zenodo.3554482) [27].
3. Because subsequent formatting and editing functions are limited after the movie is exported from IMARIS, it is easier to make any additional edits (e.g., changing the image size, cropping, or changing the playback speed) by exporting the movie from IMARIS in sequential TIFF format and then importing the image sequence into ImageJ as a multi-stack TIFF file.
4. Image processing operations involving interpolation, such as image rotation, should be kept to a minimum and not repeated because they alter the information contained in pixels. Introducing processes that modify pixel information results in data that differs from the original. To ensure reproducibility, it is essential to accurately document all image processing operations so that they can be traced.
5. During fluorescence time-lapse imaging, the fluorescence intensity can be influenced by a variety of factors. In most cases, bleaching is unavoidable. Fluctuations in excitation light intensity and slight deviations in the mechanical operating speed of the device can influence the intensity of fluorescence signals. When rapid imaging is repeated using short exposure

times, as is the case for LLSM, even slight deviations can have a significant impact.

6. The top-hat transformation is an image-processing technique for extracting small elements from images with a variable background [26]. It is effective when the objects of interest are distributed across an uneven background. In the given example, the signal from the red channel (H2B-tagRFP) bleeds through to the green (EB1-GFP) channel. Top-hat transformation is effective in reducing this bleed-through and enhancing the visibility of EB1-GFP comet around chromosomes.
7. Many errors occur during automated tracking, and the higher the density of EB1-GFP comets, the more errors are likely to occur. Since the number of microtubules is larger during late cell division but the speed of EB1-GFP comets is lower, filtering conditions must be adjusted between early and late cell division. For detailed analysis parameters and evaluation of tracking results, see ref. [8]. For tracking, one could also use u-track3D, a recently described software package that provides improved algorithms and workflows for 3D particle tracking and is competitive with both traditional and deep learning-based approaches [28].
8. In Fig. 10, it can clearly be seen that many trajectories began their growth at positions far from the centrosomes, an observation that is inconsistent with the established notion that the centrosome is the main microtubule nucleating structure. Notably, in the anaphase cell, large numbers of trajectories are seen to be co-aligned at the midpoint between the two centrosomes, and most of their start points are found in zones 4 and 5 (an area between the centrosome and the midzone). It is possible that many of these were generated by regrowth from existing microtubules at the non-centrosomal sites.
9. EB1-GFP imaging alone cannot distinguish de novo nucleation from regrowth of existing microtubules because it visualizes only the growth phase of dynamic microtubule instability. By further analyzing the direction of comet movement relative to the centrosomal axis in vertical zones dividing the intercentrosomal space (Fig. 7), it is possible in this case to identify a subpopulation of microtubules growing at an off-axis orientation, which suggests non-centrosomal microtubule nucleation.
10. AURKA belongs to the family of serine/threonine kinases. Its activation is necessary for cell division via processes involving the regulation of mitosis and cell cycle progression. AURKA is considered to be an oncogene that enhances cancer growth, and its expression in cancerous tissues is significantly higher than that in non-cancerous tissues [29, 30].

11. Aberrant expression and mutations in AURKA cause various abnormalities that lead to errors during mitosis. Since AURKA also promotes centrosome maturation [31, 32], enhanced centrosome activity may increase the number of microtubules, which could also cause abnormal cell division.

12. This chapter described a method for spatially analyzing the positions and movements of detected EB1-GFP comets as one means of analyzing large-scale LLSM data. This method enables 3D analysis of microtubule growth dynamics, even in densely packed regions where visualization is challenging for human observers. It provides valuable insights into microtubule initiation sites and orientations, demonstrating its utility across various applications. However, its current reliance on programming tailored to specific targets or objectives presents a significant hurdle for many biologists who lack expertise in programming. To address this challenge, a proposed solution could involve developing user-friendly software or tools that automate the spatial statistical analysis of LLSM data. By abstracting the technical complexities and providing intuitive interfaces, such tools would empower biologists to utilize advanced analytical techniques without extensive programming knowledge.

    To enhance analysis, leveraging AI and machine learning offers considerable potential. These algorithms can automate analysis processes, reducing manual intervention and expediting the analysis pipeline. Moreover, they excel at discerning intricate patterns and relationships within data, unveiling valuable insights that traditional methods might overlook. However, utilizing multidimensional image data presents challenges owing to its inherently large file sizes and complexity. Processing such extensive data demands substantial computational power and memory resources. Given the computational intensity of matrix operations involved, high-performance GPUs and specialized hardware are often essential. Furthermore, many machine learning algorithms struggle to efficiently scale with increasing data size. Approaches that perform well on smaller datasets may prove impractical or inefficient when applied to large image datasets. Given the size and complexity of LLSM image data, leveraging approaches that utilize the entire image is nearly impossible with existing computational resources.

    To address this challenge, we propose utilizing object information extracted and quantified from images rather than directly using the images themselves. This approach involves converting the multidimensional image data into structured numerical data representing objects within the images (in the case of this chapter, the coordinates and trajectories of the

EB1-GFP). By quantifying relevant features and attributes of these objects, such as size, shape, intensity, and spatial relationships, we can create a comprehensive dataset that encapsulates the essential information in the images. Utilizing this quantified object information offers several advantages. By focusing on object-level information rather than raw image data, the complexity and size of the dataset can be significantly reduced. Furthermore, numerical object data requires less computational resources compared to processing entire image volumes. This enables faster analysis and facilitates the application of machine learning algorithms to large-scale datasets.

We have named this sophisticated hybrid approach, which employs advanced algorithms and machine learning to analyze complex object motion and dynamic patterns, Intelligent Complex Object Kinetic Analysis (ICOKA). This method is particularly beneficial in areas where understanding the intricate motion and interaction of objects is crucial. For example, tracking nuclei in early embryos or tissues to analyze cell lineages [33–36] falls under the scope of ICOKA. While cell division is a relatively linear process, occurring repeatedly in a predictable manner, the dynamics of intracellular structures are significantly more complex. These structures can take various forms and exhibit diverse behaviors such as appearing, disappearing, fusing, and branching. Therefore, analyzing these complex movements with AI and machine learning will require advanced technologies.

Once established, ICOKA has the potential to revolutionize the analysis of complex object motion and dynamic patterns in multidimensional image data. This hybrid strategy leverages the strengths of both automated analysis and human interpretability, offering a practical pathway to broader adoption and utility in various fields.

## Acknowledgments

LLSM image data were obtained in the Betzig laboratory at the Janelia Research Campus of the Howard Hughes Medical Institute. This work was supported by the Japan Society for the Promotion of Science–NEXT program (LS128), the Takeda Science Foundation, the Uehara Memorial Foundation, JSPS KAKENHI Grant-in-Aid for Challenging Research (Pioneering) (Grant Number 20K20379), and Japan Science and Technology Agency Core Research for Evolutional Science and Technology (no. JPMJCR1863), and an intramural grant from the RIKEN Center for Life Science Technologies and Center for Biosystems Dynamics Research to Y.M.-K.

## References


1. Horio T, Hotani H (1986) Visualization of the dynamic instability of individual microtubules by dark-field microscopy. Nature 321(6070): 605–607. https://doi.org/10.1038/321605a0
2. Mitchison T, Kirschner M (1984) Dynamic instability of microtubule growth. Nature 312(5991):237–242
3. Mimori-Kiyosue Y (2011) Shaping microtubules into diverse patterns: molecular connections for setting up both ends. Cytoskeleton 68(11):603–618. https://doi.org/10.1002/cm.20540
4. Flemming W (1879) Beiträge zur Kenntnis der Zelle und ihrer Lebenserscheinungen. I. Arch Mikrosk Anat 16:302–436
5. Hyman AA, Karsenti E (1996) Morphogenetic properties of microtubules and mitotic spindle assembly. Cell 84(3):401–410. https://doi.org/10.1016/s0092-8674(00)81285-4
6. Inoué S, Sato H (1967) Cell motility by labile association of molecules. The nature of mitotic spindle fibers and their role in chromosome movement. J Gen Physiol 50(6 Suppl):259–292
7. Wittmann T, Hyman A, Desai A (2001) The spindle: a dynamic assembly of microtubules and motors. Nat Cell Biol 3(1):E28–E34. https://doi.org/10.1038/35050669
8. Yamashita N, Morita M, Legant WR et al (2015) Three-dimensional tracking of plus-tips by lattice light-sheet microscopy permits the quantification of microtubule growth trajectories within the mitotic apparatus. J Biomed Opt 20(10):101206. https://doi.org/10.1117/1.JBO.20.10.101206
9. Chen BC, Legant WR, Wang K et al (2014) Lattice light-sheet microscopy: imaging molecules to embryos at high spatiotemporal resolution. Science 346(6208):1257998. https://doi.org/10.1126/science.1257998
10. Mimori-Kiyosue Y (2021) Imaging mitotic processes in three dimensions with lattice light-sheet microscopy. Chromosom Res 29(1):37–50. https://doi.org/10.1007/s10577-021-09656-3
11. Keller PJ, Schmidt AD, Wittbrodt J et al (2008) Reconstruction of zebrafish early embryonic development by scanned light sheet microscopy. Science 322(5904): 1065–1069. https://doi.org/10.1126/science.1162493
12. Durnin J, Miceli JJ Jr, Eberly JH (1988) Comparison of Bessel and Gaussian beams. Opt Lett 13(2):79. https://doi.org/10.1364/ol.13.000079
13. Betzig E (2005) Excitation strategies for optical lattice microscopy. Opt Express 13(8): 3021–3036. https://doi.org/10.1364/opex.13.003021
14. Valm AM, Cohen S, Legant WR et al (2017) Applying systems-level spectral imaging and analysis to reveal the organelle interactome. Nature 546(7656):162–167. https://doi.org/10.1038/nature22369
15. Mimori-Kiyosue Y, Shiina N, Tsukita S (2000) The dynamic behavior of the APC-binding protein EB1 on the distal ends of microtubules. Curr Biol 10(14):865–868. https://doi.org/10.1016/S0960-9822(00)00600-X
16. Stepanova T, Slemmer J, Hoogenraad CC et al (2003) Visualization of microtubule growth in cultured neurons via the use of EB3-GFP (end-binding protein 3-green fluorescent protein). J Neurosci 23(7):2655–2664
17. Srayko M, Kaya A, Stamford J et al (2005) Identification and characterization of factors required for microtubule growth and nucleation in the early C. elegans embryo. Dev Cell 9(2):223–236. https://doi.org/10.1016/j.devcel.2005.07.003
18. Matov A, Applegate K, Kumar P et al (2010) Analysis of microtubule dynamic instability using a plus-end growth marker. Nat Methods 7(9):761–U134. https://doi.org/10.1038/Nmeth.1493
19. Abe T, Kiyonari H, Shioi G et al (2011) Establishment of conditional reporter mouse lines at ROSA26 locus for live cell imaging. Genesis 49(7):579–590. https://doi.org/10.1002/dvg.20753
20. Muroyama A, Lechler T (2017) A transgenic toolkit for visualizing and perturbing microtubules reveals unexpected functions in the epidermis. elife:6. https://doi.org/10.7554/eLife.29834
21. Applegate KT, Besson S, Matov A et al (2011) plusTipTracker: quantitative image analysis software for the measurement of microtubule dynamics. J Struct Biol 176(2):168–184. https://doi.org/10.1016/j.jsb.2011.07.009
22. Kamasaki T, O'Toole E, Kita S et al (2013) Augmin-dependent microtubule nucleation at microtubule walls in the spindle. J Cell Biol 202(1):25–33. https://doi.org/10.1083/jcb.201304031
23. Nakamura S, Grigoriev I, Nogi T et al (2012) Dissecting the nanoscale distributions and functions of microtubule-end-binding proteins

EB1 and ch-TOG in interphase HeLa cells. PLoS One 7(12):e51442. https://doi.org/10.1371/journal.pone.0051442
24. Boehm JS, Zhao JJ, Yao J et al (2007) Integrative genomic approaches identify IKBKE as a breast cancer oncogene. Cell 129(6): 1065–1079. https://doi.org/10.1016/j.cell.2007.03.052
25. Morita M, Tawara T, Nishimura M et al (2014) Communication platform for image analysis and sharing in biology. Int J Netw Comput 4(2):369–391. https://doi.org/10.15803/ijnc.4.2_369
26. Woods E, Gonzalez C (2008) Digital image processing. 3, Pearson Education Ltd
27. Lambert T, Shao L (2019) tlambert03/LLSpy: v0.4.8. https://doi.org/10.5281/zenodo.3554482
28. Roudot P, Legant WR, Zou Q et al (2023) u-track3D: Measuring, navigating, and validating dense particle trajectories in three dimensions. Cell Rep Methods 3(12):100655. https://doi.org/10.1016/j.crmeth.2023.100655
29. Vader G, Lens SM (2008) The Aurora kinase family in cell division and cancer. Biochim Biophys Acta 1786(1):60–72. https://doi.org/10.1016/j.bbcan.2008.07.003
30. Marumoto T, Zhang D, Saya H (2005) Aurora-A—a guardian of poles. Nat Rev Cancer 5(1):42–50. https://doi.org/10.1038/nrc1526
31. Joukov V, De Nicolo A (2018) Aurora-PLK1 cascades as key signaling modules in the regulation of mitosis. Sci Signal 11(543). https://doi.org/10.1126/scisignal.aar4195
32. Nikonova AS, Astsaturov I, Serebriiskii IG et al (2013) Aurora A kinase (AURKA) in normal and pathological cell division. Cell Mol Life Sci 70(4):661–687. https://doi.org/10.1007/s00018-012-1073-7
33. Malin-Mayor C, Hirsch P, Guignard L et al (2023) Automated reconstruction of whole-embryo cell lineages by learning from sparse annotations. Nat Biotechnol 41(1):44–49. https://doi.org/10.1038/s41587-022-01427-7
34. McDole K, Guignard L, Amat F et al (2018) In toto imaging and reconstruction of post-implantation mouse development at the single-cell level. Cell 175(3):859–876 e833. https://doi.org/10.1016/j.cell.2018.09.031
35. de Medeiros G, Ortiz R, Strnad P et al (2022) Multiscale light-sheet organoid imaging framework. Nat Commun 13(1):4864. https://doi.org/10.1038/s41467-022-32465-z
36. Sugawara K, Cevrim C, Averof M (2022) Tracking cell lineages in 3D by incremental deep learning. elife 11. https://doi.org/10.7554/eLife.69380

# Part II

## Spatiotemporal Manipulation of the Spindle and Target Molecules

# Chapter 5

# A Centrifuge Polarizing Microscope (CPM) That Enables the Visualization of Intracellular Structures Under High Centrifugal Forces

**Makoto Goda and Akatsuki Kimura**

## Abstract

The mitotic spindle consists of aligned filaments of dynamic microtubules that faithfully segregate mitotic chromosomes. This view of the mitotic spindle was initially established by detecting weak birefringence of the aligned filaments, which was realized by developing polarized light microscopy (Inoué S, Chromosoma 5:487–500. https://doi.org/10.1007/bf01271498, 1953). Inoué et al. developed a centrifuge polarizing microscope (CPM) (Inoué S et al., J Microsc 201:341–356. https://doi.org/10.1046/j.1365-2818.2001.00850.x, 2001; Inoué S et al., Centrifuge microscope capable of realizing polarized light observation (US Patent: US1907803A), 1999) to further understand the structural and physical basis of the functional positioning of centrosome-based structures (including the mitotic spindle) inside living cells. This chapter describes the procedure for observing biological specimens using CPM by focusing on *Caenorhabditis elegans* embryos.

**Key words** Centrifuge polarizing microscope (CPM), Mass density, Force, *Caenorhabditis elegans* embryo, Nuclear positioning

## 1 Introduction

Centrifugation is a classical method used to apply forces to living cells and investigate their physiology (reviewed in [2]). Centrifuge microscopes were designed to observe specimens during centrifugation (reviewed in [2, 4, 5]). In 1930, the Harvey-Loomis centrifuge microscope was used [6], wherein the specimen, objective lens, and prism were rotated using a rotor. An image was acquired through an eyepiece fixed outside the rotor when the specimen was passed through a light source that was also fixed outside the rotor. This microscope was used to observe cells up to 1,000× *g* of centrifugal acceleration with a 62× objective lens. Harvey et al. [7] used the microscope to quantify the surface tension of sea urchin eggs. Later, Harvey [8, 9] developed a centrifuge microscope with

Tomomi Kiyomitsu (ed.), *The Mitotic Spindle: Methods and Protocols*, Methods in Molecular Biology, vol. 2872, https://doi.org/10.1007/978-1-0716-4224-5_5,

the same principle but adopted an ultracentrifuge to obtain an image at 84,000× *g* with a low-magnification objective lens. The Beams-type microscope designed by Pickels [10] enabled Beams and King [11] to observe a cell under 150,000× *g* with a low-magnification objective lens. The specimen is rotated in a high-speed rotor and is viewed with an objective lens and a light source outside the rotor. An image was acquired without extensive blurring using a 45° mirror positioned on the rotor. Kamiya and Kuroda developed a video centrifuge microscope [12, 13] wherein the specimen and all optical devices (including a video camera, except for the light source) rotated. The design was passed through a modern centrifuge microscope, wherein all optical devices were rotated to enable continuous visualization under centrifugation up to ~30× *g* [14]. Kamitsubo et al. [15, 16] developed a centrifuge microscope wherein the rotating specimen was observed by illuminating the sample with stroboscopic light each time the rotating specimen passed under the objective lens; thus, the specimen was observed as if it was stopped with a little blur. In this microscope, only the specimen is rotated, and images are acquired using optical devices fixed outside the rotor. The basic design of the stroboscopic-type centrifuge microscope was adopted for the development of the centrifuge polarizing microscope (CPM) [2, 3] introduced in this study.

The CPM can observe samples under a centrifugal force of up to 11,500× *g* with a 40× objective lens with a spatial resolution below 1 μm. In the past, Shinya Inoué, the lead developer of the CPM, proved that the mitotic spindle consists of aligned filaments by developing polarized light microscopy and detecting the weak birefringence of the aligned filaments inside a living cell [1]. Along with this line, one of the original motivations for developing the CPM was to further understand the structural and physical basis of the functional positioning of centrosome-based structures, including the mitotic spindle inside a living cell [2]. Inoué and colleagues collaborated with Hamamatsu Photonics (Hamamatsu, Japan) and Olympus (Tokyo, Japan) to develop the CPM. The CPM was designed as a polarizing microscope that can visualize intracellular structures with a high centrifuge speed to visualize birefringent structures (including the mitotic spindle) and to move cellular structures with a small difference in mass density using centrifugation. To realize these requirements, only the sample chamber was rotated, and all the other optical components were placed outside the centrifuge rotor (Figs. 1 and 2). The optical components between the polarizer and analyzer were aligned along a straight single axis to avoid interference with the weak birefringence of the specimen (Fig. 2, green line). Normal high-performance microscope systems such as birefringent (polarizing microscope), fluorescence (532-nm excitation), and Nomarski differential interference contrast (DIC) imaging were realized in the CPM since the optical

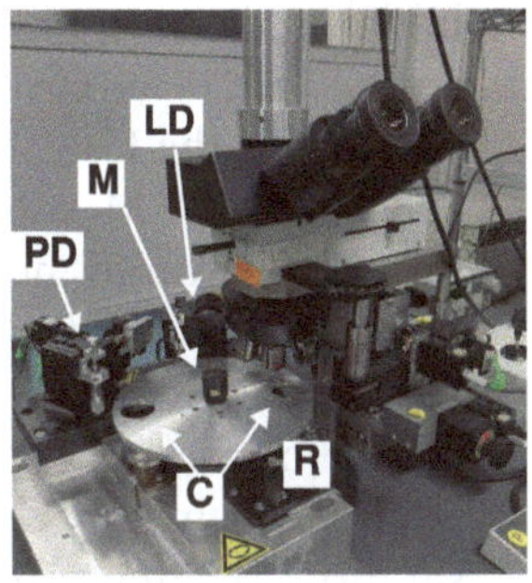


**Fig. 1** The current CPM system. The specimen is mounted in the chamber (C, empty in this photo; *see* Fig. 4), located 7.5 cm from the axis of the rotor (R). The timing mirror (M) located at the axis of the rotor reflects the laser from the laser diode (LD). The pulsed laser illuminates the chamber only when the laser is detected at the photodetector (PD); as a result, the specimen is illuminated by the laser when it passes the identical angle under the objective lens

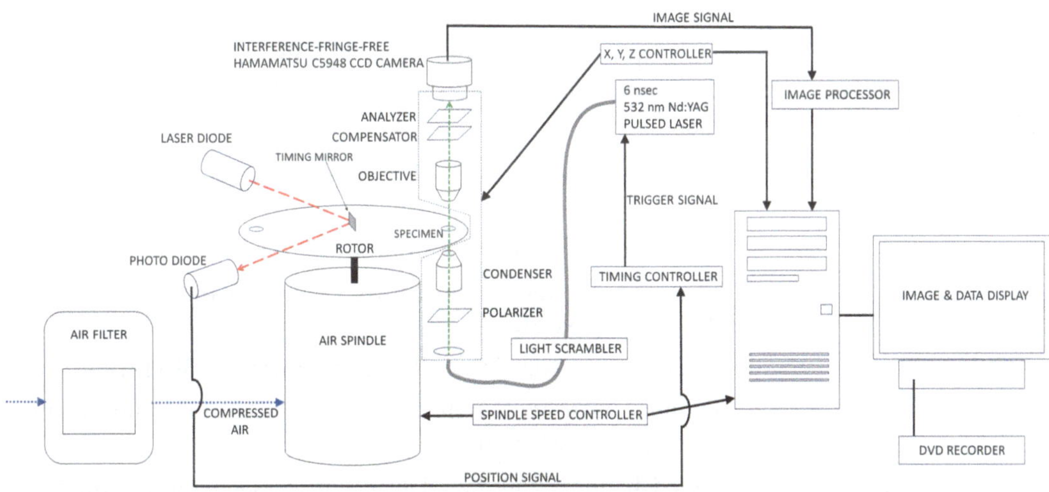


**Fig. 2** The schematic of the current CPM system. The scheme was drawn based on Fig. 4 of the Ref. [2] and updated to reflect the current system. The illumination optics for the specimen is drawn in green. The path of the laser to detect the position of the rotor is drawn in red

system does not rotate. 4×, 10×, 20×, and 40× objective lenses were used in the CPM. A high-intensity pulsed Nd:YAG laser (532 nm) irradiates the specimen for 6 ns per irradiation to achieve a spatial resolution of 1 μm of the specimen located at a radius of 7.5 cm in the rotor spinning at ~10,000 rpm. A birefringence of under 1 nm can be detected using the CPM.

Centrifuge microscopes were used to characterize the mechanical properties of cells and cell motility, such as the tension at the cell surface, the consistency of the cytoplasm, and the driving forces for cytoplasmic streaming (reviewed in [4]). They are used to analyze the cellular mechanisms of plant sensing [14, 17]. The initial application of CPM included the observation of the transient appearance of birefringent material in sea urchin eggs upon fertilization and the $Ca^{2+}$-dependent activation of marine worm oocytes [18]. The CPM was used to observe mouse oocytes [18, 19], fibroblast wound healing [18], amoeboid movement [18, 20], hemoglobin crystals [18, 21], and viruses [18].

Centrifugation was classically used to obtain insights into the nature of the mitotic spindle ([22] reviewed in [2]). Centrifugal microscopy was used to characterize the force exerted by the kinesin motor proteins [23, 24]. One of the motivations for CPM is to characterize the forces that position the mitotic spindle and centrosomes in a living cell [2]. Accordingly, we recently characterized the forces that position centrosomes in the *Caenorhabditis elegans* embryo using CPM [25].

## 2 Materials

### 2.1 Centrifuge Polarizing Microscope (CPM) System

A photograph of the current system is shown in Fig. 1. A schematic of the optical components is shown in Fig. 2 (drawn based on Fig. 4 of the Ref. [2]).

1. Microscope system including XYZ controller (Olympus, custom-made)
2. Camera (Hamamatsu Photonics, C5948-based custom-made)
3. High-speed spindle (Dover, 315-RF) and motion controller (Dover, DMM-2001)
4. Laser system (New Wave Research, Polaris II-30Hz)
5. Timing controller for the synchronization of the illumination (Olympus, custom-made)
6. Four-Channel Digital Delay/Pulse Generator (Stanford Research Systems Inc., Model DG535)
7. Personal computer (PC) (Dell OptiPlex GXMT 5133), CPM spindle driver (custom-made), CPM software (custom-made)
8. Image Processor (Hamamatsu, Argus box)
9. Digital Scan Converter (Sony, DSC-1024G)
10. DVD recorder (Panasonic, DMR-BRW1010)

### 2.2 Reagents

1. Percoll (Sigma)
2. 1× buffer: The buffer appropriate for observing the specimen in usual imaging (*see* **Note 1**).
3. 4× buffer: 4-fold concentration of the 1× buffer.
4. 1× buffer with 75% Percoll: 4× buffer is mixed with Percoll at a 1:3 ratio; as a result, the resultant solution is 1× buffer with 75% Percoll.

**2.3 Cleaning the Chamber**

1. Air duster
2. 100% Ethanol

**2.4 Biological Specimens and Mounting**

1. *C. elegans* embryo (for example): *C. elegans*, N2 strain is cultured using a conventional method [26].
2. Dissection microscope equipped with transmitted light to dissect *C. elegans*
3. Pipette: Gel loader

## 3 Methods

**3.1 Preparing the CPM System**

1. Powering up the CPM system.
2. Setting the laser: set the laser power (PFN) to 60% or greater, the mode to "continuous," the repetition rate to 30, and the video mode to "N."
3. Microscope settings: remove the compensator, and adjust the angle of the analyzer and rectifier on the condenser lens (if a rectified condenser lens is used) to the dark background (Fig. 3b). After putting the compensator back to the position, search the sample in focus using the focus knob of the XYZ position controller. Adjust the Z position of the condenser lens (*see* **Note 2**). Center the position of the field diaphragm using the XY adjustment for the field diaphragm (Fig. 3a). For fluorescent microscopy, *see* **Note 3**. For DIC imaging, *see* **Note 4** (Fig. 3c, d). Optimize the Köhler illumination (Fig. 3e).
4. (Optional) The CPM system can illuminate the specimen with a mercury lamp under non-centrifugation conditions (*see* **Note 5**). This function is mainly used to adjust microscope systems, such as the optical axis.

**3.2 Mounting the Sample**

1. The sample chamber consists of a plastic window, a glass window, two screws, a silicone spacer, a top holder, and a bottom holder (Fig. 4a) [2]. Two chambers (*see* **Note 6**) are placed at two positions of the rotating stage (Fig. 1, "C").
2. Assemble the chamber (*see* **Note 7**). First, place the two windows on each side of the silicone spacer (Fig. 4b (i)). Second, place the spacer with the windows on the bottom holder (Fig. 4b (ii); *see* **Note 8**). Third, place the top holder on top of the spacer (Fig. 4b (iii)). Finally, fix the chamber with the screws.
3. Prepare the specimen: in the case of a *C. elegans* embryo, dissect an adult worm under the dissecting microscope in 1× buffer.
4. Load the specimen with the buffer into the chamber through the loading hole at the top of the chamber holder using a

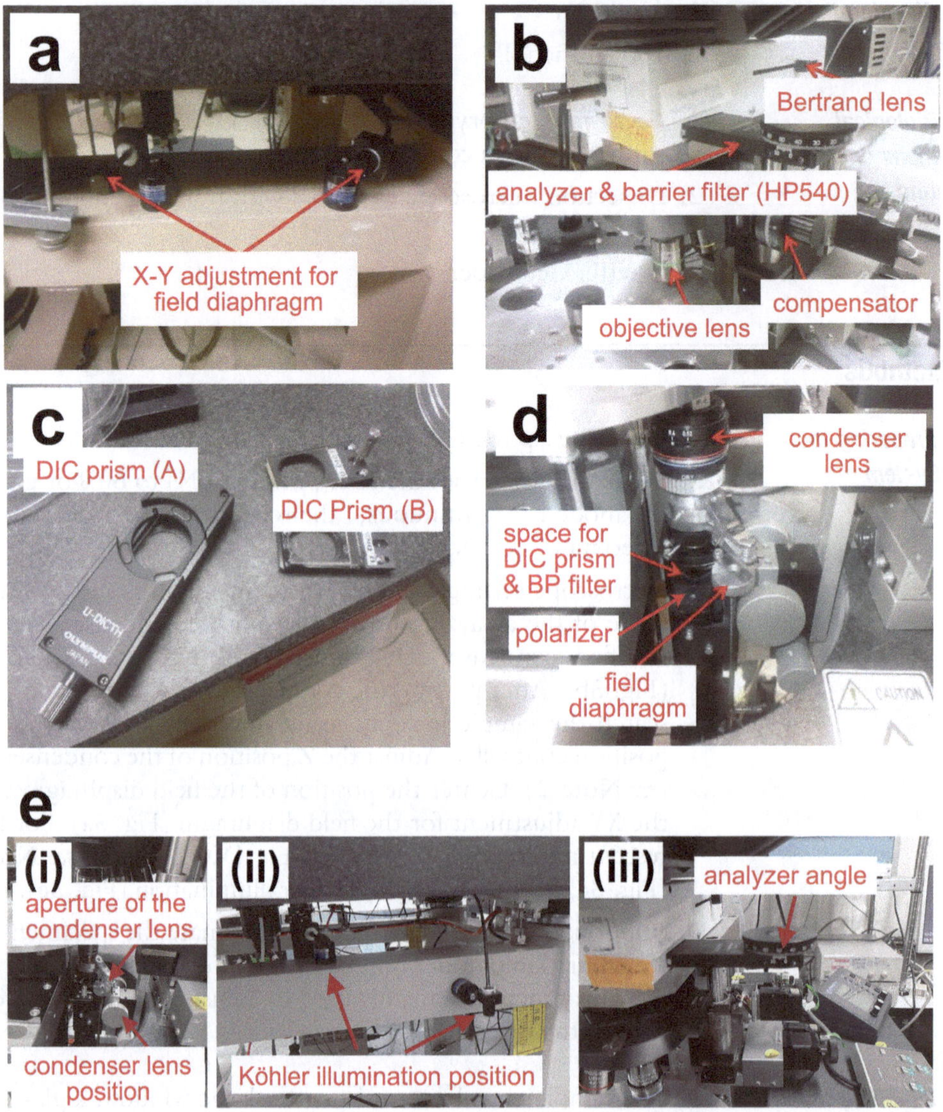


**Fig. 3** The position of the parts to be adjusted upon set up. (**a**) X-Y adjustment for the field diaphragm. (**b**) The compensator. (**c**) DIC prism. (**d**, **e** (i)) Condenser lens. (**e** (ii)) Köhler illumination position. (**e** (iii)) Angle of the analyzer

micropipette with a gel loader (Fig. 4c (i); *see* **Note 9**). The tip of the loader will enter the chamber smoothly (Fig. 4c (ii)) if the holes in the top holder and spacer are properly aligned (*see* **Note 8**). Typically, we initially load 8 μL of 75% Percoll solution and then carefully (to avoid disturbing the layer) load 4 μL of 1× buffer with the specimen on top of the Percoll layer (*see* **Note 10**). Place the same amount of solution in the other chamber (with or without a specimen); as a result, the weights of the two chambers will be identical.

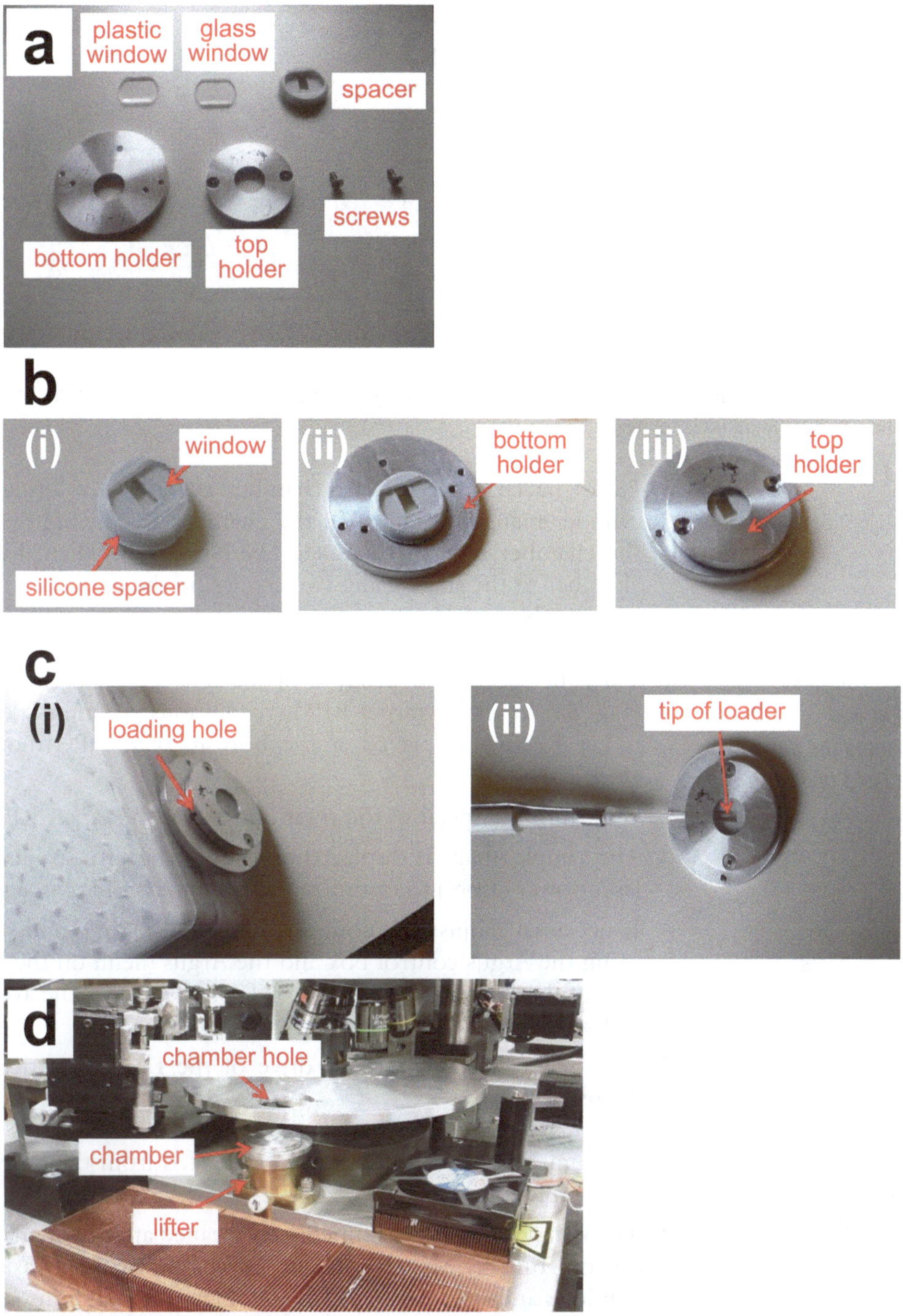


**Fig. 4** The specimen chamber. (**a**) The parts of the chamber. (**b**) Assembly procedure. (**c**) Mounting the sample. (**d**) Mounting the chamber on the rotor

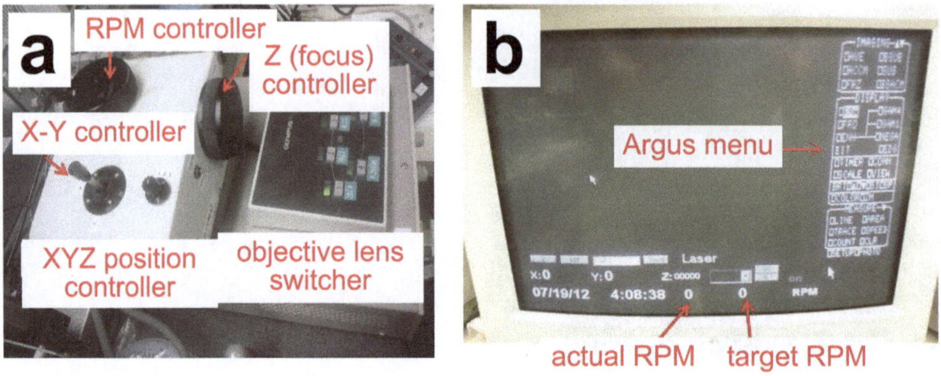


**Fig. 5** Manipulation of the XYZ position, objective lens, and rotation speed. (**a**) Switches and knobs of the controller. (**b**) Appearance of the display

5. Place the chamber on the chamber cylinder. By lifting the cylinder manually (Fig. 4d), the chamber will fit tightly into the chamber hole of the stage (*see* **Note 11**). Then, fix the chamber to the stage using two screws on the two sides of the hole (*see* **Note 12**).

### 3.3 Starting the Rotation and Positioning the Sample in the Field of View

1. Turn the speed-control knob of the XYZ position controller (Fig. 5a) to set the target RPM (revolutions per minute). The target RPM is displayed at the bottom of the PC monitor (Fig. 5b).
2. Wait until the speed of the rotation reaches the target RPM.
3. The sample image is displayed on the PC monitor. Adjust the laser power (PFN) to optimize the brightness of the image.
4. If necessary, adjust the contrast, range, averaging, and so on using the Argus control box and the Argus menu on the right side of the PC monitor (Fig. 5b). Typically, we set the average of four frames.
5. Start recording to the hard drive of the DVD recorder (*see* **Note 13**).

### 3.4 Cleaning the Chamber and Shutting Down the System

1. After imaging, the sample can be collected from the chamber with a micropipette.
2. The sample chamber is immediately cleaned after each imaging session. Initially, the chamber is divided into parts. Wash the windows and spacer using a stream of tap water and rinse them with distilled water. The windows are rinsed further with 100% ethanol. Remove distilled water or ethanol from the windows and spacers using Kimwipes and air duster.
3. Shut off the CPM system by reversing the procedure to start the system.

## 4 Notes

1. For the imaging of the *C. elegans* embryo, we use the following buffer as the 1× buffer: 88.5 mM NaCl, 30 mM KCl, 2.55 mM $CaCl_2$, 2.55 mM $MgCl_2$, and 3.75 mM HEPES-NaOH, pH 7.2.
2. Pay attention to the distance between the condenser lens and the rotating stage to avoid collisions.
3. The fluorescent microscopy of the CPM is a transmission type. For fluorescent imaging, replace the analyzer for a 540 nm-high-pass filter. The filter serves as the emission filter.
4. For DIC, place the DIC prism (A in Fig. 3c) along the optical axis instead of the compensator. Insert the DIC prism (B in Fig. 3c) into the space just before the condenser lens (Fig. 3d).
5. Observations under non-centrifugation conditions with mercury (Hg) lamp illumination can be conducted using the CPM system as follows: First, the power on the Hg power supply. Position the sample under the objective lens by rotating the stage. Fix the position of the stage using a stage stopper. Switch the Hg-Laser knob to "Hg." Finally, pull the camera-binocular switch for visual observation.
6. The difference between the total weights of the two chambers should be under 1 mg.
7. Do not touch the center region of the windows, which are illuminated by the laser throughout the preparation. Remove the dust from the surfaces of the windows using an air duster before creating the chambers.
8. When placing the silicone spacer on the bottom holder, adjust the angle of the spacer such that the loading hole of the spacer aligns with that of the top holder, which will be placed in the next step. If the positions of the two holes do not match, the sample will not be loaded into the chamber (Fig. 4c (ii)).
9. During sample loading, keep the chamber vertically against gravity as much as possible (Fig. 4c (i)); as a result, the Percoll layer will be stably positioned as the lower layer.
10. The volume of the sample must be under 15 μL.
11. If the chamber does not fit into the hole, its top holder should be replaced with a new one. Note that the weight of the two chamber sets should be always balanced.
12. Do not drive screws with excessive strength because the screw thread is not strong.
13. The recorded movies in the current CPM system are in "m2ts" format. We convert the movies into "mov" format using a Windows software and open the mov file using the FFMPEG plug-in for ImageJ software.

## Acknowledgments

We thank the late Shinya Inoué for inventing the CPM and for discussions on its use. We thank Shinya Inoué, Rudolf Oldenbourg, and the Marine Biological Laboratory for allowing us to use the CPM and later permitting us to transfer the CPM to the National Institute of Genetics, Japan. We thank Naobumi Okada, Tomomi Tani, and Hajime Takahashi for their support in transporting the CPM to Japan. We thank Michael Shribak, Zenki Ikeda, and Gohta Goshima for their input into our collaborative research using CPM. We thank Naoki Noda, Hiroaki Ishikawa, and Mari Yoshida for their comments on this manuscript. The study using CPM was supported by grants from the Marine Biological Laboratory, JSPS KAKENHI, and NIGMS/NIH.

### References

1. Inoué S (1953) Polarization optical studies of the mitotic spindle: I. The demonstration of spindle fibers in living cells. Chromosoma 5: 487–500. https://doi.org/10.1007/bf01271498
2. Inoué S, Knudson RA, Goda M et al (2001) Centrifuge polarizing microscope. I. Rationale, design and instrument performance. J Microsc 201:341–356. https://doi.org/10.1046/j.1365-2818.2001.00850.x
3. Inoué S, Suzuki K, Ichie K et al (1999) Centrifuge microscope capable of realizing polarized light observation (US Patent: US1907803A)
4. Hiramoto Y, Kamitsubo E (1995) Centrifuge microscope as a tool in the study of cell motility. Int Rev Cytol 157:99–128. https://doi.org/10.1016/s0074-7696(08)62157-9
5. Toyota M, Terao-Morita M, Ikeda N et al (2012) Live-cell imaging of plant gravity sensing by using a vertical-stage confocal microscope and a centrifuge microscope. Plant Morphol 24:23–32. https://doi.org/10.5685/plmorphol.24.23
6. Harvey EN, Loomis AL (1930) A microscope-centrifuge. Science 72:42–44. https://doi.org/10.1126/science.72.1854.42
7. Harvey EN (1931) The tension at the surface of marine eggs, especially those of the sea urchin, Arbacia. Biol Bull 61:273–279. https://doi.org/10.2307/1536947
8. Harvey EN (1932) The centrifuge-microscope for super-centrifugal forces. Science 75:267–268. https://doi.org/10.1126/science.75.1940.267.b
9. Harvey EN (1934) The air turbine for high speed centrifuging of biological material, together with some observations on centrifuged eggs. Biol Bull 66:48–54. https://doi.org/10.2307/1537461
10. Pickels EG (1936) Optical designs for observing objects in centrifugal fields of force. Science 83:471–472. https://doi.org/10.1126/science.83.2159.471
11. Beams HW, King RL (1937) The suppression of cleavage in Ascaris eggs by ultracentrifuging. Biol Bull 73:99–111. https://doi.org/10.2307/1537872
12. Kuroda K, Kamiya N (1981) Behavior of cytoplasmic streaming in Nitella during centrifugation as revealed by the television centrifuge-microscope. Biorheology 18:633–641. https://doi.org/10.3233/bir-1981-183-625
13. Kuroda K, Kamiya N (1989) Propulsive force of Paramecium as revealed by the video centrifuge microscope. Exp Cell Res 184:268–272. https://doi.org/10.1016/0014-4827(89)90386-8
14. Toyota M, Ikeda N, Sawai-Toyota S et al (2013) Amyloplast displacement is necessary for gravisensing in Arabidopsis shoots as revealed by a centrifuge microscope. Plant J 76:648–660. https://doi.org/10.1111/tpj.12324
15. Kamitsubo E, Kikuyama M, Kaneda I (1988) Apparent viscosity of the endoplasm of Characean internodal cells measured by the centrifuge method. Protoplasma Suppl 1:10–14. https://doi.org/10.1007/978-3-7091-9008-1_2
16. Kamitsubo E, Ohashi Y, Kikuyama M (1989) Cytoplasmic streaming in internodal cells of Nitella under centrifugal acceleration: a study

done with a newly constructed centrifuge microscope. Protoplasma 152:148–155. https://doi.org/10.1007/bf01323074

17. Nakamura M, Toyota M, Tasaka M et al (2015) Live cell imaging of cytoskeletal and organelle dynamics in gravity-sensing cells in plant gravitropism. Methods Mol Biol 1309:57–69. https://doi.org/10.1007/978-1-4939-2697-8_6
18. Inoué S, Goda M, Knudson RA (2001) Centrifuge polarizing microscope. II. Sample biological applications. J Microsc 201:357–367. https://doi.org/10.1046/j.1365-2818.2001.00851.x
19. Liu L, Hammar K, Smith PJS et al (2001) Mitochondrial modulation of calcium signaling at the initiation of development. Cell Calcium 30:423–433. https://doi.org/10.1054/ceca.2001.0251
20. Fukui Y, Uyeda TQ, Kitayama C et al (2000) How well can an amoeba climb? Proc Natl Acad Sci USA 97:10020–10025. https://doi.org/10.1073/pnas.97.18.10020
21. Hoffman JF, Inoué S (2006) Directly observed reversible shape changes and hemoglobin stratification during centrifugation of human and Amphiuma red blood cells. Proc Natl Acad Sci USA 103:2971–2976. https://doi.org/10.1073/pnas.0510884103
22. Lillie FR (1909) Karyokinetic figures of centrifuged eggs; an experimental test of the center of force hypothesis. Biol Bull 17:101–119. https://doi.org/10.2307/1536032
23. Hall K, Cole DG, Yeh Y et al (1993) Force-velocity relationships in kinesin-driven motility. Nature 364:457–459. https://doi.org/10.1038/364457a0
24. Hall K, Cole D, Yeh Y et al (1996) Kinesin force generation measured using a centrifuge microscope sperm-gliding motility assay. Biophys J 71:3467–3476. https://doi.org/10.1016/s0006-3495(96)79542-5
25. Goda M, Shribak M, Ikeda Z et al (2024) Live-cell imaging under centrifugation characterized the cellular force for nuclear centration in the Caenorhabditis elegans embryo. Proc Natl Acad Sci USA, in press
26. Brenner S (1974) The genetics of Caenorhabditis elegans. Genetics 77:71–94. https://doi.org/10.1093/genetics/77.1.71

# Chapter 6

# Manipulation of Spindle Position Using Magnetic Tweezers in Sea Urchin Embryos

**Aude Nommick, Jing Xie, and Nicolas Minc**

## Abstract

The regulation of mitotic spindle position and orientation, and consequent division plane specification, is critical for early embryo development, tissue architecture and stem cells. Accordingly, defects in spindle positioning have been associated with the emergence of tissue disorders and mis-specification of cell fates, causing the formation of tumors, for instance, in stem cell populations. In such context, methods to physically manipulate mitotic spindle position or orientation in living cells bear the promise of dissecting the causal impact of spindle mis-positioning on cell or tissue behavior. Here, we describe a method to directly modulate mitotic spindle position and orientation with in vivo magnetic tweezers in developing sea urchin embryos. This method allows for monitoring the impact of spindle orientation on embryo development and quantifying forces and torques needed to move or rotate spindles. It may be transposable to many other cell, embryo or tissue types.

**Key words** Sea urchin, Embryogenesis, Morphogenesis, Cell division, Mitosis, Spindle, Magnetic tweezers, Forces, Microinjection

## 1 Introduction

The mitotic spindle is a microtubule (MTs) based assembly that carries the fundamental functions of segregating chromosomes and specifying the site of cytokinetic ring assembly [1–3]. Therefore, its position and orientation are critical to dictate the size and arrangement of daughter cells in embryos and tissues. Spindles may be positioned in the cell center for symmetric divisions or asymmetrically, such as during asymmetric stem cell divisions. They may also orient with respect to different cues, including cell polarity, cell shape and tissue tension [4–6]. These cues ultimately impact the spatial distribution of forces exerted by astral microtubules, which radiate from centrosomes to the cell cortex. Molecular mechanisms that regulate spindle position have been studied for decades (for recent reviews, *see* [7–9]). In contrast, the exact magnitude of forces and torques that move or rotate spindles in place have been

Tomomi Kiyomitsu (ed.), *The Mitotic Spindle: Methods and Protocols*, Methods in Molecular Biology, vol. 2872,
https://doi.org/10.1007/978-1-0716-4224-5_6,

only rarely measured [10–12], limiting our quantitative understanding of force-generating machineries orienting and positioning spindles.

The proper positioning and orientation of spindles and consequent divisions impact tissue architecture and embryo morphogenesis. For instance, early embryo development relies on a series of reductive divisions with stereotypical 3D patterns that define the position of blastomeres, cell fates, germ layers and developmental axis [9, 13]. In proliferating epithelial tissues, spindles are commonly oriented in the tissue plane, generating daughter cells that remain side by side in the epithelial layer [4, 7, 14]. Failures to control epithelial spindle orientation have been linked to cancer development [15, 16]. Indeed, such mis-regulations could drive defects in the positioning of cells in the layer, exposing them to a non-conformal environment, and lead to dysplasia, hyperplasia, epithelial to mesenchymal transition (EMT), metastasis or the emergence of cancer stem cell populations in stem cell niches. While genetic manipulations have been instrumental in demonstrating the role of spindle orientation for proper embryo and tissue development [14, 15, 17], methods to physically modulate spindle orientation bear the promise of directly monitoring the impact of a misregulated division on cell and tissue behavior [18].

Here, we propose a method based on calibrated in vivo magnetic tweezers, which allows for quantifying forces that position spindles, as well as control spindle orientation and position and consequent division plane specification in intact developing embryos. This method is based on the injection of specific magnetic beads that recruit dynein motors to self-target themselves at the centrosome, allowing external force application at mitotic spindle poles with a magnet tip. We present detailed protocols for bead preparation, injection and force application, as well as examples of its application.

## 2 Materials

### 2.1 Sea Urchin Animals Maintenance and Eggs Collection

1. Sea urchin animals (*Paracentrotus Lividus*, Roscoff Marine Station)
2. Aquariums equipped with different filters (Biobox) and skimmers. Water temperature 16–17 °C, pH 7.9–8.2
3. Artificial seawater (ASW), prepared from diluting reef ocean salt in deionized water (Reef Crystals, Instant Ocean)
4. 0.5 M KCl to stimulate gamete spawning
5. 5 mL syringe to inject sea urchins
6. Glass beakers to collect eggs
7. 1.5 mL Eppendorfs tubes to collect sperm

8. Artificial seawater filtered (ASW.F) with a vacuum filter system (Stericup Quick Release Filter 500 mL, Millipore 0.22 μm, PES Membrane)
9. Plastic dishes (MatTek P50G-0-14-F, 50mm No.O)
10. Pasteur plastic pipettes (3 mL, 158 mm)
11. Incubators to maintain eggs and embryos at 16 °C
12. Nitex mesh of 60–70 μm (Genesee Scientific) (*see* **Note 1**)

## 2.2 Magnetic Tweezers

### 2.2.1 Magnetic Beads

1. 1 μm streptavidin magnetic particles (Nanolink® SoluLink).
2. Washing solution: 1 M NaCl with 1% Tween-20
3. Biotin-Atto 488/565 (Sigma-Aldrich). 2 mg/mL stock diluted in DMSO.
4. Rotary shaker for Eppendorf tubes.
5. Sonicator.
6. Magnetic separation rack for 1.5 mL Eppendorf tubes.

### 2.2.2 Magnetic Tweezers Setup

1. Rod-shaped strong neodymium magnets (diameter 4 mm; height 10 mm; S-04-10 AN; Supermagnet) prolonged by a sharpened steel piece with a tip radius of ~50 μm to create a magnetic gradient.
2. Gold to electro-coat the surface of the steel tip to prevent oxidation.
3. Inverted microscope equipped with a 3D micromanipulator (Injectman 4, Eppendorf).

### 2.2.3 Magnetic Tweezer Calibration

1. Mono-dispersed super-paramagnetic Dynabeads 2.8 μm (280-M Streptavidin; Thermofisher) or 1 μm (MyOne Streptavidin C1; Thermofisher)
2. Nonmagnetic fluorescent tracers 0.5 μm (Molecular probes; Invitrogen).
3. Computer with TrackMate plugin (Image J) and MATLAB (Mathworks)
4. Poly-L-Lysine (1 mg/mL$^{-1}$)
5. 80% glycerol solution
6. Inverted epifluorescent microscope equipped with a micromanipulator (Injectman 4, Eppendorf), a micro-injection system (FemtoJet 4, Eppendorf), Micro-Manager (Open Imaging), and a camera, at room temperature of 18–20 °C.

## 2.3 Micro-injection

### 2.3.1 Capillaries and Loaders

1. Sigmacote siliconizing reagent.
2. Borosilicate glass capillaries TW100-4 (1 mm diameter) (World Precision Instruments, WPI) or filament-glass capillaries TW100F-4 (1 mm diameter) (WPI).

3. Needle puller (P-1000, Sutter Instrument)
4. Microforge MF2 Narishige.
5. Microgrinder EG-402 Narishige.
6. epTIPS Microloader 20 μL (Eppendorf)

*2.3.2 Injection Dishes*

1. μ-Dish 35 mm 1.5, low (IBIDI)
2. 1% protamine solution: 0.5 g Protamine sulfate salt from salmon (Sigma) + 50 mL $H_2O$

*2.3.3 mRNA Synthesis*

1. mMessageMachine Kit (Ambion) for mRNA synthesis.
2. MEGAclearKit (Ambion) for mRNA purification.
3. Poly (A) Tailing Kit (ThermoFisher) (optional)
4. Sonicator.
5. Centrifuge (e.g. Eppendorf 5425R)

*2.3.4 Imaging*

7. An inverted epifluorescence microscope. We use a TI-Eclipse microscope (Nikon) combined with a complementary metal–oxide–semiconductor (CMOS) camera (Hamamatsu), using a 20× dry objective (Apo, NA 0.75, Nikon) and a 1.5× magnifier, yielding a pixel size of 0.216 μm.

## 3 Methods

### 3.1 Gametes Collection

1. Female and male gametes are collected by intracoelomic injection of ~1 mL of 0.5 M KCl.
2. Egg laying is done in a glass beaker filled with ASW by inverting female sea urchins onto the beaker. Eggs are then washed twice in ASW.F. Sperm is collected as pure as possible with a pipet, kept at 4 °C, and used within 2–4 days (always test its efficacy before use).
3. Filter eggs 3 times through a 70-μm Nitex mesh to remove the vitelline membrane (only for injection purposes, to facilitate the adhesion to the protamine dish) (*see* **Note 1**).
4. Keep the eggs in the incubator at 16–18 °C in ASW.F until use. Eggs can be used only on the same day of collection.
5. Optional: if you want to observe DNA, incubate them in Hoechst 10 μg/mL in ASW.F until injection.

### 3.2 Preparation of Micro-injection Capillaries

*3.2.1 Capillary Coating (Only for Beads Injection)*

1. Incubate capillaries in a 15 mL tube with 1 mL of SigmaCote.
2. Place them in a 50 mL tube and incubate overnight at RT in a rotor.

3. Place capillaries under the hood to dry them (before, eliminate the bubbles inside each capillary by tapping them down on an absorbent paper).
4. Let dry overnight under the hood.

*3.2.2 Capillary Fabrication for Beads Injection (Fig. 1a)*

1. With the capillary puller, pull coated capillaries with the specific program calibrated for them.
2. With the Microforge, cut the tip of the capillary to a diameter of 10 μm (using a 10× Objective), with the temperature power at ~50%.
3. With the MicroGrinder machine, grind the tip of the capillary with an angle of 40° on the diamond grinder until observing a horizontal shape at its base.

### 3.3 Coating of Injection Dishes

1. Prepare a 1% protamine solution and let it dissolve on the rotor for 8–10 h.
2. Take non-coated dishes, put a drop of protamine in each middle circle, and remove them by aspiration after 2 min.
3. Place all the dishes in a large volume of deionized water for 30 min or overnight at RT.
4. Let the dish dry bottom-up on tissues overnight, covering them with paper.

### 3.4 Magnetic Beads Preparation

Streptavidin magnetic particles, which exhibit spontaneous microtubule-minus-end-directed motion in sea urchin embryos, are used to apply magnetic forces on the mitotic spindle [10, 19, 20] (Fig. 1b). To prepare and fluorescently label beads for injection:

1. A solution of 10 μL of undiluted streptavidin beads is first washed in the shaker for 40 min in 100 μL of washing solution and then sonicated for 5 min.
2. Washed beads are then incubated for 15 min in 100 μL of 2 μg/mL Atto565-biotin, and rinsed twice in 100 μL PBS 10 min. At each washing step, remove the supernatant using the magnetic separation rack.
3. Finally, beads are resuspended in 20 μL PBS (or protein/mRNA mix, *see* Subheading 3.5) and kept at 4 °C until use (within 2–3 days).

### 3.5 mRNA/Protein Mix Preparation

To visualize intracellular structures such as microtubules or DNA while performing spindle-pulling experiments, one can inject purified fluorescent proteins that are directly visible after injection, with a signal that is maintained for several hours, depending on the probe. In addition, if the purpose is to perform experiments after the 4–8 cell stage, one may inject mRNA probes, since a protein probe will be already too diluted at this stage, or degraded.

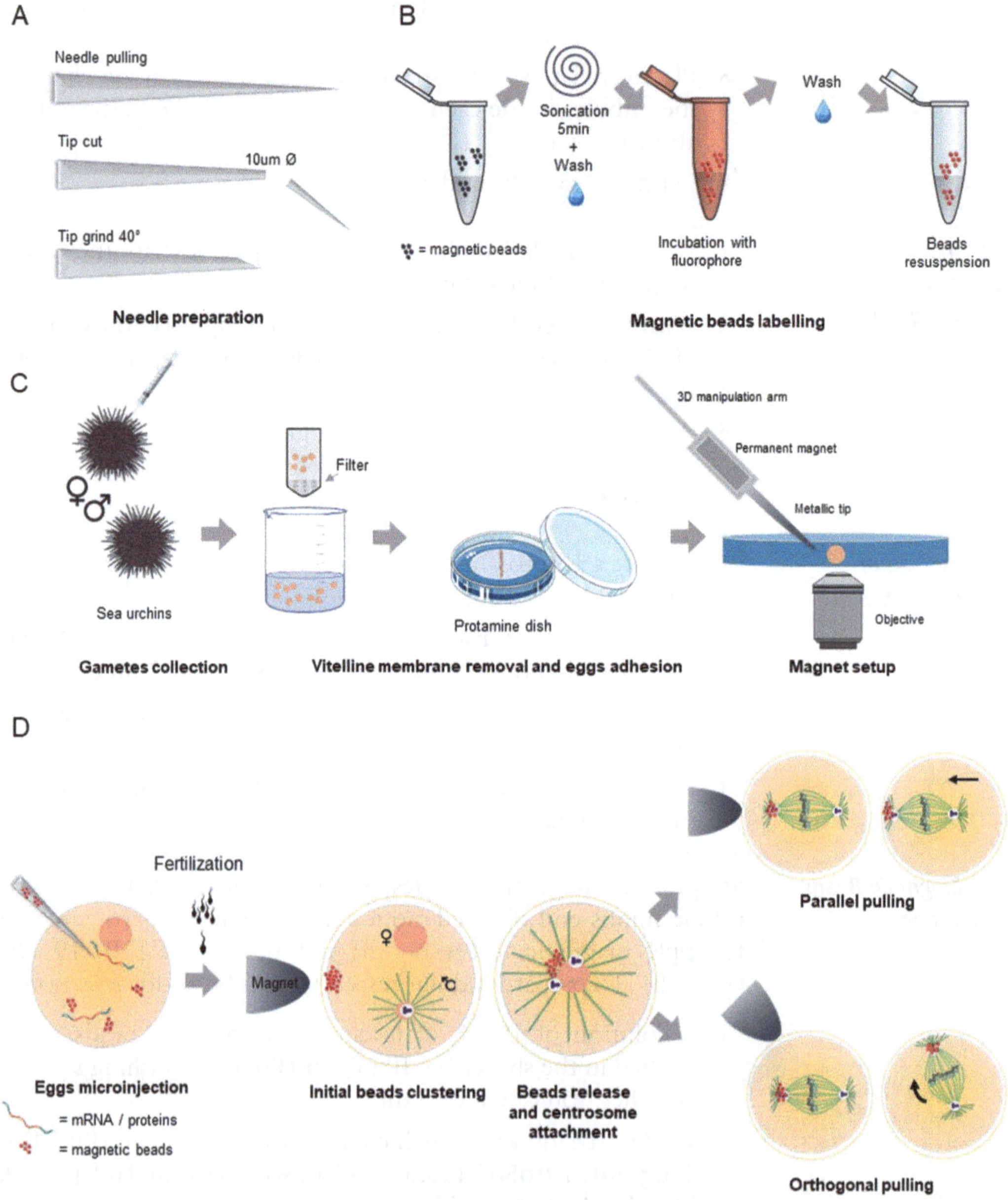


**Fig. 1** Procedures to manipulate spindle position using magnetic tweezers in sea urchin embryos. (**a**) Microinjection capillaries preparation (*see* Subheading 3.2). (**b**) Magnetic beads labeling with fluorophores. Magnetic beads in black, Biotin-Atto-labelled beads in red (*see* Subheading 3.4) (**c**) Gametes collection by intracelomic sea urchins injection. Vitelline membrane removal by filtering eggs through a 70 μm mesh, and eggs adhesion on 1% protamine dish with ASW.F to inject them. Microscope magnet setup composed of a 3D manipulation arm (controlled by a 3D micromanipulator), a permanent magnet, and a metallic gold-coated tip that creates a magnetic gradient (*see* Subheadings 3.1, 3.3, and 3.7) (**d**) Eggs microinjection with magnetic beads, mRNA and/or proteins. Fertilization. Initial beads clustering by approaching the magnet tip before aster centration, followed by magnet removal for beads release and their attachment to centrosomes. On the right side, the two types of possible spindle pulling: Parallel pulling, to generate spindle displacement along its initial long axis (black arrow, direction of displacement). Orthogonal pulling to generate spindle displacement and rotation (black arrow, rotation direction). In green, microtubules, in red, fluorescent beads; in orange pronuclei and nuclei; in purple, centrosomes and pericentriolar material; in blue, chromosomes. The dashed orange line surrounding the embryo represents the fertilization membrane (*see* Subheadings 3.6 and 3.7)

In contrast, mRNA probes are not useful for experiments that require a fluorescent signal before the 4–8 cell stage, as they need several hours to be expressed. It is also possible to combine in the same injection mRNA, proteins, and beads.

### 3.6 Protein Mix for Beads Injection

1. Take the concentrated and purified protein stock aliquot at −80 °C.
2. Resuspend the protein in PBS at the desired concentration and centrifuge 10 min at 16,000 *g* RT (to remove potential contaminants or aggregates), and only keep the supernatant for the next step.
3. At **step 3** of magnetic beads preparation (*see* Subheading 3.4), use 20 μL of this protein solution to resuspend beads instead of PBS alone. Prepare this mix just before injection to avoid unspecific binding of proteins to magnetic beads and keep it on ice during the experiment.

#### 3.6.1 mRNA Mix for Beads Injection

Capped-mRNA are synthetized with the kit mMessage mMachine and purified with the associated kit MEGAClearPURIF kit (*see* **Note 2d**).

Optional: If you have problems obtaining a good signal, you can increase mRNA stability and translation efficiency by adding a Poly(A) sequence with the Poly(A) Tailing kit.

1. Take the mRNA stock aliquot at −80 °C.
2. Centrifuge the aliquot 1 h at 19,000 rpm at 4 °C (*see* **Note 2d**)
3. Dissolve the desired volume of mRNA with the appropriate volume of beads + PBS mix prepared before (or beads + proteins mix, *see* Subheading 3.5).
4. Mix vigorously by pipetting to dissolve the mRNA properly and obtain a homogeneous bead suspension (just before filling the injection capillary). Keep the solution on ice during the experiment.

### 3.7 Eggs Injection

1. Before injection, 20 μL of unfertilized eggs are adhered on protamine-coated glass-bottom dishes along a vertical line (Fig. 1c) (*see* **Note 1**).
2. Install the dish on the inverted epifluorescent microscope, equipped with a micromanipulator and a micro-injection system.
3. Fill the injection capillary with 2–3 μL of injection mix, using epTIPS Microloader (*see* **Note 2a**).
4. Mount the capillary on the micro-injection system. Find the focus of the capillary tip using a 10× objective. Make sure that there is a constant visible flow of beads in your liquid medium,

and no blockage inside the capillary. Before starting, expel the air from the capillary to push the liquid at the tip extremity (*see* **Note 2b, c**).

5. Once the flow has stabilized by adjusting the injection pressure parameters (for this type of needle and beads, the injection pressure is around 12 psi with an injection time of 0.12 s), start injecting using the 20× objective. Try to inject eggs apically: apply light pressure to their surface, then insert the capillary on the egg by lowering it slightly. If the injection is done directly on the side of the embryo, there is a risk of damaging it too much, and the beads may come out directly through the injection hole. Do not leave unfertilized eggs on the protamine dish for more than 20–30 min without fertilizing them, as they will become too debilitated.
6. Once eggs have been injected, fertilize them. For that, dilute 1 μl of pure sperm in 2 mL of ASW.F and quickly mix strongly by pipetting to activate the sperm. Directly after, add one drop of sperm mix on eggs (*see* **Note 3**).

### 3.8 Pulling Forces Application on Mitotic Spindles

After fertilization, beads are spontaneously transported toward the centrosomes and form a large aggregate at the center of the aster. This aggregate remains stable and occasionally splits during centrosome duplication. Thus, beads sometimes end up on one or two spindle poles at metaphase. The targeting of beads towards the centrosomes presumably occurs in a dynein- and microtubule-dependent manner [19, 20].

1. Just after injection, approach the magnet tip to all the fertilized eggs that have beads inside to attract all the beads and form a "magnetic cluster" at the egg cortex. This will avoid bead interference with pronuclei fusion and nuclei centration mechanisms. Be careful; if you form the magnetic cluster after more than ~2 min of fertilization, you could alter aster centration. Then, remove the magnet, releasing the beads free in the embryo and wait until metaphase (Fig. 1d).
2. While waiting for metaphase, using the 20× objective, place the magnet tip on the side of the field of view (over a zone of the dish far from eggs to avoid force application on beads), and draw a circle (using ImageJ circle tool) to save the position. Then, remove the magnet using the micromanipulator option that saves the magnet position (*see* **Note 4**).
3. At metaphase, chromosomes marked with Hoechst, line up along the metaphase plate, and the spindle is easily visible as a dumbbell-shaped smooth region in DIC (Differential Interference Contrast), presumably because of its association with the endoplasmic reticulum [10]. This initial visualization allows

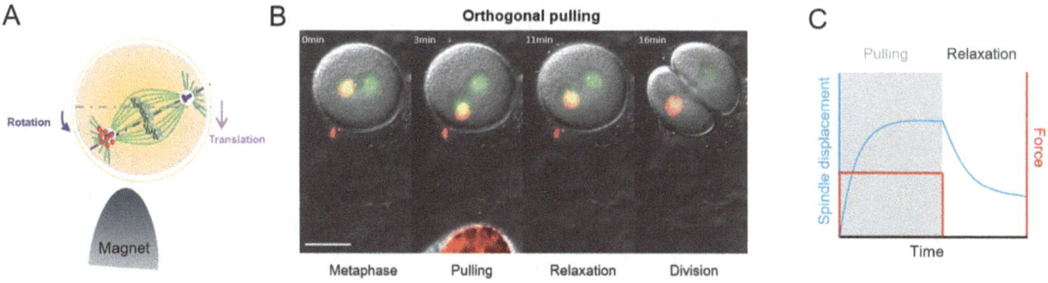


**Fig. 2** Spindle rotation by magnetic tweezers in sea urchin embryos. (**a**) Scheme of the pulling experiment performed in **b**. In green, microtubules, in red, fluorescent beads; in purple, centrosomes and pericentriolar material; in blue, chromosomes. The dashed orange line surrounding the embryo represents the fertilization membrane. (**b**) Time-lapse of an orthogonal pulling experiment at the 1cell stage (*see* Subheading 3.7). From left to right: [1] urchin egg in metaphase, [2] magnet approach and spindle rotation, [3] spindle relaxation to the cell center, [4] Cytokinesis and cell division. In green, Tau-Venus protein (microtubule marker); in red, fluorescent magnetic beads. Scale bar, 50 μm (**c**) Simplified graph representing the evolution of the magnetic force applied (red) related to spindle displacement (blue) over time. The gray area corresponds to the period of force application, and the white area to the magnet release, leading to spindle relaxation

for selecting spindles that are properly oriented along the axis of the magnet and defining the parallel and orthogonal axis directions for force applications.

4. Once the embryo is selected, place it near the circle previously drawn with ImageJ (that corresponds to the magnet tip saved position) and move down the magnet. Leave the magnet at a fixed position, either in the same axis of the spindle (for parallel pulls and spindle translation displacement) or orthogonal to the spindle (for orthogonal pulls and spindle rotation + translation displacements) (Figs. 1 and 2). Leave the magnet in that position or remove it depending on the purpose of the experiment (long or short force application). For experiments implicating long force application, the magnet can be progressively moved away from the egg using the automatic stage of the microscope.

### *3.9 Imaging*

Time-lapses of multiple fluorescence (depending on the probes used) are recorded on an inverted epifluorescence microscope (*see* **Note 5**). Time-lapses are typically performed at a rate of 1 image every 10 or 20 s, but this may vary depending on the purpose of the experiment.

### *3.10 Magnetic Tweezers Force Calibration*

In order to retrieve the force applied on spindles, the magnet and magnetic beads must be a priori calibrated [19, 21]. This is achieved in vitro using a test viscous solution fluid.

1. Prepare a diluted suspension of 2.8 μm (280-M Streptavidin) or 1 μm (MyOne Streptavidin C1) mono-dispersed super-paramagnetic beads.

2. To correct for the drift and slight movement of the fluid, mix the magnetic beads suspension with a small fraction of 0.5 μm nonmagnetic fluorescent tracers.
3. Place a large droplet of 80% glycerol (viscosity $8.0 \times 10^{-2}$ Pa sec at 22 °C) on the cover glass and put a few μL of the bead's suspension on top of the droplet. Mix the glycerol and beads very gently, close to the surface, and add more beads if needed.
4. Under the microscope, place the magnet tip close to a region with a good amount of beads away from the boundaries and film the movement of magnetic beads and fluorescent tracers at a rate of 3 frames per second.
5. Using the TrackMate plugin in Image J [22], find the position of beads in each frame. Employ MATLAB or Excel to compute the average velocity of nonmagnetic tracers and subtract it from the velocity of the magnetic beads. Then, find the corrected speed of the magnetic beads projected along the magnet tip axis, as a function of their distance to the magnet tip.
6. Using Stokes's law; $F = 6\pi\eta RV$, one can retrieve the magnet force as a function of the distance to the magnet, which is usually fitted using a double exponential function [19].
7. To compute the dependence of the force on aggregate size, use the same setup as above.
8. Prepare a solution of beads aggregate by mixing the beads suspension with 100 μL Poly-L-Lysine (1 mg/mL$^{-1}$) and incubating for 3 min, and place them in the test viscous solution.
9. Monitor the speed of aggregates at a fixed distance from the magnet tip (usually 100 μm, as in in vivo experiments).
10. Using Image J, compute the size of aggregates by measuring its two principal length axes, L1 and L2, and use an effective radius, $R = (\sqrt{L1 L2})/2$ to apply Stoke's law.
11. From the velocity of moving aggregates and Stoke's law, one obtains the size dependence of the magnetic force, which is usually best fitted with a cubic function [19].
12. Combine distance dependence and size dependence to calculate the exact force applied on spindles in vivo.

### 3.11 Analysis of Spindle Position and Orientation

1. Load time-lapses of spindles moving under external magnetic force on Fiji and use the TrackMate plugin in ImageJ [22] for analysis. Movies need to be checked and corrected manually in some cases.
2. Crop and rotate time-lapses to align the initial spindle axis to the horizontal X-axis.

3. Track the size of bead aggregates and their distance to the magnet tip and correct for the displacement of the egg in case it has moved during force application. Magnet tip position is tracked at each time point in the DIC channel and the position of bead aggregates are tracked from their fluorescence channel. The size of beads aggregate is also recorded at different time points to calculate a mean-size if this one changes during the time-lapse.
4. Spindle position and orientation are tracked manually in Fiji by tracking the centroids of two smooth disks, which correspond to spindle poles in the DIC channel. This allows computing spindle length, position, and angles with the magnetic force axis. Spindle displacements and magnetic forces are then projected along the horizontal X-axis for parallel pulling or along the Y-axis for orthogonal pulling.
5. Spindle position may also be recorded using the chromosome plate position in the Hoechst signal, but we often noticed some small delay in the motion of chromosomes as compared to that of spindle poles, presumably caused by some time-scales associated with internal viscoelastic structures that hold chromosomes.
6. Compute the complete spindle displacement as a function of time. This curve is usually renormalized by the instantaneous force application to compensate for force variations during each pull. Depending on the context, this renormalized curve can be fitted with different types of viscoelastic models to extract spindle viscous drag or spindle centering stiffness [10–12].

## 4 Notes

1. *Urchin eggs do not adhere to the protamine dish.* The diameter of eggs can vary slightly so that the vitelline membrane does not disappear even after passing through the 70-µm Nitex mesh. In such case, pass the eggs through a 60-µm mesh 3 times and place them in a new protamine dish.
2. *The capillary is clogged.* If there is no flux of beads from the capillary, even when the pressure is increased, the capillary may be clogged. This could happen for several reasons:
   (a) When filling the capillary with the injection mix, it is very important to avoid air bubbles that will prevent the liquid from flowing through the capillary tip properly. For that, remove the microloader tip from the capillary very gently.
   (b) Before finding the focus of the capillary to inject, it is important to activate the pressure inside the needle before

immersing it in ASW. This will avoid water penetration inside it, which may cause rapid irreversible protein aggregations at the needle tip and block it.

(c) Beads, protein clusters, debris in the medium or fragments of urchin egg membrane can clog the tip of the needle. In all these cases, the first solution is to use the "clean" option of the injection system to eject the clogging element by pressure or to wash the capillary tip thoroughly with a paper. If the flow does not return, the capillary must be changed. If the clogging is just occurring at the tip of the capillary, another possibility is to cut the tip using the border of the dish (be careful; tips with diameters higher than 10 μm may damage the egg) or forceps.

(d) Too large a volume of mRNA clog capillaries. If the departure concentration of mRNA stock is too low and you have to add a high volume for your final injection mix, the viscosity of the solution will increase too much. Because of that, it is important to centrifuge the mRNA aliquot 1 h at 19,000 rpm at 4 °C before using it. This step will limit the increase in viscosity of RNA solution that may block the injection capillary, and remove potential aggregates. If the mRNA solution viscosity still blocks the capillary, there are two more possibilities: (i) Synthetize a new stock of mRNA to obtain a higher concentration in order to use less volume (even with the same mRNA synthesis kit, final concentrations vary a lot between experiments), (ii) inject first your eggs with the beads and then change the capillary to inject the same eggs with mRNA separately. To inject mRNA alone without beads, non-coated and special capillaries should be used, with an inner filament inside (TW100F-4), which are more adapted to a viscous solution.

3. *Urchin eggs do not develop properly.* To evaluate eggs and sperm quality, it is recommended to perform fertilization tests before injection on non-injected eggs. This will allow the sperm concentration to be calibrated to add to the eggs to avoid multiple fertilization (polyspermy) of the same egg, which will impact embryo development.

4. *The spindle does not move with the magnet force.* If the spindle does not move after applying forces, it could be due to insufficient bead aggregate size. On the contrary, an overly large amount of injected beads could cause damage to the embryo. It is thus very important at the moment of bead injection to stabilize properly the beads flux. In addition, one can modulate the magnet-beads distance to increase or decrease force magnitude.

5. *High-resolution microscopy.* If the purpose is to visualize fine structures such as microtubules or F-actin, live imaging and pulling experiments can be performed on a spinning-disk confocal microscope, e.g., TI-Eclipse equipped with a CSU-X1FW spinning head, and an EM-CCD camera, using a 40× or 60× Oil-immersion objective equipped with a micromanipulator for magnetic tweezers.

## Acknowledgments

We thank all members of the Minc team for the discussion and technical help. J.X. acknowledges the "École Doctorale FIRE—Program Bettencourt," a fellowship from the Chinese Scholarship Council (201708070046) and from the labex "Who Am I." A.N. is supported by a Post-doc fellowship from the ARC Foundation (PDF2021070004082). This work was supported by the Centre National de la Recherche Scientifique (CNRS), the Université de Paris, and grants from La Ligue Contre le Cancer (EL2021. LNCC/ NiM), the Agence Nationale pour la Recherche (ANR, "TiMecaDev"), the Fondation Bettencourt Schueller ("Coup d'élan"), a scientific subvention from the foundation "Simone and Cino Del Duca" and the European Research Council (ERC CoG "Forcaster" no. 647073 and ERC PoC "MagMech" No. 101069173) to N.M.

### References

1. Green RA, Paluch E, Oegema K (2012) Cytokinesis in animal cells. Annu Rev Cell Dev Biol 28:29–58
2. Scholey JM, Brust-Mascher I, Mogilner A (2003) Cell division. Nature 422:746–752
3. Walczak CE, Heald R (2008) Chapter three—Mechanisms of mitotic spindle assembly and function. In: Jeon KW (ed) International review of cytology, a survey of cell biology. Academic, London, pp 111–158
4. Bosveld F, Markova O, Guirao B et al (2016) Epithelial tricellular junctions act as interphase cell shape sensors to orient mitosis. Nature 530:495–498
5. Minc N, Burgess D, Chang F (2011) Influence of cell geometry on division-plane positioning. Cell 144:414–426
6. Campinho P, Behrndt M, Ranft J et al (2013) Tension-oriented cell divisions limit anisotropic tissue tension in epithelial spreading during zebrafish epiboly. Nat Cell Biol 15:1405–1414
7. di Pietro F, Echard A, Morin X (2016) Regulation of mitotic spindle orientation: an integrated view. EMBO Rep 17:1106–1130
8. Lechler T, Mapelli M (2021) Spindle positioning and its impact on vertebrate tissue architecture and cell fate. Nat Rev Mol Cell Biol 22:691–708
9. Sallé J, Minc N (2021) Cell division geometries as central organizers of early embryo development. Semin Cell Dev Biol S1084-9521(21):00208-1
10. Xie J, Najafi J, Le Borgne R et al (2022) Contribution of cytoplasm viscoelastic properties to mitotic spindle positioning. Proc Natl Acad Sci USA 119:e2115593119
11. Garzon-Coral C, Fantana HA, Howard J (2016) A force-generating machinery maintains the spindle at the cell center during mitosis. Science 352:1124–1127
12. Anjur-Dietrich MI, Hererra VG, Farhadifar R et al. (2023) Clustering of cortical dynein

regulates the mechanics of spindle orientation in human mitotic cells. 2023.09.11.557210

13. Pierre A, Salle J, Wuhr M et al (2016) Generic theoretical models to predict division patterns of cleaving embryos. Dev Cell 39:667–682
14. Saleh J, Fardin MA, Barai A et al (2023) Length limitation of astral microtubules orients cell divisions in murine intestinal crypts. Dev Cell 58:1519–1533.e6
15. Nakajima Y, Meyer EJ, Kroesen A et al (2013) Epithelial junctions maintain tissue architecture by directing planar spindle orientation. Nature 500:359–362
16. Caussinus E, Gonzalez C (2005) Induction of tumor growth by altered stem-cell asymmetric division in Drosophila melanogaster. Nat Genet 37:1125–1129
17. Jankele R, Jelier R, Gönczy P (2021) Physically asymmetric division of the C. elegans zygote ensures invariably successful embryogenesis. elife:e61714
18. Mukherjee RN, Salle J, Dmitrieff S et al (2020) The perinuclear ER scales nuclear size independently of cell size in early embryos. Dev Cell 54:395–409 e7
19. Tanimoto H, Salle J, Dodin L et al (2018) Physical forces determining the persistency and centering precision of microtubule asters. Nat Phys 14:848–854
20. Salle J, Xie J, Ershov D et al (2019) Asymmetric division through a reduction of microtubule centering forces. J Cell Biol 218:771–782
21. Najafi J, Dmitrieff S, Minc N (2023) Size- and position-dependent cytoplasm viscoelasticity through hydrodynamic interactions with the cell surface. Proc Natl Acad Sci 120:e2216839120
22. Tinevez JY, Perry N, Schindelin J et al (2017) TrackMate: an open and extensible platform for single-particle tracking. Methods 115:80–90

# Chapter 7

# Photochemical Control of Cell Division Using a Photoswitchable CENP-E Inhibitor

**Akira Matsura, Shota Hiruma, Nobuyuki Tamaoki, and Ryota Uehara**

## Abstract

Photoswitchable compounds are potent tools for elucidating molecular functions in dynamic cellular processes. Photoswitchable inhibitors targeting various mitotic spindle factors have been developed. In this chapter, we describe experimental methods for photo-controlling mitotic chromosome dynamics using a recently developed photoswitchable inhibitor of mitotic kinesin, CENP-E. This inhibitor undergoes reversible photoisomerization to a more inhibitory *trans* or less inhibitory *cis* state by visible or UV light irradiation, respectively, enabling photoswitching of CENP-E function both in vitro and in vivo. First, we explain the procedures used to optimize the experimental condition for efficient photoswitching of CENP-E functionality in cultured cells. We then describe how to conduct de novo photo-control of mitotic chromosome motion using the inhibitor under a microscope.

**Key words** Mitotic kinesin, Photoswitch, Azobenzene, Chromosome segregation

## 1 Introduction

Cell division is a dynamic process in which diverse events, such as the reorganization of complex cellular structures and the rearrangement of organelles, proceed stepwise in a matter of minutes. Spatiotemporal regulation of these events is governed by numerous cell division-associated proteins whose exact functions are unidentified. Therefore, a valuable approach to understanding cell division mechanisms is to perturb the activities of cell division-associated proteins with high spatiotemporal resolution and examine their effects on cell division events. Optogenetic approaches using light-inducible hetero-dimerization systems for spatiotemporal perturbation of different mitotic factors have successfully elucidated their new functions in cell division control [1–3]. However, optogenetic approaches require genetic engineering of the target genes, limiting their utility and applications in different biological systems.

Tomomi Kiyomitsu (ed.), *The Mitotic Spindle: Methods and Protocols*, Methods in Molecular Biology, vol. 2872,
https://doi.org/10.1007/978-1-0716-4224-5_7,

As complementary to optogenetics, optochemistry using photo-controllable compounds is becoming a new powerful option for spatiotemporal perturbation of cell division proteins. Unlike optogenetics approaches, optochemistry allows photo-controlled protein perturbations in wild-type (genetically unmodified) cells or organisms. This feature enables applications of optochemical techniques to a wide range of biological materials and potentially to clinical therapy [4, 5]. There are older examples of perturbation of cell division progression by photoconversion of anti-mitotic compounds, such as studies using UV-inactivation of colcemid [6]. However, many sophisticated designs of photoswitchable molecules have been developed in recent years, facilitating reversible photocontrol of different enzymatic activities and cellular processes [7, 8]. For example, the development of azobenzene-derived photoswitches enabled optical control of microtubule or F-actin reorganization in living cells [9, 10].

This chapter explains experimental procedures of optochemical control of cell division in human cultured cells using a photoswitchable CENP-E inhibitor (PCEI-HU) recently developed by our group [11]. CENP-E is a mitotic kinesin motor protein essential for proper chromosome segregation [12, 13]. CENP-E binds to the kinetochores and transports mitotic chromosomes along spindle microtubules from spindle poles to the cell equatorial plane [12]. We developed PCEI-HU by replacing the core imidazopyridine ring in GSK923295, a preexisting allosteric inhibitor of CENP-E ATPase activity that locks CENP-E in a rigor microtubule-bound state [14], with a pyrazolylazo group (an azobenzene derivative). The introduced pyrazolylazo group undergoes reversible photoisomerization between more bulky *cis* configuration and elongated *trans* configuration upon UV or visible light irradiation, respectively (Fig. 1a). The *cis* or *trans* configuration restricts and permits the CENP-E inhibitory effect of PCEI-HU, respectively [11]. As detailed in the following sections, the fast photoswitching with sufficient thermal stability of each photoisomer makes PCEI-HU an ideal tool for spatiotemporal modulations of CENP-E activity within a dividing cell. The alternating UV/visible light irradiation of PCEI-HU-treated cells enabled repeatable photoswitching of mitotic chromosome congression during prometaphase in HeLa cells. This technique was also utilized to specify the exact mitotic stage at which the CENP-E function is required for chromosome alignment [11, 15]. We recently achieved photocontrol of mitotic progression in zebrafish embryos using PCEI-HU (Matsura et al., unpublished), suggesting its utility in wide variety of biological materials.

Here, we describe the experimental system and procedure we use for photo-controlling experiments in PCEI-HU-treated human cultured cells, emphasizing precautions and the knack particularly required for experiments using optochemical tools.

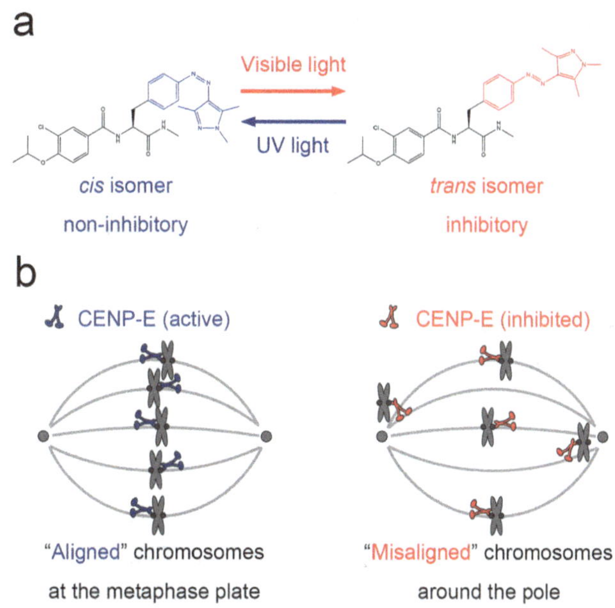


**Fig. 1** (**a**) Reversible photoisomerization between *cis* (non-inhibitory) and *trans* (inhibitory) state of PCEI-HU. (**b**) Schematic images of PCEI-HU-mediated photoswitching of CENP-E function to regulate mitotic chromosome alignment in a dividing cell

## 2 Materials

### 2.1 Compounds

1. Dimethyl sulfoxide (DMSO, cell culture grade).
2. Photoswitchable CENP-E inhibitor (PCEI-HU, FDV-0037, Funakoshi, Tokyo, Japan, *see* **Note 1**).
3. 10 mM MG-132 in DMSO. Store at −20 °C.
4. 1 mM SiR-DNA (CY-SC007, Cytoskeleton Inc., Denver, CO, USA) in DMSO. Store at −20 °C.

### 2.2 Cell Culture

1. Cell culture medium: Dulbecco's modified Eagles medium (DMEM) supplemented with 10% fetal bovine serum (FBS) and 1× antibiotic-antimycotic solution (A5955, Sigma-Aldrich, St. Louis, MO, USA).
2. Live-imaging medium: Phenol red-free DMEM supplemented with 10% fetal bovine serum (FBS) and 1× antibiotic-antimycotic solution
3. 24-well cell culture dish (662160, Greiner Bio-One GmbH, Frickenhausen, Germany)
4. Multi-well glass bottom dish (D141400, Matsunami, Osaka, Japan)
5. HeLa Kyoto cells (CVCL_1922)

### 2.3 Cell Fixation and Immunostaining

1. Phosphate-buffered saline (PBS).
2. Cell fixation buffer: 3.2% Paraformaldehyde (PFA) in PBS (1.04005.1000, Merck KGaA, Darmstadt, Germany).
3. Cell permeabilization buffer: 0.5% Triton X-100 in PBS
4. 4′,6-diamidino-2-phenylindole (DAPI)
5. Mounting media (Fluoromount, K024, Diagnostic BioSystems, Pleasanton, CA, USA)

### 2.4 Light Irradiation and Observation

1. LED controller unit (CL-1501, Asahi Spectra, Tokyo, Japan).
2. UV LED head unit (365 nm LED, CL-H1-365-9-1-B, Asahi Spectra, Tokyo, Japan, *see* **Note 2**).
3. Visible light LED head unit (505 nm LED, CL-H1-505-9-1-B, Asahi Spectra, Tokyo, Japan, *see* **Note 2**).
4. UV filter (WINCOS1905UH, Yamahira, Saitama, Japan, *see* **Note 3** and Fig. 2a).
5. Epifluorescence microscope (we use Ti2 from Nikon, Tokyo, Japan) with ×20 0.75 NA Plan-Apochromatic, a filter cube for far-red fluorescence (Nikon, *see* **Note 4**), Zyla4.2 sCMOS camera (Andor, Belfast, UK), and a stage top incubator (STXG-WSKMX-SET, Tokai Hit, Shizuoka, Japan). In the case of transparent light observation, *trans*-to-*cis* photoisomerization can be avoided by putting a UV filter in a transparent light path (*see* Fig. 4a).
6. Micro-manager (https://micro-manager.org/).
7. Image J (https://imagej.net/ij/).

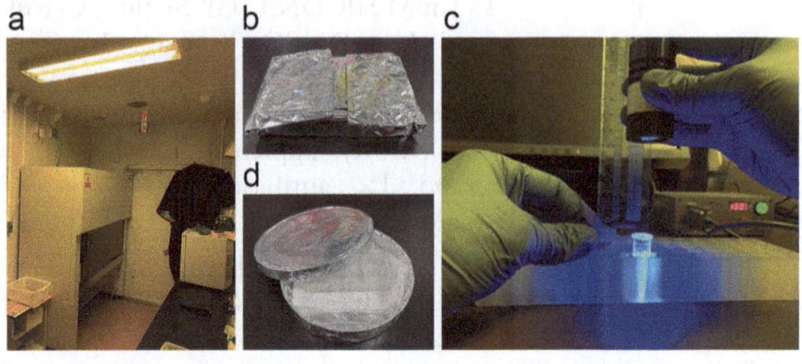

**Fig. 2** (**a**) A yellow room setup. (**b**, **d**) Light protection for PCEI-HU-treated cell culture using aluminum foil cover. (**c**) A visual presentation showing how to conduct light irradiation for photoisomerization of PCEI-HU in a 1.5 mL microtube

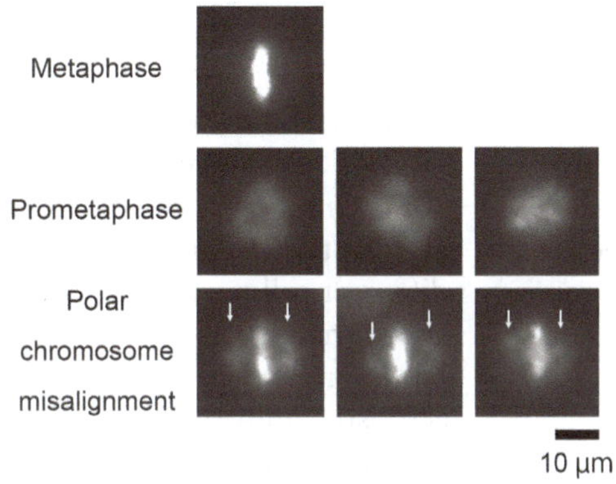


**Fig. 3** Examples of mitotic chromosome distributions in normal prometaphase, metaphase, or typical "polar chromosome misalignment" defects in CENP-E-inhibited cells. Arrows indicate misaligned chromosomes trapped at the spindle pole area

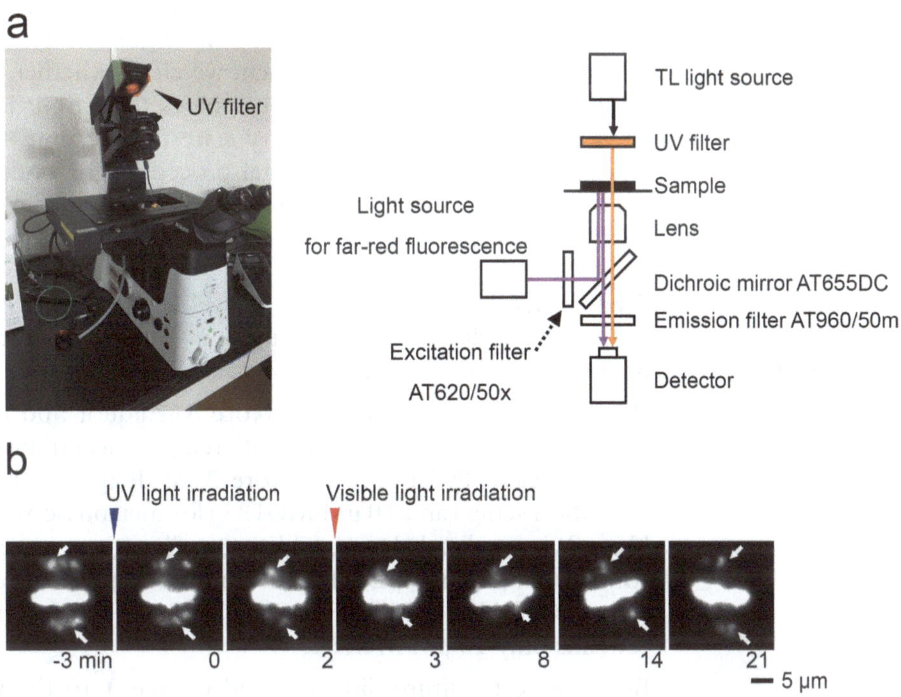


**Fig. 4** (**a**) Microscope configuration for de novo optical control of mitotic chromosome movement. To avoid photoconversion of PCEI-HU by observation light, one needs to use a filter cube for far-red fluorescence imaging or a UV filter in a transparent light path for transparent light observation. The arrowhead in the photo indicates the UV filter inserted into the transparent light path. (**b**) Live imaging of a PCEI-HU-treated HeLa Kyoto cell. PCEI-HU was introduced to cell culture at a non-irradiated state, and cell culture was irradiated with UV or visible light during live imaging at the timepoint indicated by arrowheads. DNA was stained using SiR-DNA. Arrows indicate the reversible motion of misaligned chromosomes during the UV/visible light irradiation cycle

## 3 Methods

### 3.1 Optimization of PCEI-HU Treatment Condition for Photoswitching Manipulation

To achieve efficient photoswitching manipulation, one needs to treat PCEI-HU at a concentration where CENP-E is blocked in a *trans*-enriched photostationary state but not *in a cis*-enriched state (*see* **Note 5**). Since the efficacy of PCEI-HU can differ among different cell types or experimental settings, it is crucial to find an appropriate condition of PCEI-HU treatment for each experimental design. When CENP-E function is blocked, cells suffer a characteristic mitotic defect with a small proportion of mitotic chromosomes mislocated at the spindle poles, while most other chromosomes are aligned on metaphase plate (polar chromosome misalignment) [14, 16] (Fig. 1b, see also Figs. 3 and 4b). Therefore, the effect of PCEI-HU on in vivo CENP-E function can be evaluated by observing mitotic chromosome configuration in DAPI-stained samples. Usually, we first specify the minimum concentration of non-irradiated PCEI-HU (100% *trans* isomer) that sufficiently induces polar chromosome misalignment phenotype (described in Subheading 3.1.1). Then, we check whether PCEI-HU at the same concentration does not block the CENP-E function in a *cis*-enriched photostationary state (Subheading 3.1.2). This section explains the experimental procedure for optimizing PCEI-HU concentration for in vivo photoswitching manipulation in HeLa Kyoto cells.

#### 3.1.1 Specification of Minimum PCEI-HU Concentration for Blocking CENP-E Function in a trans Photo-Isomeric State

1. Seed HeLa Kyoto cells on a 24-well cell culture dish at $1.0 \times 10^5$ cells/mL (*see* **Note 6**) in 500 μL cell culture medium. Incubate for 22 h at 37 °C with 5% $CO_2$.
2. (Perform in a yellow room, *see* **Note 3**) Aliquot 500 μL cell culture medium containing different concentrations of non-irradiated PCEI-HU (*see* **Note 7** for designing the concentration series) and 10 μM MG132 (for metaphase arrest, *see* **Note 8**) immediately before adding to cell culture. As negative controls, aliquot 500 μL culture medium that does not contain either PCEI-HU or MG-132 (DMSO vehicle control) or that contains only 10 μM MG132.
3. Bring the cell culture dish prepared in **step 1** to the yellow room (*see* **Note 9**). Replace culture media with the ones prepared in **step 2** (apply 500 μL for each well). Cover the cell culture dish with aluminum foil to avoid light exposure and return it to the incubator (Fig. 2b). Incubate for 2 h at 37 °C with 5% $CO_2$.
4. Prepare 3.2% PFA in PBS while incubating cells after PCEI-HU administration.
5. Bring the cell culture dish to the yellow room. Remove the culture medium, rinse once with 500 μL PBS, and apply

500 μL 3.2% PFA in PBS for each well. Fix cells for 10 min at 25 °C. All procedures hereafter can be conducted under regular environmental lights.

6. Rinse the fixed cells three times with 500 μL PBS for each well.
7. Remove PBS and apply 500 μl 0.5% Triton in PBS for each well. Permeabilize the fixed cells for 10 min at 25 °C.
8. Rinse the permeabilized cells three times with 500 μL PBS for each well.
9. Remove PBS and apply 100 μL PBS containing 1 μg/mL DAPI. Stain the samples for 30 min at 25 °C.
10. Rinse the stained cells three times with 500 μL PBS for each well. Remove PBS and apply 100 μL of mounting media for each well.
11. Image mitotic chromosome configuration in the DAPI-stained cells on an epifluorescence microscope. We typically use a 20× objective. Optimize imaging conditions by imaging a few mitotic cells (*see* **Note 10**). Once the imaging condition is determined, keep it unchanged throughout all samples.
12. Manually scan the partial area of each well and acquire images of all mitotic cells found, including prometaphase, metaphase, and anaphase cells. We typically image 100 mitotic cells for each condition in each experimental trial.
13. Open the acquired images using Image J. Categorize mitotic cells into three groups, which are "metaphase," "prometaphase," and "polar chromosome misalignment" (Fig. 3, *see* **Note 11**), based on their mitotic configurations. Calculate the frequency of polar chromosome misalignment in each condition using spreadsheet software and determine the minimum concentration of PCEI-HU (in *trans* isomer) sufficient for inducting polar chromosome misalignment.

#### *3.1.2 Comparison of PCEI-HU Effect on Mitotic Chromosome Alignment Between* cis *and* trans *Photo-Isomeric States*

1. Seed cells as in Subheading 3.1.1, **step 1**.
2. (Perform in a yellow room, *see* **Note 3**) Aliquot 500 μL culture medium containing PCEI-HU at the concentration determined in Subheading 3.1.1 into three 1.5 mL microtubes. As negative controls, aliquot 500 μL medium without PCEI-HU into three 1.5 mL microtubes.
3. To achieve *a trans-* or *cis-*enriched photo-stationary state, subject these tubes to three different patterns of light irradiation as follows.
   (a) Non-irradiated condition (100% trans isomer).
   (b) UV-irradiated condition (enriched with *cis* isomers). Irradiate PCEI-HU solution with 365 nm light at 100% for 60 s (*see* **Note 12**).

(c) Visible light-irradiated condition (enriched with *trans* isomers). Irradiate PCEI-HU solution with 505 nm light at 100% for 60 s (*see* **Note 12**).

For light irradiation, stand each 1.5 mL microtube upright on a tube rack, open the microtube lid, and irradiate the tube with light from 5 cm above the sample using LED lights (Fig. 2c, *see* **Note 2**). Cover each tube separately with aluminum foil to avoid cross-irradiation by the light for other tubes. Complete light irradiation immediately before adding to cell culture.

4. Add MG132 to each tube at a final concentration of 10 μM.
5. Treat cells prepared at **step 1** with the compound prepared at **steps 2–4** (apply 500 μL cultured medium with compound for each well). Incubate for 2 h at 37 °C with 5% $CO_2$ (*see* **Note 13**).
6. Conduct cell fixation, DAPI-staining, and microscopic observation as described in Subheading 3.1.1, **steps 4–13**. Once one confirms that PCEI-HU blocks CENP-E function in a *trans*- but not *cis*-enriched photostationary state at the optimized concentration, one can treat PCEI-HU at the concentration for the following photoswitching experiment.

### 3.2 Optical Control of Mitotic Chromosome Alignment Using PCEI-HU

This section explains a protocol for de novo optical manipulation of mitotic chromosome alignment using PCEI-HU in HeLa Kyoto cells. To conduct live imaging of mitotic chromosomes during optical manipulation, one must use infrared fluorescent DNA probes that do not interfere with the light for PCEI-HU conversion. Here, we describe a method using SiR-DNA probe (*see* **Note 14**).

1. Seed cells on multi-well glass bottom dish at $1.5 \times 10^5$ cells/mL in 100 μL culture medium. Incubate for 20 h at 37 °C with 5% $CO_2$.
2. Replace culture media with 100 μL live-imaging medium containing 0.5 μM SiR-DNA. Incubate for 2 h at 37 °C with 5% $CO_2$.
3. Replace culture media with 100 μL live-imaging medium to remove excess SiR-DNA from cell culture.
4. Bring the cell culture dish to the yellow room. Replace culture media with 100 μL live-imaging medium containing 10 μM MG132 with or without non-irradiated PCEI-HU at the concentration determined in Subheading 3.1.1 (*see* **Note 15**). Cover the cell culture dish with aluminum foil to avoid light exposure and return it to the incubator (Fig. 2d). Incubate for 2 h at 37 °C with 5% $CO_2$.

5. While incubating cells at **step 4**, prewarm the stage incubator on the epifluorescence microscope.
6. Set the cell culture dish prepared in **step 4** on the stage incubator on the microscope.
7. Optimize live imaging conditions by imaging a few mitotic cells. Once the imaging condition is determined, keep it unchanged throughout all samples. We typically use the multi-acquisition mode of Micro-manger software to image cells at 1-min time intervals with 3 z-slices of 4-μm steps (*see* **Note 16**). Be careful not to observe samples in visible fluorescence channels: a single irradiation of UV or visible excitation light would result in substantial photoisomerization of PCEI-HU.
8. For de novo photoconversion of PCEI-HU photostationary state, irradiate the cell culture dish from the top of the stage top incubator (we typically irradiate from 3.2 cm above the stage) within the interval of image acquisition using 365 nm or 505 nm LED light source (for *trans*-to-*cis* or *cis*-to-*trans* conversion, respectively). Sufficient photoswitching is achieved by irradiation of 365 nm light at 5% for 25 s or 505 nm light at 100% for 35 s (*see* **Notes 2** and **17**).
9. By alternating 365 nm and 505 nm light irradiation, one may achieve an on-off manipulation of chromosome alignment from the spindle pole to the cell equatorial plane in a reversible manner [11] (*see* **Note 18**). An example of alternating light irradiation is shown in Fig. 4b.

## 4 Notes

1. PCEI-HU is commercially available from Funakoshi, Tokyo, Japan. We store PCEI-HU as 10 mM stock in DMSO. The compound can be kept at −20 °C for at least 6 months. PCEI-HU is synthesized as a 100% *trans* isomer and keeps *trans* configuration unless irradiated with UV light. Exposure to environmental light, including room light and sunlight, can cause *cis* isomerization of PCEI-HU. Therefore, PCEI-HU, whether stored as powder or solution, should be kept with light shedding. We usually cover the stock tubes with aluminum foil. Handling of PCEI-HU, particularly the stock for long-term storage, should be performed in a yellow room, where radiation of <500 nm wavelength light is blocked (*see* **Note 3**). If one is inevitably handling PCEI-HU in a laboratory without a yellow room, cover all windows and turn off room lights to minimize environmental light irradiation.

2. By the measurement in our experimental setting, the intensity of the 365 or 505 nm LED light was 416 or 141 mW·cm$^{-2}$ at 100%, respectively, when the detector of a light meter was irradiated from 1 cm above. Note that there is an optimal wavelength window for efficient photoisomerization: we have found that either 505 nm or 510 nm light enables sufficient *cis*-to-*trans* photoisomerization, whereas 436 nm light results in 10%-reduced photoisomerization.
3. An instant yellow room can be set up by simply covering all light sources in a room with a UV filter (Fig. 2a). We use the WINCOS1905UH filter, which shields <500 nm wavelength light to less than 1%.
4. We use the AT620/50× excitation light filter for far-red fluorescence imaging, AT655DC dichroic mirror, and AT960/50m emission light filter (Chroma, VT, USA). The excitation filter shield <590 nm light more than 99%.
5. We found that PCEI-HU at the *cis*-enriched photostationary state was about 10 times less efficient in blocking CENP-E function than at the *trans*-enriched photostationary state both in vitro and in vivo [11]. Therefore, when introduced at excessively high concentrations (e.g., 10 times higher than the appropriate concentration range), PCEI-HU at the *cis*-enriched photostationary state also blocks the CENP-E function.
6. Cell seeding density should be optimized when using different cell lines.
7. We recommend initially trying a PCEI-HU concentration range from 0.2 to 2 μM. We have achieved efficient photo-switching of CENP-E function with 0.25–1 μM PCEI-HU in all mammalian cell lines tested so far (including HeLa Kyoto, HAP1, LLC-PK1, and MDCK).
8. We recommend co-treating PCEI-HU with MG132 to facilitate the evaluation of CENP-E function in mitotic cells. In asynchronous cell populations, a substantial proportion of mitotic cells are at early prometaphase, when most mitotic chromosomes remain unaligned. At this phase, it is not easy to evaluate the functionality of CENP-E based on mitotic chromosome configuration. MG132 blocks the transition from metaphase to anaphase by inhibiting APC/C-mediated proteolysis. Therefore, MG132 synchronizes mitotic cells at a stage where most chromosomes align at the metaphase plate, enabling efficient detection of polar chromosome misalignment upon inhibition of CENP-E function.
9. In case the yellow room is not equipped with cell culture hoods, this process may be performed on a pre-cleaned regular bench in the yellow room. Alternatively, if one prefers to handle

cell culture in a hood, one may perform this process in a regular laboratory (i.e., non-filtered room) with minimized environmental lights (*see* also **Note 1**).

10. Try different levels of ND filter and excitation light exposure time to obtain clear mitotic chromosome images without saturated pixels. Optimize the number and interval of z-stacks to cover the entire volume of mitotic cells.

11. By definition of prometaphase (i.e., mitotic cells with at least one misaligned chromosome), cells with polar chromosome misalignment are generally categorized as prometaphase cells. In this experiment, however, it is crucial to distinguish cells with polar chromosome misalignment from other types of prometaphase cells. This is because CENP-E inhibition induces the former but not the latter mitotic chromosome configuration. By quantifying the frequency of polar chromosome misalignment, we can monitor the inhibitory state of CENP-E in mitotic cells.

12. In aqueous environments, UV or visible light irradiation makes PCEI-HU reach a photostationary state (PSS) containing >90% *cis* (non-inhibitory) or >85% *trans* (inhibitory) isomer, respectively [11]. This *trans*-to-*cis* or *cis*-to-*trans* conversion of PCEI-HU occurs within 20 or 10 s, respectively [11]. The efficiency and dynamics of photoisomerization can be checked using reversed-phase high-performance liquid chromatography (RP-HPLC) or UV-vis absorption spectra analysis [11].

13. The thermal relaxation lifetime for *cis* PCEI-HU isomer is > 24 h [11]. Therefore, once converted to *trans* or *cis* isomer by -pre-irradiation at **step 3** in Subheading 3.1.2, most PCEI-HU molecules keep their isomeric state throughout the 2-h incubation.

14. Alternatively, genetically coded near-infrared fluorescent probes for chromosomes, such as histone H2B-emiRFP670 (a gift from Vladislav Verkhusha, Addgene plasmid# 136571 http://n2t.net/addgene:136571; RRID: Addgene_136571) [17] can be used.

15. In this protocol, we initially treat cells with non-irradiated PCEI-HU to conduct a photo-switching experiment with mitotic cells that initially suffer polar chromosome misalignment. In the case of performing an experiment with mitotic cells initially treated with PCEI-HU at a *cis*-enriched photostationary state, pre-irradiate PCEI-HU with 365 nm light as described in Subheading 3.1.2, **step 3. (b)** (use live-imaging medium instead of cell culture medium).

16. We recommend using a confocal microscope with higher magnifications to trace the behavior of single chromatids.

17. For the de novo *cis* isomerization experiment, finding an appropriate range of light amount within which sufficient photoisomerization is achieved with minimum phototoxic effects on cellular processes is important. We recommend comparing the mitotic progression of non-treated cells with or without UV irradiation to evaluate potential phototoxicity in the designed UV irradiation condition. Consider reducing light power or irradiation time when detecting phototoxic effects.
18. CENP-E inhibition by PCEI-HU does not induce re-misalignment of chromosomes once they are entirely aligned at the metaphase plate. Therefore, cells should be irradiated with visible light before all chromosomes align to achieve an on-off manipulation of chromosome motion with UV/visible light irradiation cycles.

## Acknowledgments

Study in the Uehara laboratory is supported by JSPS KAKENHI (Grant Numbers JP19KK0181, JP19H05413, JP19H03219, JPJSBP120193801, JP21K19244, and JP24K02017 to Ryota Uehara), the Princess Takamatsu Cancer Research Fund, the Orange Foundation, the Smoking Research Foundation, Daiichi Sankyo Foundation of Life Science, the Terumo Life Science Foundation, the Hoansha Foundation, and the Akiyama Life Science Foundation. Akira Matsura is supported by JST SPRING, Grant Number JPMJSP2119. We thank Kazuya Matsuo for advice on experimental design. The authors declare no competing financial interests.

## References

1. Okumura M, Natsume T, Kanemaki MT et al (2018) Dynein-Dynactin-NuMA clusters generate cortical spindle-pulling forces as a multi-arm ensemble. elife 7. https://doi.org/10.7554/eLife.36559
2. Wagner E, Glotzer M (2016) Local RhoA activation induces cytokinetic furrows independent of spindle position and cell cycle stage. J Cell Biol 213(6):641–649. https://doi.org/10.1083/jcb.201603025
3. Zhang H, Aonbangkhen C, Tarasovetc EV et al (2017) Optogenetic control of kinetochore function. Nat Chem Biol 13(10):1096–1101. https://doi.org/10.1038/nchembio.2456
4. Dunkel P, Ilaš J (2021) Targeted cancer therapy using compounds activated by light. Cancers 13(13):3237
5. Fuchter MJ (2020) On the promise of photopharmacology using photoswitches: a medicinal chemist's perspective. J Med Chem 63(20): 11436–11447. https://doi.org/10.1021/acs.jmedchem.0c00629
6. Aronson J, Inoué S (1970) Reversal by light of the action of N-methyl N-desacetyl colchicine on mitosis. J Cell Biol 45(2):470–477. https://doi.org/10.1083/jcb.45.2.470
7. Welleman IM, Hoorens MWH, Feringa BL et al (2020) Photoresponsive molecular tools for emerging applications of light in medicine. Chem Sci 11(43):11672–11691. https://doi.org/10.1039/d0sc04187d
8. Volarić J, Szymanski W, Simeth NA et al (2021) Molecular photoswitches in aqueous

environments. Chem Soc Rev 50(22): 12377–12449. https://doi.org/10.1039/D0CS00547A

9. Borowiak M, Küllmer F, Gegenfurtner F et al (2020) Optical manipulation of F-actin with photoswitchable small molecules. J Am Chem Soc 142(20):9240–9249. https://doi.org/10.1021/jacs.9b12898
10. Borowiak M, Nahaboo W, Reynders M et al (2015) Photoswitchable inhibitors of microtubule dynamics optically control mitosis and cell death. Cell 162(2):403–411. https://doi.org/10.1016/j.cell.2015.06.049
11. Mafy NN, Matsuo K, Hiruma S et al (2020) Photoswitchable CENP-E inhibitor enabling the dynamic control of chromosome movement and mitotic progression. J Am Chem Soc 142(4):1763–1767. https://doi.org/10.1021/jacs.9b12782
12. Wood KW, Sakowicz R, Goldstein LS et al (1997) CENP-E is a plus end-directed kinetochore motor required for metaphase chromosome alignment. Cell 91(3):357–366. https://doi.org/10.1016/s0092-8674(00)80419-5
13. Putkey FR, Cramer T, Morphew MK et al (2002) Unstable kinetochore-microtubule capture and chromosomal instability following deletion of CENP-E. Dev Cell 3(3):351–365. https://doi.org/10.1016/s1534-5807(02)00255-1
14. Wood KW, Lad L, Luo L et al (2010) Antitumor activity of an allosteric inhibitor of centromere-associated protein-E. Proc Natl Acad Sci USA 107(13):5839–5844. https://doi.org/10.1073/pnas.0915068107
15. Yoshizawa K, Matsura A, Shimada M et al (2023) Tetraploidy-linked sensitization to CENP-E inhibition in human cells. Mol Oncol 17(6):1148–1166. https://doi.org/10.1002/1878-0261.13379
16. Yao X, Abrieu A, Zheng Y et al (2000) CENP-E forms a link between attachment of spindle microtubules to kinetochores and the mitotic checkpoint. Nat Cell Biol 2(8):484–491. https://doi.org/10.1038/35019518
17. Matlashov ME, Shcherbakova DM, Alvelid J et al (2020) A set of monomeric near-infrared fluorescent proteins for multicolor imaging across scales. Nat Commun 11(1):239. https://doi.org/10.1038/s41467-019-13897-6

# Part III

## Quantitative Live Imaging

# Chapter 8

# Quantitatively Analyzing the Morphological Growth Dynamics of Bipolar and Non-bipolar Spindles Using Xenopus Egg Extract

**Akihiro Tanaka, Tatsuya Fukuyama, Yusuke T. Maeda, and Yuta Shimamoto**

## Abstract

The bipolar shape of the microtubule-based spindle is a pivotal morphological phenotype for accurate chromosome segregation during cell division. However, existing descriptions of spindle morphogenesis remain largely qualitative. Here, we introduce a method that provides a quantitative description of the morphological growth dynamics of spindles using Xenopus egg cytoplasmic extract and a computational image analysis pipeline. Our method allows for determining the characteristic growth dynamics of bipolar and non-bipolar spindles, comparing their spatiotemporal differences and changes in response to physical and biochemical perturbations.

**Key words** Xenopus egg extract, Spindle assembly, Morphodynamics, Large-field confocal live imaging, Quantitative shape analysis, Machine learning

## 1 Introduction

The spindle, comprised of microtubules and microtubule-associated proteins, plays a crucial role in the segregation of chromosomes during cell division. The accuracy of segregation depends on the bipolar configuration of the spindle. However, aberrant morphological phenotypes, such as multipolar spindles, sporadically arise in a cell population, leading to chromosome segregation errors. The incidence of spindle assembly failure varies among cell types and species, with some ranging from 1 in 1000–10,000 in somatic cells to 1 in 10 in oocytes [1, 2].

---

**Supplementary Information** The online version contains supplementary material available at https://doi.org/10.1007/978-1-0716-4224-5_8.

Tomomi Kiyomitsu (ed.), *The Mitotic Spindle: Methods and Protocols*, Methods in Molecular Biology, vol. 2872, https://doi.org/10.1007/978-1-0716-4224-5_8,

Xenopus egg extract is a nearly undiluted cytoplasm derived from unfertilized eggs of African clawed frogs, offering a variety of unique assays owing to its cell-free nature. The extract can be prepared by mechanical crushing and fractionation of laid eggs, while the cell cycle is arrested at the metaphase of meiosis II by cytostatic factor (CSF). The addition of $Ca^{2+}$ releases the cell-cycle arrest and drives the extract into interphase, initiating nuclear assembly around added sperm. Subsequent restoration of CSF activity cycles the extract back into the M-phase, initiating nuclear envelope breakdown and spindle assembly. In the homogenized bulk extract cytoplasm, hundreds of spindles can be simultaneously assembled around replicated chromosomes. Among these spindles, 70–80% typically exhibit a bipolar configuration, while the remaining 20–30% display non-bipolar shape phenotypes, including multipolar structures [3–5]. The extract allows for controlled biochemical and physical perturbations, rendering the system invaluable for studying spindle assembly mechanics. However, existing analyses are largely limited to the steady-state dynamics of spindles that had already reached metaphase or the 'frozen' snapshots of spindle assembly intermediates.

Here, we present a method for quantitatively analyzing the temporal growth dynamics of spindles from nuclear envelope breakdown to steady-state metaphase (Fig. 1) [6]. Using cytoplasmic extracts and an in-house developed image analysis pipeline, our approach captures the evolving sequence of multiple spindle assemblies that simultaneously take place within the bulk cytoplasm. The acquired timelapse data are subjected to an image analysis pipeline that integrates Fourier-based shape decomposition and machine learning-based phenotype sorting. Originally developed for studying the shape dynamics of moving amoeba cells [7], our computational approach provides a quantitative platform for determining the growth dynamics of spindles. The outcome of the analysis includes the growth trajectories of individual spindles traveling across a two-dimensional "morphospace," which visually illustrates the characteristic assembly path associated with each spindle shape phenotype (e.g., bipolar, multipolar).

## 2 Materials

### 2.1 Solutions

All solutions are prepared using ultrapure water and research-grade reagents and sterilized using a membrane filter. The prepared solutions are stored at 4 °C unless otherwise noted.

1. *MMR (25×):* 125 mM HEPES, 2.5 mM EDTA, 2.5 M NaCl, 50 mM KCl, 25 mM $MgCl_2$ and 50 mM $CaCl_2$; pH 7.85 adjusted using NaOH.
2. *XB salt (20×)*: 2 M KCl, 20 mM $MgCl_2$ and 2 mM $CaCl_2$.

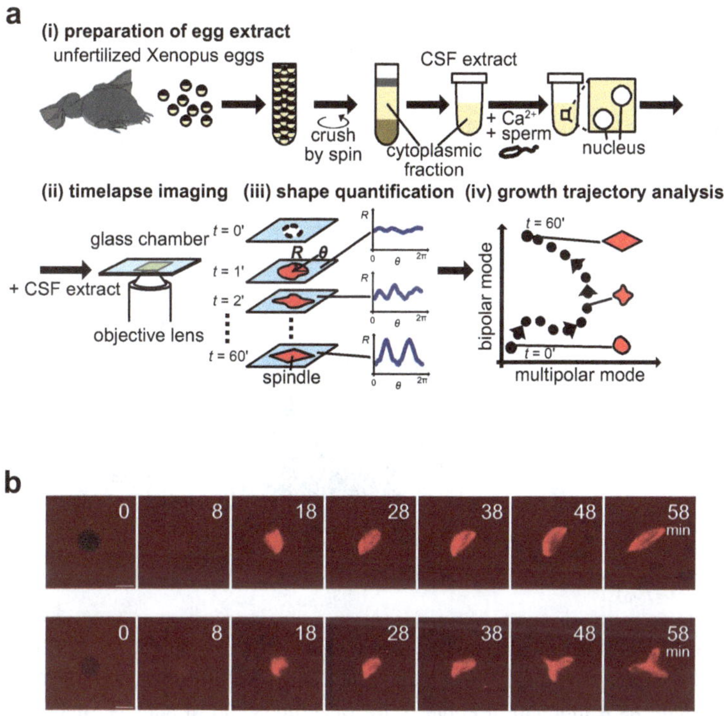


**Fig. 1** Quantitative analysis of spindle-growth dynamics in frog egg extract. (**a**) Assay workflow. (i) Unfertilized eggs of *Xenopus laevis* are crushed and fractionated by centrifugation. The cytoplasmic fraction, which maintains the M-phase arrest by CSF activity, is retrieved and added with $Ca^{2+}$ and sperm to drive the cell cycle into interphase. Subsequent dilution of the interphase extract with the CSF extract drives the cell cycle back to M-phase and spindle assembly starts. (ii) The spindle assembly process is observed by immediately sealing the reaction in a glass chamber and visualizing microtubules using a spinning-disk confocal fluorescence microscope. Multiple spindle assemblies can be recorded by raster-scanning the sample stage while performing timelapse. (iii) The acquired timelapse images are used to quantify spindle shape dynamics over time by tracing the spindle contour at each time frame using a rotational vector $R(\theta)$ drawn from the structure's center (*see* Fig. 2a). (iv) The spindle shape profile is then decomposed into discrete Fourier modes, including bipolar and multipolar shape modes. The calculated bipolar and multipolar shape modes are plotted on a two-dimensional morphospace to generate the spindle's growth trajectory from nuclear envelope breakdown to steady-state metaphase. (**b**) Example timelapse images of growing spindles. Bipolar (upper) and multipolar (lower) spindle assembly cases are shown. Scale bars, 20 μm

3. *2 M sucrose*
4. *1 M K-HEPES*: pH 7.7 adjusted using KOH.
5. *0.5 M EGTA*: pH 7.7 adjusted using KOH.
6. *1 M* $MgCl_2$
7. *5%(w/v) Gelatin*: bovine skin type B in water. Heat to dissolve and store at −20 °C.
8. *Cyto-D*: 10 mg/mL cytochalasin D dissolved in DMSO. Store at −20 °C.

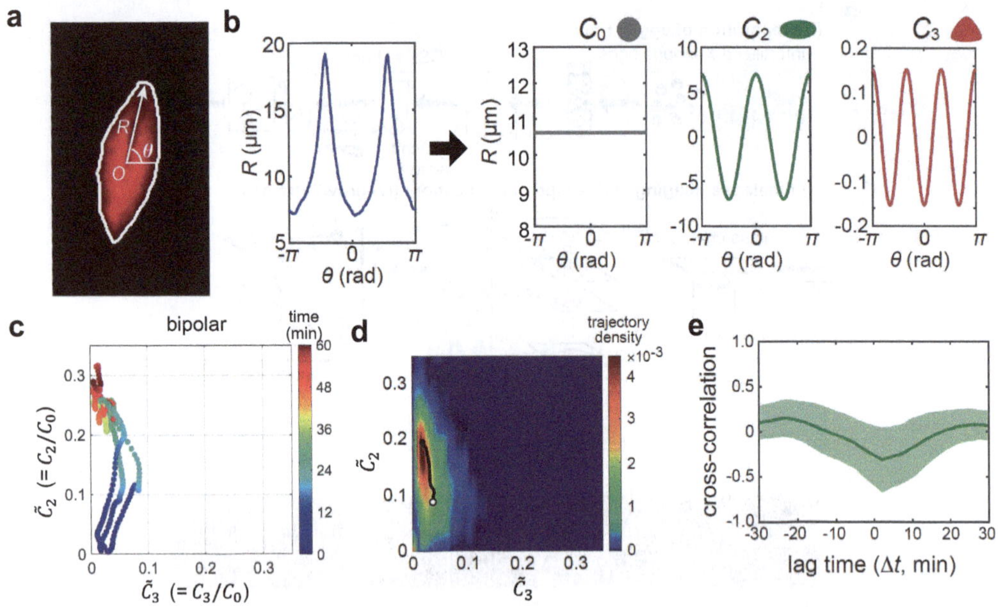

**Fig. 2** Morphospace analysis of spindle-growth dynamics. (**a**) Detection of spindle contour in the microtubule channel image after denoising. The rotational vector, ***R***, drawn from the structure's centroid, *O*, is used to trace the spindle outline as a function of angle $\theta$. (**b**) Decomposition of spindle shape using Fourier-based analysis. The length profile of ***R*** as a function of $\theta$ (blue-colored trace) is decomposed into discrete Fourier modes, $C_n$ (traces in other colors). (**c**) Time evolution of the normalized Fourier modes, $\tilde{C}_2$ and $\tilde{C}_3$ , revealing spindle growth trajectories that travel across the morphospace. Three examples of bipolar assembly cases are shown. Colors indicate the elapsed time. (**d**) Trajectory landscape. The growth trajectory data from multiple bipolar spindles ($n = 87$) are projected onto the single plane and the probability density of the trajectories at each coordinate point is calculated. The warmer color indicates a higher probability of trajectory passage. The solid line indicates the mean trajectory line; the start and end points are shown in open and filled circles. (**e**) An example cross-correlation plot of $\tilde{C}_2$ and $\tilde{C}_3$ for bipolar assembly case. For multipolar assembly case, see Fukuyama et al. [6]

9. *Protease Inhibitors (LPC)*: 10 mg/mL each of Leupeptin, Pepstatin A and Chymostatin dissolved in DMSO. Store at −20 °C.
10. *Energy mix (50×)*: 190 mM Creatin phosphate, 25 mM ATP, 25 mM $MgCl_2$ and 5 mM EGTA; adjust pH 7.7 using KOH and store at −80 °C.
11. *Pregnant mare serum gonadotropin (PMSG)*: 100 U/mL in water. Store at −20 °C.
12. *Human chorionic gonadotropin (hCG)*: 1000 U/mL in water. Store at −20 °C.
13. *Sperm dilution buffer (5×)*: 50 mM K-HEPES (pH 7.7), 5 mM $MgCl_2$, 500 mM KCl, 750 mM sucrose and 50 μg/mL cytochalasin D. Store at −20 °C.

14. *10 mM $Ca^{2+}$*: 10 mM $CaCl_2$ dissolved in 1× Sperm Dilution Buffer. Store at −20 °C.
15. *XB:* 1× XB salt with 50 mM sucrose and 10 mM K-HEPES. Adjust pH to 7.7 using 10 N KOH. Prepare 1 L for each extract preparation. Adjust the temperature to 18 °C.
16. *CSF-XB:* 5 mM EGTA and 1 mM $MgCl_2$ in XB. Prepare 250 mL for each extract preparation. Adjust the temperature to 18 °C.
17. *CSF-XB+PI:* 1/1000× LPC stock in CSF-XB. Prepare 90 mL for each extract preparation. Adjust the temperature to 18 °C.
18. *Cysteine solution:* 165 mM L-cysteine and 1× XB salt. Adjust pH to 7.7 using 10 N NaOH. Prepare 400 mL for each extract preparation. Adjust the temperature to 18 °C.

### 2.2 Reagents

1. *Demembranated sperm:* A working stock of ~$2 \times 10^4$ sperm/μL can be flash frozen and stored at −80 °C; prepare according to Murray. [8].
2. *X-rhodamine-labeled tubulin*: A working stock of ~1 mg/mL diluted in CSF extract can be flash frozen and stored at −80 °C; prepare according to Hyman et al. [9].
3. *SYTOX green DNA dye*: A working stock can be prepared at the final concentration of 100 nM in CSF-XB. Choose a proper fluorescent dye depending on your microscope setting. The dye concentration should be optimized to achieve a sufficient signal-to-noise ratio and minimized phototoxicity.
4. *VaLaP*: used to seal the imaging chamber. Mix equal weights of Vaseline, Lanolin and Paraffin in a beaker and heat the mixture to ~80 °C until thoroughly blended. Store at room temperature and warm to melt when required.

### 2.3 Equipment

1. Disposable 1-mL syringes
2. Syringe needles (27 gauge for hormone injection; 16 gauge for extract retrieval)
3. Pasteur glass pipette with a wide-bored, fire-polished tip (~5 mm in inner diameter)
4. Cylinders (100 mL, 250 mL, 1000 mL and 2000 mL)
5. Ultracentrifuge (e.g., Optima XE-90, Beckman)
6. Swing rotor (e.g., SW55 Ti, Beckman)
7. Thin-wall, ultra-clear centrifuge tubes (No.344057, Beckman)
8. Clinical centrifuge (e.g., 5702R, Eppendorf)
9. Glass beaker (500 mL)
10. Glass petri dish (15 cm diameter and 7.5 cm height)
11. Aspirator

12. Sample tubes (1.5 mL)
13. Glass coverslips (18 × 18 mm; thickness No.1)
14. Glass slides (24 × 60 mm)
15. Fluorescence imaging microscope equipped with a spinning-disk confocal unit (e.g., Yokogawa CSU-X1), excitation lasers (e.g., 488 nm and 561 nm, OBIS, Coherent), a medium power, high numerical aperture objective lens (e.g., 20× and 0.8 NA, MDR70270, Nikon), a motorized two-axis sample stage (e.g., MS-2000, Applied Scientific Instrumentation) and a high sensitivity camera (e.g., scientific CMOS, Neo, Andor).
16. Image acquisition software (e.g., NIS-Elements, Nikon) equipped with timelapse and raster-scan modes. The software controls the timing of laser excitation, image acquisition, and the movement of the sample stage.
17. Image analysis software (MATLAB and ImageJ)

### 2.4 Animals

1. Frogs (*Xenopus laevis*). Refer to Xenbase.org (https://www.xenbase.org/xenbase/) for the supply company near your location. Female frogs are used for egg collection and can be re-used after 3–4 months of a resting period. Male frogs are used for the preparation of demembranated sperm. Animal-related experiments must be conducted in strict accordance with the protocol approved by an institutional committee.

## 3 Methods

Extracts are prepared by mechanically crushing unfertilized eggs according to the established method [10]. These eggs are naturally arrested at metaphase of meiosis II by cytostatic factor (CSF) and maintained the arrest throughout the preparation using buffers containing a $Ca^{2+}$ chelator. To initiate spindle assembly, the prepared extract is first cycled to interphase by the addition of $Ca^{2+}$ along with demembranated sperm. Following the formation of nuclei, the extract is cycled back into metaphase by diluting in CSF-arrested extract. Immediately after, the reaction is sandwiched between a clean glass slide and a coverslip and is subjected to time-lapse imaging. The imaging is performed at minute time intervals with minimized phototoxicity. Employing raster scanning of the sample stage, multiple spindles that grow and mature into the steady-state structures can be imaged nearly simultaneously. Finally, the acquired spindle image data are subjected to computational analysis using an in-house MATLAB program, which quantitatively determines the characteristic growth dynamics of bipolar and non-bipolar spindles. The entire procedure is summarized in Fig. 1a.

### *3.1 Preparation of Egg Extract*

1. Prime female frogs with the injection of PMSG on Day 1 and Day 3 (50 U per frog per injection). Complete the second injection within 2 weeks before extract preparation (*see* **Note 1**).
2. One day before the extract preparation, inject the primed frogs with hCG (500 IU per frog). Place the frogs in separate containers filled with 2 L of 1× MMR. Maintain the frogs overnight in the dark in a 16 °C incubator.
3. Collect the laid eggs following ~16 h post hCG injection and remove eggs of abnormal morphology (e.g., puffy-shaped, smaller-sized, non-spherical or pigmented ones). Use a wide-bore glass pipette to handle eggs. Any debris and dirt should also be removed throughout the washing steps.
4. Wash the remaining eggs with 1× MMR. Keep harvesting eggs of abnormal morphology.
5. Transfer the washed eggs to a glass beaker pre-coated with gelatin (*see* **Note 2**).
6. Dejelly the eggs. Add cysteine solution (~200 mL) to the beaker and gently swirl. After 2–3 min, discard the solution and add the remaining cysteine solution. Repeat until the jelly coat is removed. Do not exceed 9 min as prolonged exposure to cysteine may damage eggs (*see* **Note 3**).
7. Place XB (~250 mL) in a petri dish pre-coated with gelatin. Transfer the eggs to the petri dish.
8. Wash the eggs with buffers in the following order: XB, CSF-XB and CSF-XB+PI. Gently swirl the petri dish and "puff" the eggs using a wide-bore pipette to thoroughly remove residual cysteine (*see* **Note 4**).
9. Place 1 mL CSF-XB+PI and 10 μL Cyto-D into thin-wall centrifuge tubes and transfer the eggs to the tubes. Fill each tube as much as possible with eggs.
10. Spin to pack the eggs in a clinical centrifuge at 600× *g* for 1 min. Aspirate off the buffer.
11. Spin again at 600× *g* for 1 min and then at 1400× *g* for 30 s. Aspirate off the buffer.
12. Transfer the tubes to a swing bucket pre-chilled to 16 °C.
13. Spin at 10,000 rpm for 15 min at 16 °C using a SW55 rotor. Eggs are crushed and separated into layers. After ultracentrifugation, place the tubes on ice.
14. Collect the cytoplasmic fraction from each tube using a 1-mL syringe and a 16-gauge needle. Punctate the tube wall just above the black/gray layer sedimented at the tube bottom and carefully syringe out the golden-yellow fraction.

15. Transfer the extracted fraction to pre-chilled 1.5-mL test tubes and add Cyto-D (to 10 μg/mL), LPC (to 10 μg/mL each) and Energy Mix (to 1×). Place the tubes on ice until required.

### 3.2 Glass Cleaning

The steps in this section can be carried out before the extract preparation.

1. Place glass coverslips and slides in racks. Place the racks in beakers.
2. Sonicate with water, then 1 N KOH, and then water each for 10 min. Between each sonication, discard the solution and rinse the equipment with water several times (*see* **Note 5**).
3. Dry the cleaned glass surface using a stream of nitrogen and store it in a dust-free box with desiccant. These slides and coverslips are used to assemble an imaging chamber in Subheading 3.3.

### 3.3 Spindle Assembly Imaging

For spindle assembly imaging, freshly prepared extracts are placed in a test tube and cycled into interphase to assemble nuclei. The reaction is then cycled back into the M-phase immediately before sealing in the imaging chamber. To handle extracts, wide-bore pipette tips should be used to avoid excess mechanical shearing. These tips can be prepared by chopping off 1–2 mm of the pointed end of commercial pipette tips using a laser blade. Do not vortex the extract.

1. In a clean 1.5-mL test tube, place 30 μL of the fresh CSF extract prepared as in Subheading 3.1 and add sperm at 1000–-2000 nuclei/μL.
2. Add the calcium stock solution at the final concentration of 0.4 mM. Briefly flick the tube to mix the reaction and incubate for 80 min at 18 °C. Following the 80-min incubation, check nuclear assembly using DAPI staining (*see* **Note 6**).
3. Add X-rhodamine tubulin and SYTOX-green to the reaction mix for visualization of spindles and chromosomes, respectively. The optimal dye concentrations depend on the microscope setting and labeling stoichiometry. A good starting point would be 100 nM and 1 nM, respectively (*see* **Note 7**).
4. Add an equal volume (30 μL) of fresh CSF extract that has been kept on ice to the test tube containing extract with pre-assembled nuclei. Mix gently and thoroughly by tapping.
5. Pick a cleaned slide glass prepared in Subheading 3.2 and place 5 μL of the extract reaction on the glass surface.
6. Mount a clean coverslip onto the reaction to make an extract sandwich. Avoid trapping of air bubbles. Avoid excess mechanical shear.

7. Seal the coverslip edges using VaLaP and place the chamber on a microscope's sample stage.
8. Turn on a fluorescence light source to visualize nuclei. Adjust the focus of the microscope to nuclei and start timelapse imaging. Spindle assembly typically starts at 10–15 min following the addition of CSF extract in Subheading 3.3, **step 4**, associated with nuclear envelope breakdown. The spindle assembly typically completes within 45–60 min (*see* **Notes 8** and **9**).
9. The imaging chamber can be raster-scanned by moving the motorized sample stage for simultaneous recording of multiple spindle assemblies. We typically use a 20× objective lens and scan 5 × 6 squares (2.3 × 2.7 mm) at 1-min intervals. This allows for capturing 50–100 spindles with negligible photodamage. The raster scanning is also beneficial for seamless tracking of individual spindles, which drift across the cytoplasm and often cross the border between adjacent image areas.

### 3.4 Data Analysis

Imaging analysis is performed according to the following steps: (1) cropping spindle images, (2) tracing spindle shape outline, (3) decomposing the shape outline into discrete Fourier modes, and (4) plotting the Fourier mode amplitudes onto a two-dimensional "morphospace." The image analysis is performed across the entire timelapse sequence so that the dynamics of spindle growth can be plotted as single-line trajectories. The analysis can be performed using ImageJ plug-ins and in-house developed MATLAB codes. The MATLAB codes are available in Electronic Supplementary Materials.

#### 3.4.1 Spindle Cropping

1. In ImageJ, open a timelapse image sequence file as acquired in Subheading 3.3.
2. In the initial image of the timelapse (time = 0), target a nucleus and place a square ROI (300 × 300 pixels) that covers the entire structure. Then, crop the image.
3. Proceed on to the next image. Adjust the ROI position to center the targeted structure. Repeat cropping until the timelapse sequence ends. Examples of images are shown in Fig. 1b.
4. Save the cropped image stack in TIFF format.

#### 3.4.2 Shape Tracing

1. Open the cropped image stack in ImageJ.
2. Process each image for noise reduction using the ImageJ toolbox (CSBDeep; https://csbdeep.bioimagecomputing.com).
3. Open the MATLAB code "Fourier_Cn_analysis.m." Load the image stack and execute the code. The images are automatically processed according to the following steps.

(a) Detecting the spindle area based on image binarization.

(b) Tracing the contour of the highlighted area (Fig. 2a). The centroid position of the highlighted area is calculated, and then a tracing vector ($\boldsymbol{R}$) is drawn from the centroid to the contour at angle $\theta$. The vector is rotated over 360° along the contour, generating a spindle shape profile, $R(\theta)$ ($0° \leq \theta < 360°$) (blue trace, Fig. 2b).

(c) Fourier-based decomposition (Fig. 2b). The acquired spindle shape profile is decomposed into discrete Fourier modes according to the following equation:

$$R(\theta) = \sum_{n=0}^{+\infty} C_n e^{-in\theta} \ (n = 0, 1, 2, \ldots) \tag{1}$$

Here, the amplitude of each Fourier coefficient, $C_n$, can be calculated as follows:

$$C_n = \frac{1}{2\pi} \int_{-\pi}^{\pi} R(\theta) e^{in\theta} d\theta \tag{2}$$

The 0-th order Fourier mode, $C_0$, represents the spindle's size. The 2nd-order Fourier mode, $C_2$, represents the degree of second-order rotational symmetry and is the measure of bipolarity. The 3rd-order Fourier mode, $C_3$, measures the degree of multipolarity. The value of $C_1$ should be zero because the origin of the tracing vector $\boldsymbol{R}$ is set to the structure's centroid (*see* **Note 10**).

(d) Normalization of $C_2$ and $C_3$. The coefficient values of $C_2$ and $C_3$, obtained in **step c** of this section, are divided by the value of $C_0$. The normalized amplitudes of each mode are defined as $\tilde{C}_2$ ($= C_2/C_0$) and $\tilde{C}_3$ ($= C_3/C_0$), respectively (*see* **Note 11**).

(e) Graph plotting. The obtained values of ($\tilde{C}_2, \tilde{C}_3$) are plotted onto a two-dimensional plane composed of $\tilde{C}_2, \tilde{C}_3$ coordinates. We call this plane the "morphospace."

(f) Repeat **steps a–e** of this section across all timelapse frames and generate a trajectory of spindle growth in the morphospace (Fig. 2c).

#### *3.4.3 AI-Based Phenotype Sorting*

1. Open the MATLAB code "classification.m" and create the MATLAB date file named "classification.mat."
2. Train the machine learning algorithm. First, prepare still images of spindles that had reached metaphase. The images should be acquired independently of the timelapse assay described in Subheading 3.3. This can be done by performing

a spindle assembly reaction in a test tube and squashing the reaction between a slide and a coverslip after incubating >60 min to attain steady-state metaphase and supplementing with X-rhodamine tubulin. Acquire multiple spindle images from each reaction in a non-biased manner. Collect data using at least three independent extract preparations. Then, manually annotate each spindle image, either as "circular (monopolar)," "bipolar," or "multipolar." Following the annotation, calculate the values of $\tilde{C}_2$ and $\tilde{C}_3$ according to Eq. 2. Finally, map all the annotated spindle data onto a single morphospace based on the calculated values of $(\tilde{C}_2, \tilde{C}_3)$ (*see* **Note 12**). The algorithm creates a phase map based on the training image data and defines the area of each morphological phenotype [6].

3. Test the machine learning model. Load a sample spindle image to the program and execute the program. Check if the correct spindle phenotype is produced as output. Perform a blind test using multiple spindle images and with several participants. If the performance appears low (<95%), go back to **step 2** of this section and re-train the algorithm with more spindle images (*see* **Note 13**).
4. Sort the timelapse image data using the trained algorithm. Open the last image of each timelapse sequence analyzed in Subheading 3.4.2 and load it onto the program. Execute the program to classify the data based on the final spindle shape. Repeat this step for all available timelapse data. In the following section, the growth dynamics of spindles can be quantitatively analyzed using dataset classified to each assembly phenotype (e.g., bipolar, multipolar).

#### 3.4.4 Landscape Mapping

1. In MATLAB, pool all the spindle trajectory data classified into the same phenotypic category and map them onto a single morphospace of $(\tilde{C}_2, \tilde{C}_3)$.
2. Calculate the density of the pooled trajectories at each coordinate point, using a MATLAB function "ksdensity." If there is any characteristic growth dynamics associated with a specific assembly phenotype, a cluster of the trajectories should be seen as color coding (Fig. 2d).
3. To further verify if the characteristic growth dynamics exists for the specific assembly phenotype, the following correlation analysis can be performed using the MATLAB code "crosscorr_analysis.m." This code calculates the cross-correlation between $\tilde{C}_2$ and $\tilde{C}_3$ and generate a profile, $C_n$, as a function of time shift, $\Delta t$, according to the following equation:

$$C_r(\Delta t) = \frac{\langle \tilde{C}_2(t) - \langle \tilde{C}_2 \rangle \rangle \cdot \langle \tilde{C}_3(t + \Delta t) - \langle \tilde{C}_3 \rangle \rangle}{\sqrt{\left\langle (\tilde{C}_2(t) - \langle \tilde{C}_2 \rangle)^2 \right\rangle \cdot \left\langle (\tilde{C}_3(t + \Delta t) - \langle \tilde{C}_3 \rangle)^2 \right\rangle}} \quad (3)$$

Here, $\langle \cdot \rangle$ denotes the time average. Perform this calculation for each spindle timelapse data. Then, pool and average all the data within each phenotype class.

4. Generate a two-dimensional graph of $C_r$ against $\Delta t$ (Fig. 2e). If there are any characteristic growth dynamics associated with a specific assembly phenotype, the cross-correlation profile yields a peak. The positive correlation indicates that bipolar and multipolar growths occur concomitantly. The negative correlation indicates that bipolar and multipolar growth modes are in mutual inhibition (*see* **Note 14**).

### *3.5 Molecular and Mechanical Perturbations*

1. Molecular perturbation experiments can be performed by the addition of reagents (e.g., chemical inhibitors, recombinant proteins) at the timing of **step 3** in Subheading 3.3, just before the initiation of spindle assembly. The reagents are diluted in CSF-XB buffer and added at a volume that should not exceed 5% of the total extract reaction. For reagents of low solubility in aqueous solution, DMSO can be added up to 0.5% (v/v) to the final concentration in the reaction. A buffer control should be run side-by-side with the reaction supplemented with the reagent. Image acquisition and analyses can be performed using the identical image analysis pipeline described above; the resulting spindle growth characteristics can be examined by generating the morphospace trajectory and cross-correlation profile (Subheading 3.4).
2. Mechanical perturbation experiments can be performed using an open-top imaging chamber and a pair of microneedles. First, an extract reaction (~5 μL) is spread onto a coverslip surface, constructing the bottom of the chamber and covered with mineral oil. Using a pair of glass microneedles, whose shafts are held using micromanipulators, spindles in the extract can be captured and mechanically perturbed. For example, a focused pole of a bipolar spindle can be captured by inserting the tips of microneedles into the structure and then split by moving the two microneedles apart, transforming into a multipolar configuration [6]. The following changes in spindle shape morphology can be timelapse imaged and then analyzed using the image analysis pipeline. The detailed protocol for the spindle micromanipulation assay is provided in Shimamoto and Kapoor [11].

## 4 Notes

1. Eggs obtained from frogs of the first-time oviposition often yield low-quality extracts; we routinely use eggs from the second and subsequent times of oviposition.
2. Eggs of ~50 mL in total volume are sufficient to fill one ultracentrifuge tube used in the subsequent step.
3. As the dejellying proceeds, eggs start to pack with each other. It is also expected to observe a greasy layer coming off of the eggs' surface. Remove cysteine as quickly and thoroughly as possible when the dejellying step is completed.
4. Dejellied eggs are extremely fragile. Handle them carefully and always keep them soaked in buffer. Avoid excess mechanical stress when transferring and washing the eggs.
5. We prefer Matsunami glass washed with KOH over other types of glasses (e.g., PEG or lipid-grafted glasses) as we find it minimally perturbs spindle assembly.
6. The assembled nuclei may start to clump with each other as the time elapses. A gentle finger-tapping of the tube every 10–15 min during the assembly can avoid clumping.
7. Each dye concentration must be adjusted to minimize influences on the intrinsic spindle dynamics. This can be examined by comparing spindle assembly reactions performed in the imaging chamber and in a test tube. At the concentrations of 200 nM for X-rhodamine tubulin and 1 nM for SYTOX green, we routinely observe spindles with similar size and shape and comparable assembly ratios (70–80% of spindles are bipolar).
8. Go back to Subheading 3.3, **step 4** and prepare a new assembly reaction if nuclear envelope breakdown had been already started. It is recommended that several interphase reactions be set up side-by-side.
9. It is also recommended to set up other interphase reactions following the first cycling reaction with tens of minutes time delay. This allows for acquiring multiple timelapse data from multiple cycling reactions in a single day. The assay must be completed before the activity of prepared extracts dampens. This typically occurs within 6–9 h after extract preparation.
10. In our experimental setting, the higher-order Fourier modes ($n \geq 4$) are typically small and can be omitted.
11. This normalization procedure scales the calculated shape parameters to the structure's size. For example, a larger triangle yields a larger $C_3$ amplitude than a smaller triangle does in the analysis, and the size-dependency is cancelled by the normalization to $C_0$.

12. The data of $(\tilde{C}_2, \tilde{C}_3)$ and associated annotation ("monopolar," "bipolar," or "multipolar") can be saved in the MATLAB file "classification.mat." The saved MATLAB file can be loaded and run using the MATLAB code "classification.m" to call the machine learning model when necessary.
13. In our experimental setting, two hundreds of still spindle images are sufficient to train the algorithm with >95% confidence accuracy.
14. In our experimental settings, at least 30 spindle data is required to conclude the existence of characteristic growth dynamics in the correlation analysis.

## Acknowledgments

This work was supported by JSPS Grant-in-Aid for Scientific Research JP23K19357 (to A.T.), JP22K14014 (to T.F.), JP18H05427, JP20H01872 and JP21K18605 (to Y.T.M.), JP22K18362, JP22H02590, JP23K23854, Takeda Science Foundation and the Uehara Memorial Foundation (to Y.S.), and NIG JOINT 91A2019 and 58A2020 (to Y.T.M. and Y.S.). A.T. is the recipient of the National Institute of Genetics and JSPS postdoctoral research fellowships (JP24KJ0225), and the Research fellowship by Nagahisa Science Foundation.

## References

1. Ganem NJ, Godinho SA, Pellman D (2009) A mechanism linking extra centrosomes to chromosomal instability. Nature 460(7252): 278–282
2. So C, Menelaou K, Uraji J et al (2022) Mechanism of spindle pole organization and instability in human oocytes. Science 375(6581): eabj3944
3. Kronja I, Kruljac-Letunic A, Caudron-Herger M et al (2009) XMAP215-EB1 interaction is required for proper spindle assembly and chromosome segregation in Xenopus egg extract. Mol Biol Cell 20(11):2684–2696
4. Halpin D, Kalab P, Wang J et al (2011) Mitotic spindle assembly around RCC1-coated beads in Xenopus egg extracts. PLoS Biol 9(12): e1001225
5. Takagi J, Shimamoto Y (2017) High-quality frozen extracts of Xenopus laevis eggs reveal size-dependent control of metaphase spindle micromechanics. Mol Biol Cell 28(16): 2170–2177
6. Fukuyama T, Yan L, Tanaka M et al (2022) Morphological growth dynamics, mechanical stability, and active microtubule mechanics underlying spindle self-organization. Proc Natl Acad Sci USA 119(44):e2209053119
7. Maeda YT, Inose J, Matsuo MY et al (2008) Ordered patterns of cell shape and orientational correlation during spontaneous cell migration. PLoS ONE 3(11):e3734
8. Murray AW (1991) Cell cycle extracts. Methods Cell Biol 36:581–605
9. Hyman A, Drechsel D, Kellogg D et al (1991) Preparation of modified tubulins. Methods Enzymol 196:478–485
10. Desai A, Murray A, Mitchison TJ et al (1998) The use of Xenopus egg extracts to study mitotic spindle assembly and function in vitro. Methods Cell Biol 61:385–412
11. Shimamoto Y, Kapoor TM (2018) Analyzing the micromechanics of the cell division apparatus. Methods Cell Biol 145:173–190

# Chapter 9

# Visualization and Analyses of Cytoplasmic Streaming in *C. elegans* Zygotes

**Kenji Kimura and Fumio Motegi**

## Abstract

Cytoplasmic streaming is the bulk flow of cytoplasm observed, not only in plants but also in animal oocytes and embryos. The flow of viscous fluid within the cytoplasm generates forces that re-arrange intracellular organelles, such as mitotic spindles and nuclei, to regulate cell growth, migration, and polarity. Cytoplasmic streaming is established by motor proteins and the viscoelastic cytoskeleton, including the actin filaments and microtubules. The mode of action can be changed in response to cell fate and developmental stage; however, its regulatory mechanism is unclear. The nematode *Caenorhabditis elegans* is a valuable model organism for analyzing cytoplasmic streaming because actin- and microtubule-dependent flows occur during oogenesis and embryogenesis, respectively. In this chapter, we describe methods for visualizing and quantitatively analyzing cytoplasmic streaming in *C. elegans* oocytes and embryos.

**Key words** Cytoplasmic streaming, Fluid flow, Actin, Microtubules, Oocyte, Zygote, *C. elegans*

## 1 Introduction

The cytoplasm is a viscous fluid composed of macromolecules, such as large protein complexes, organelles, and the cytoskeletons. In this crowded environment, the movement of organelles is limited to short distances. A collective flow of the gel-like cytoplasm, known as cytoplasmic streaming, occurs in many larger animal cell types, particularly oocytes and embryos [1]. A force generated by the unidirectional flow of the cytoplasm displaces cytoplasmic organelles through hydrodynamic coupling, which overcomes the slow rate of diffusion. This results in the movement of maternally deposited macromolecules across longer distances to specific locations within the cell. Recent studies of cytoplasmic-streaming mechanisms in various animal models have emphasized its important roles in patterning oocytes and embryos. Cytoplasmic streaming displaces the meiotic spindle toward the subcortical region, ensuring successful meiosis in mouse oocytes [2]. It also plays an important

Tomomi Kiyomitsu (ed.), *The Mitotic Spindle: Methods and Protocols*, Methods in Molecular Biology, vol. 2872,
https://doi.org/10.1007/978-1-0716-4224-5_9,

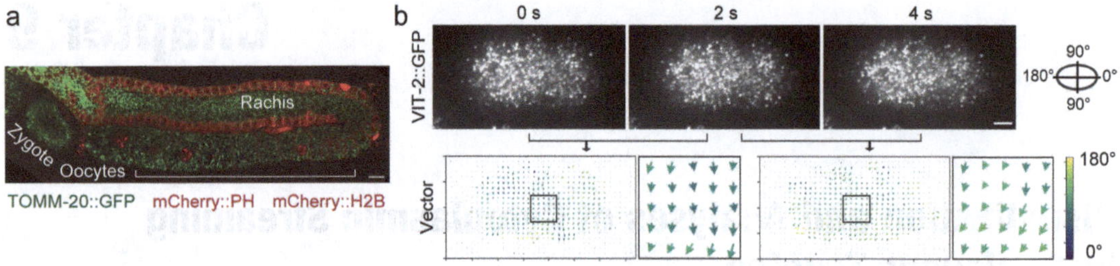


**Fig. 1** Cytoplasmic streaming in *C. elegans* syncytial germ cells and the zygote. (**a**) A confocal fluorescence image of mitochondria labeled with TOMM-20::GFP, the plasma membrane labeled with mCherry::PH, and chromosome labeled with mCherry::H2B in the gonad. (**b**) Confocal fluorescence images of yolk granules labeled with VIT-2::GFP in the zygote immediately after fertilization. Flow vectors of yolk granules for 2 s are shown as arrows in the bottom panels. The flow direction is indicated by the color bar. The boxed areas are magnified on the right. Scale bars, 5 μm

role in establishing embryonic polarity, which, in turn, affects the positioning of the mitotic spindle during asymmetric cell division [3]. Therefore, cytoplasmic streaming is closely associated with the rearrangement of intracellular structures during development.

Cytoplasmic streaming is primarily driven by motor proteins. ATP-dependent motor proteins can exert forces on the cytoskeletal structures, and their movement along cytoskeletal filaments generates friction-based forces in the viscous cytoplasm. The type of cytoskeleton varies depending on the species and their developmental stage. In mouse oocytes, cytoplasmic streaming is induced by actin polymerization at the animal pole [2]. In zebrafish oocytes and ascidian *Ciona* zygotes, cytoplasmic streaming is mediated by contraction of the cytoplasmic actin meshwork near the animal pole [4] and at the vegetal hemisphere [5], respectively. Conversely, *Drosophila* oocytes (after stage 10) induce cytoplasmic streaming through microtubule sliding generated by kinesin-1 motor proteins [6, 7]. Although recent studies have characterized various types of cytoplasmic streaming, it remains unclear how these various processes are regulated and utilized during different stages of development.

The nematode *C. elegans* is a suitable model organism for examining the mechanism underlying cytoplasmic streaming. It exhibits both actin- and microtubule-dependent cytoplasmic streaming at different developmental stages. During early oogenesis, the cytoplasm of syncytial germ cells within the gonadal tissue establishes a collective flow along the distal–proximal axis in an actin-dependent manner (Fig. 1a) [8]. Following release from meiotic arrest, a fertilized zygote initiates the kinesin-1-dependent transport of intracellular organelles along cortical microtubules, which results in the formation of a circular pattern of cytoplasmic streaming (Fig. 1b) [9–11]. After the completion of meiosis, the cortical actin network retracts from the posterior pole (near the sperm-derived centrosomes) to the anterior pole, which in turn, generates cytoplasmic flow along the anteroposterior axis [12, 13]. Thus, different mechanisms generate cytoplasmic streaming during oogenesis and embryogenesis. To further examine the mechanisms and functions of cytoplasmic streaming, it is necessary to develop techniques that enable time-lapse imaging of intracellular flow and its quantitative analysis. Here, we describe methods for visualizing, observing, and quantifying cytoplasmic streaming in *C. elegans* oocytes and embryos by confocal microscopy and computer-assisted image analysis.

## 2 Materials

### *2.1 Preparation and Mounting of* C. elegans

1. *C. elegans* strains expressing cytoplasmic organelle markers tagged with fluorescent proteins (*see* **Note 1**). Strains useful for monitoring cytoplasmic streaming are listed in Table 1. These strains are available from the *Caenorhabditis* Genetics Center (https://cgc.umn.edu).
2. Stereomicroscope equipped with a transmitted light source for handling *C. elegans.*
3. A worm picker can be prepared by inserting a 1-inch piece of 32-gauge platinum wire into the tip of a Pasture pipette. See WormBook "Maintenance of *C. elegans*" [14].

**Table 1**
***C. elegans* strains available from CGC**

| Strain name | Genotype | Labeled organelle | Observable cytoplasmic streaming* | Ref. |
|---|---|---|---|---|
| WH327 | *unc-119(ed3) III; ojIs23 [pie-1p::GFP:: C34B2.10].* | Endoplasmic reticulum | 1, 2, 3 | [29] |
| DH1033 | *sqt-1(sc103) II; bIs1 [vit-2::GFP + rol-6(su1006)] X.* | Yolk granule | 2, 3 | [30] |
| JA1403 | *unc-119(e2498) III; weIs15 [pie-1p::GFP::eea-1 (FYVEx2) + unc-119(+)].* | Early endosome | 1, 2, 3 | [31] |
| RT688 | *unc-119(ed3) III; pwIs28[unc-119(+); pie-1p:: cav-1::GFP]* | Cortical granule | 2 | [32] |
| EGD623 | *unc-119(ed3) III; egxSi152 [mex-5p: tomm-20:: gfp::pie-1 3UTR + unc-119(+)] II.* | Mitochondria | 1, 2, 3 | [33] |
| EGD629 | *unc-119(ed3) III; egxSi155 [mex-5p::tomm-20:: mKate2::pie-1 3UTR + unc-119(+)] II.* | Mitochondria | 1, 2, 3 | [33] |

*1: Cytoplasmic streaming in the gonad
*2: Meiotic cytoplasmic streaming in the zygote immediately after fertilization
*3: Cortical and cytoplasmic flow in the zygote

4. An alcohol lamp or a gas burner for sterilizing a platinum wire.
5. M9 buffer: 22 mM $KH_2PO_4$, 42 mM, $Na_2HPO_4$, 86 mM NaCl, 1 mM $MgSO_4$.
6. Anesthesia solution: 1–4 mM levamisole in M9 buffer.
7. 2% agarose in ultrapure water for making agarose pads.
8. Two glass slides with two pieces of vinyl tape (0.2 mm thickness) attached as spacers on top.
9. Coverslips (22 × 22 mm) and glass slides (76 × 26 mm).

### 2.2 Image Acquisition and Analysis

1. Confocal laser microscopy: We used a spinning-disk confocal system (CSU-X1 or CSU-W1; Yokogawa) mounted onto a microscope (Ti2-E; Nikon) equipped with a 60× or 100× objective lens and an EM-CCD or CMOS camera controlled by Metamorph software (Molecular Devices).
2. Image processing software, such as Fiji/ImageJ.
3. Software using Python or R for quantitative analyses, such as Spyder (Version 5.4.3, https://www.spyder-ide.org/) or RStudio (Version 2023.12.1, https://posit.co/products/open-source/rstudio/).

## 3 Methods

In a gravid *C. elegans* hermaphrodite at an early adult stage, cytoplasmic streaming occurs prominently within a central lumen of the syncytial germ cells, known as rachis (Fig. 1a). It also occurs transiently in the zygotes immediately after fertilization (Fig. 1b). These cytoplasmic streaming events can be observed in live animals in a healthy state because the gonad and newly fertilized zygotes are osmotically sensitive and fragile. Therefore, it is necessary to immobilize live worms without applying high pressure. An effective method for worm immobilization is treatment with low dose (1–4 mM) levamisole [15]. Genetically-encoded fluorescent proteins or fluorescent dyes that are injected or penetrate into the rachis are used to visualize the flow of cytoplasmic organelles, including endosomes, yolk granules, cortical granules, endoplasmic reticulum (ER), and mitochondria. Both vesicular organelles and the tubular ER network establish flows at similar velocity and direction, indicating a collective motion of the viscous cytoplasm [9]. Transgenic *C. elegans* strains that label various organelles are available from the *Caenorhabditis* Genetics Center. Alternatively, the desired strain may be generated using standard methods, such as CRISPR-Cas9-based genome editing [16, 17]. Staining of organelles in live animals can be achieved through the preincubation of the worms with fluorescent dyes, such as LysoTracker or MitoTracker for a couple of days [18, 19]. Even without fluorescent proteins and dyes, the movement of cytoplasmic granules is visible to some extent using differential interference contrast microscopy; however, these granules are dense and difficult to quantitate. Cytoplasmic streaming can be observed by introducing small fluorophore particles, such as FluoSpheres and quantum dots. Although these particles introduced into the cytoplasm behave as cytoplasmic probes, it is necessary to carefully inject the particle solution into the rachis to avoid excessive damage [8]. In this chapter, we described a simple method for visualizing and analyzing cytoplasmic streaming using *C. elegans* zygotes that express fluorescent proteins in specific organelles.

### 3.1 Mounting Worms

1. To prepare agarose pads for mounting worms, melt a 2% agarose solution in a microwave oven. The melted agarose may be maintained on a heat block at 75 °C.
2. To prevent the worms from moving during live imaging, transfer gravid worms into a drop of anesthesia solution (10 μL) for approximately 5 min at room temperature around 20–25 °C (Fig. 2a) (*see* **Notes 2** and **3**).
3. To prepare an agarose pad, place a glass slide between two additional glass slides with a layer of tape that defines the desired thickness of the pad. A drop of melted agarose (approximately 100–150 μL) is then applied to a glass slide using a Pasteur pipet or micropipette. Immediately press on the agarose drop with another glass slide onto the taped slides (Fig. 2b) (*see* **Note 4**).
4. Remove the upper glass slide by sliding and add 10 μL of anesthesia solution onto the center of the agarose pad. Transfer the anesthetized worms onto the pad (Fig. 2c).
5. Place the coverslip gently onto the worms, but avoid applying high pressure.

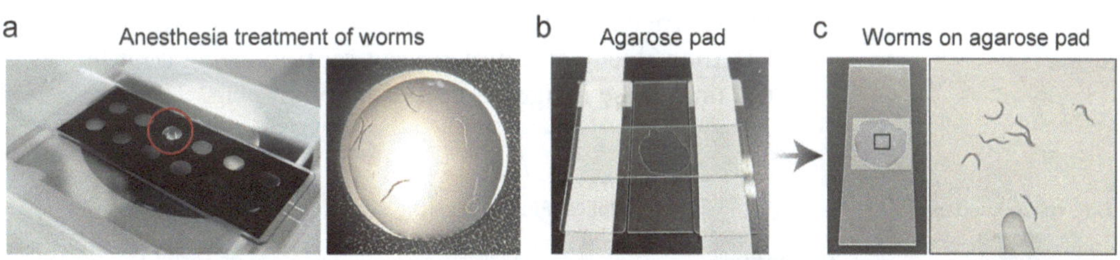


**Fig. 2** Immobilization of worms on an agarose pad. (**a**) Worms are soaked in a drop of the anesthesia solution (2 mM levamisole) in a well on a coated slideglass. The area of the red circle is magnified on the right. (**b**) An agarose pad is prepared on the middle glass slide. The glass slides on each side of the middle glass slide have two pieces of vinyl tape attached and serve as spacers. (**c**) Worms mounted on an agarose pad are shown. The boxed area is magnified on the right

### 3.2 Live Cell Imaging of Cytoplasmic Streaming

1. Place the agarose pad with worms on the confocal microscope stage at room temperature.
2. Adjust the focal plane and stage position until the worm is centered in the field of view using transmitted light and a 10× or 20× objective lens.
3. Adjust the field of view to the location where the cytoplasmic streaming is occurring (the rachis in the gonad or the oocyte just before fertilization) by changing the objective lens to 60× or 100×.
4. Acquire confocal images of fluorescently labeled organelles in the gonad or fertilized oocyte every 2–5 s and record the cytoplasmic streaming. It is recommended to keep the exposure time to a minimum to avoid photodamage.

### 3.3 Image Processing, Quantitation, and Data Analysis of Cytoplasmic Streaming

To quantify and estimate the velocity and direction of flow from acquired time-lapse imaging data, optical flow analysis methods can be used. The Farnebäck algorithm is a method for optical flow analysis [20], which calculates the velocity and direction of moving objects in all pixels for a series of images. The algorithm is implemented in Python using the OpenCV library (cv2.calcOpticalFlowFarneback); however, other programming languages may also be used.

1. Open time-lapse images clearly capture the fluorescence signal using Fiji/ImageJ software (*see* **Note 5**).
2. Rotate the images so that the long axis of the gonad or zygote is parallel to the horizontal axis using Fiji/ImageJ (Image > Transform > Rotate) (Fig. 3, *see* **Note 6**).
3. Crop the region of interest in the images where cytoplasmic streaming occurs and save it as an AVI file in Fiji/ImageJ (File > Save As).
4. Import the saved AVI file into Python using the OpenCV library function (cv2.VideoCapture).
5. Execute the Farnebäck algorithm function (cv2.calcOpticalFlowFarneback) from the OpenCV library in Python to calculate the velocity and direction of motion for all coordinates in the time-lapse images at each time point (Fig. 3).

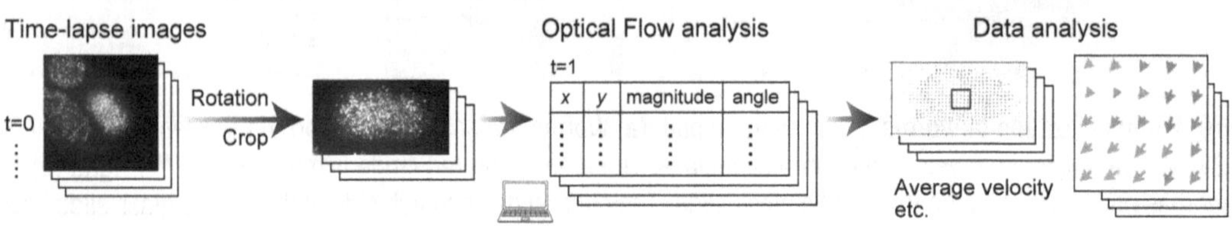


**Fig. 3** Workflow for quantitative analysis of cytoplasmic streaming. See the Subheading 3 for details

6. Export the data as a CSV file using the function in the Python library, such as to csv.
7. Calculate and analyze the flow characteristics (average of velocity, direction, etc.) from the obtained numerical data using appropriate software or programming languages (*see* **Note 7**). We use R to analyze numerical data.
8. Flow vectors may be displayed as arrows using functions in programming languages, such as Arrows in R or Matplotlib (plt.quiver) in Python.

## 4 Notes

1. *C. elegans* strains are maintained by the standard method [21] and should not be starved before imaging of cytoplasmic streaming.
2. Glass slides with 5–6 mm diameter holes, such as Matsunami TF1006/TF1205 or MP Biomedicals Multitest slides, are recommended for anesthetizing the worms.
3. There are various methods of immobilizing worms besides anesthesia, including the use of microbeads, microfluidic devices, and photo-crosslinked polyethylene glycol [22–27]. One method we have used is mounting worms on a 10% agarose pad with a suspension of 0.1 μm polystyrene microbeads; however, this method tends to physically compress the worms, which prevents cytoplasmic streaming in the gonad. Anesthesia is useful for immobilizing worms to image cytoplasmic streaming, but it is not recommended for prolonged periods (more than a few hours) because of the gradual weakening of the worm.
4. We recommend that agarose pads be prepared just before use to prevent them from drying and hardening. Otherwise, high pressure can occur which affects the worms by preventing cytoplasmic streaming. It is important to ensure that the agarose pads are not too thin or too hard.
5. To ensure an accurate analysis, it is recommended to avoid data in which the intestine overlaps the gonad or zygote, as gut auto-fluorescence can mask and blur the cytoplasmic marker signals. In addition, if the worm position changes during live imaging or the background signal is uneven, the resulting data should not be used for analysis. If necessary, image correction processing can be performed.
6. Fluorescence signals at each time point may be displayed in different colors to show the trajectories of organelles during cytoplasmic streaming using Fiji software with the Color Footprint Rainbow macro developed by Y. Hiratsuka (Japan

Advanced Institute of Science and Technology). To enhance the trajectories, the Difference of Gaussians approach [28] may be used to improve the fluorescence signals in the time-lapse images before using the macro. Briefly, the fluorescence signals are enhanced by subtracting the images obtained with a larger Gaussian blur (e.g., $\sigma = 4$) from the images obtained with a smaller blur (e.g., $\sigma = 1$). Both the Gaussian blur process and the image subtraction may be done using the tools in Fiji/ ImageJ (Process > Filters > Gaussian Blur and Process > Image Calculator, respectively).

7. Flow quantification data may be extracted from a specific region by specifying the coordinate range of the image. Outliers can be excluded by specifying the velocity range. Large image sizes or longer time lapses may require more time for calculation.

## References


1. Kimura K, Motegi F (2021) Fluid flow dynamics in cellular patterning. Semin Cell Dev Biol 120:3–9
2. Yi K, Unruh JR, Deng M et al (2011) Dynamic maintenance of asymmetric meiotic spindle position through Arp2/3-complex-driven cytoplasmic streaming in mouse oocytes. Nat Cell Biol 13(10):1252–1258
3. Kimura K, Kimura A (2020) Cytoplasmic streaming drifts the polarity cue and enables posteriorization of the *Caenorhabditis elegans* zygote at the side opposite of sperm entry. Mol Biol Cell 31(16):1765–1773
4. Shamipour S, Kardos R, Xue SL et al (2019) Bulk actin dynamics drive phase segregation in *Zebrafish* oocytes. Cell 177(6):1463–1479
5. Ishii H, Tani T (2021) Dynamic organization of cortical actin filaments during the ooplasmic segregation of ascidian *Ciona* eggs. Mol Biol Cell 32(3):274–288
6. Serbus LR, Cha BJ, Theurkauf WE et al (2005) Dynein and the actin cytoskeleton control kinesin-driven cytoplasmic streaming in *Drosophila* oocytes. Development 132(16): 3743–3752
7. Lu W, Winding M, Lakonishok M et al (2016) Microtubule-microtubule sliding by kinesin-1 is essential for normal cytoplasmic streaming in Drosophila oocytes. Proc Natl Acad Sci USA 113(34):E4995–E5004
8. Wolke U, Jezuit EA, Priess JR (2007) Actin-dependent cytoplasmic streaming in *C. elegans* oogenesis. Development 134(12):2227–2236
9. Kimura K, Mamane A, Sasaki T et al (2017) Endoplasmic-reticulum-mediated microtubule alignment governs cytoplasmic streaming. Nat Cell Biol 19(4):399–406
10. McNally KL, Martin JL, Ellefson M et al (2010) Kinesin-dependent transport results in polarized migration of the nucleus in oocytes and inward movement of yolk granules in meiotic embryos. Dev Biol 339(1):126–140
11. Yang HY, McNally K, McNally FJ (2003) MEI-1/katanin is required for translocation of the meiosis I spindle to the oocyte cortex in *C. elegans*. Dev Biol 260(1):245–259
12. Munro E, Nance J, Priess JR (2004) Cortical flows powered by asymmetrical contraction transport PAR proteins to establish and maintain anterior-posterior polarity in the early *C. elegans* embryo. Dev Cell 7(3):413–424
13. Niwayama R, Shinohara K, Kimura A (2011) Hydrodynamic property of the cytoplasm is sufficient to mediate cytoplasmic streaming in the *Caenorhabditis elegans* embryo. Proc Natl Acad Sci USA 108(29):11900–11905
14. Stiernagle T (2006) Maintenance of *C. elegans*. WormBook:1–11
15. Chai Y, Li W, Feng G et al (2012) Live imaging of cellular dynamics during *Caenorhabditis elegans* postembryonic development. Nat Protoc 7(12):2090–2102
16. Dickinson DJ, Goldstein B (2016) CRISPR-based methods for *Caenorhabditis elegans* genome engineering. Genetics 202(3): 885–901
17. Paix A, Folkmann A, Seydoux G (2017) Precision genome editing using CRISPR-Cas9 and linear repair templates in *C. elegans*. Methods 121–122:86–93

18. Kimura K, Tanaka N, Nakamura N et al (2007) Knockdown of mitochondrial heat shock protein 70 promotes progeria-like phenotypes in *Caenorhabditis elegans*. J Biol Chem 282(8): 5910–5918
19. Kimura K, Kimura A (2011) Intracellular organelles mediate cytoplasmic pulling force for centrosome centration in the *Caenorhabditis elegans* early embryo. Proc Natl Acad Sci USA 108(1):137–142
20. Farnebäck G (2003) Two-frame motion estimation based on polynomial expansion. In: Bigun J, Gustavsson T (eds) Image analysis, vol 2749. Springer, Berlin/Heidelberg, pp 363–370
21. Brenner S (1974) The genetics of *Caenorhabditis elegans*. Genetics 77(1):71–94
22. Burnett K, Edsinger E, Albrecht DR (2018) Rapid and gentle hydrogel encapsulation of living organisms enables long-term microscopy over multiple hours. Commun Biol 1:73
23. Suzuki M, Sakashita T, Funayama T (2019) Immobilization of live *Caenorhabditis elegans* individuals using an ultra-thin polydimethylsiloxane microfluidic chip with water retention. J Vis Exp 145
24. Dong L, Cornaglia M, Krishnamani G et al (2018) Reversible and long-term immobilization in a hydrogel-microbead matrix for high-resolution imaging of *Caenorhabditis elegans* and other small organisms. PLoS One 13(3): e0193989
25. Keil W, Kutscher LM, Shaham S et al (2017) Long-term high-resolution imaging of developing *C. elegans* larvae with microfluidics. Dev Cell 40(2):202–214
26. Zellag RM, Zhao Y, Gerhold AR (2022) Live-cell imaging and analysis of germline stem cell mitosis in *Caenorhabditis elegans*. Bio Protoc 12(1):e4272
27. Niwa S (2017) Immobilization of *Caenorhabditis elegans* to analyze intracellular transport in neurons. J Vis Exp 128
28. Applegate KT, Besson S, Matov A et al (2011) plusTipTracker: quantitative image analysis software for the measurement of microtubule dynamics. J Struct Biol 176(2):168–184
29. Poteryaev D, Squirrell JM, Campbell JM et al (2005) Involvement of the actin cytoskeleton and homotypic membrane fusion in ER dynamics in *Caenorhabditis elegans*. Mol Biol Cell 16(5):2139–2153
30. Grant B, Hirsh D (1999) Receptor-mediated endocytosis in the *Caenorhabditis elegans* oocyte. Mol Biol Cell 10(12):4311–4326
31. Andrews R, Ahringer J (2007) Asymmetry of early endosome distribution in *C. elegans* embryos. PLoS One 2(6):e493
32. Sato K, Sato M, Audhya A et al (2006) Dynamic regulation of caveolin-1 trafficking in the germ line and embryo of *Caenorhabditis elegans*. Mol Biol Cell 17(7):3085–3094
33. Fan X, De Henau S, Feinstein J et al (2020) SapTrap assembly of *Caenorhabditis elegans* MosSCI transgene vectors. G3 (Bethesda) 10(2):635–644

# Chapter 10

# 4D Microscopy and Tracking of Chromosomes and the Spindle in *C. elegans* Early Embryos

**Julien Dumont and Gilliane Maton**

## Abstract

Maintaining genomic integrity throughout successive cell divisions is essential for the proper development and functioning of organisms. Chromosome alignment and segregation occur on a microtubule-based spindle originating from centrosomes. The molecular and cellular mechanisms involved in accurate chromosome segregation during early embryonic divisions are highly conserved between worms and humans. Therefore, *C. elegans* serves as a robust model for investigating mitotic cell divisions within a metazoan system. Throughout early embryonic development, filming and tracking successive cell divisions becomes progressively more challenging as the number of cells increases and cell size decreases. To address this challenge, we describe a method for preparing live samples, performing 4D time-lapse imaging, and semi-automated tracking of chromosomes and spindle poles during early mitotic divisions in *C. elegans* embryos.

**Key words** Cell division, Mitosis, Mitotic spindle, Chromosome segregation, Time-lapse imaging, Embryonic division, *C. elegans*

## 1 Introduction

The maintenance of genomic integrity is essential to ensure proper development and functioning of organisms. Chromosome segregation errors during mitosis or meiosis result in aneuploidy, which can cause spontaneous abortions, developmental abnormalities, or tumorigenesis [1, 2]. Therefore, tight regulation of the many steps involved in chromosome segregation is essential to ensure their accurate distribution to daughter cells. During cell division, chromosomes align and sister chromatids segregate within a highly sophisticated spindle-shaped structure built from microtubules [3, 4]. In animal somatic cells and spermatocytes, microtubules are primarily assembled from the two centrosomes located at each spindle pole. Centrosomes consist of two centrioles surrounded by the pericentriolar material (PCM), within which microtubule nucleation and anchoring take place [5–8].

Tomomi Kiyomitsu (ed.), *The Mitotic Spindle: Methods and Protocols*, Methods in Molecular Biology, vol. 2872,
https://doi.org/10.1007/978-1-0716-4224-5_10,

The nematode *Caenorhabditis elegans* is a powerful model organism for investigating embryonic divisions and chromosome segregation. Its suitability for laboratory research is due to its easy maintenance, a short generation time of merely 3–4 days at room temperature, and prolific reproductive capabilities [9]. While most individuals are hermaphrodites capable of self-fertilization, occasional emergence of males occurs due to spontaneous non-disjunction events of the sexual chromosome during meiosis, allowing for straightforward genetic crosses through mating with hermaphrodites. Additionally, *C. elegans* embryos are easily accessible, transparent and characterized by an invariant developmental lineage [10]. Live-cell microscopy is therefore a method of choice for studying the mitotic embryonic divisions of *C. elegans*.

The *C. elegans* zygote typically measures 50 × 30 μm and displays an ovoid shape. Its first division is asymmetric, giving rise to a large anterior AB and a smaller posterior P blastomere. A succession of asymmetric divisions leads to the formation of a first-stage larva (L1) comprised of approximately 500 cells. After hatching and progression through four larval stages (L1–L4) accompanied by additional cell divisions, adult hermaphrodites, comprised of 959 somatic cells, begin producing offspring [11].

A classical approach for analyzing cell division and sister chromatid segregation in *C. elegans* embryos is to perform time-lapse imaging from transgenic strains expressing fluorescently tagged microtubules or spindle poles and chromosome markers. This is typically achieved by capturing a restricted number of Z planes, separated by a considerable distance (usually 4–10 Z-planes spaced apart by 1–2 μm) across the dividing embryo [12, 13]. This approach is adopted to minimize the potential phototoxic effects of laser exposure. This, however, results in excessively undersampled Z-stacks, making three-dimensional reconstruction of images impossible. As a result, two-dimensional Z-projected images are typically utilized for image analysis [14, 15]. While this technique is efficient for visualizing the spindle and chromosomes in single-cell embryos, its limitation lies in the relatively thin Z slice it acquires, making it unsuitable for analyzing more mature multicellular embryos. Specifically, the spindles and chromosomes could be located anywhere within the embryo's full thickness in these older embryos, necessitating a different imaging approach [16]. Here, we introduce a 4D (XYZ and time) confocal microscopy approach complemented by semi-automated tracking of chromosomes and spindle poles. This novel approach is adapted from a method we recently developed to analyze *C. elegans* chromosome segregation during oocyte meiosis [17]. In this chapter, we outline the practical setup required for the preparation of the acquisition system, dissection of embryos, time-lapse imaging, and semi-automated analyses using the Imaris software to track the spindle poles and chromosomes during early embryogenesis in *C. elegans*.

## 2 Materials

### 2.1 Worm Strains

For imaging mitotic divisions, we use transgenic *C. elegans* strains expressing fluorescent markers for chromatin (mCherry::HIS-11$^{H2B}$ or mCherry::HIS-58$^{H2B}$), and either microtubules (GFP::TBB-2$^{\beta\text{-tubulin}}$), or centrosomes (GFP::TBG-1$^{\gamma\text{-tubulin}}$). Additionally, we may utilize the AuroraB kinase homolog (GFP::AIR-2$^{\text{AuroraB}}$) that marks chromosomes from prophase to metaphase and translocates to the central spindle at anaphase. Although GFP::AIR-2$^{\text{AuroraB}}$ is not tracked or used for distance measurement, it serves as a valuable indicator of anaphase onset and central spindle integrity during anaphase. These marker combinations are particularly adapted to monitor the distances between chromosomes and spindle poles during the early embryonic divisions, typically up to the 64-cell stage. Depending on specific experimental needs, alternative fluorescent markers can obviously be employed.

### 2.2 Materials and Solutions

1. High quality borosilicate glass coverslips No. 1.5 (0.13–0.17 mm) 24 × 60 mm.
2. No. 15 blade scalpel (Bard-Parker 371615) or 23G × 1″ needle (No. 16, Terumo Neolus NN-2613R).
3. A pair of superfine eyelashes with a handle (Ted Pella, Inc, Prod No.113) (*see* **Note 1**).
4. Dumostar Dumont Tweezers #5, 11 cm, straight, 0.1 × 0.06 mm tips.
5. Meiosis Medium (*see* **Note 2**): 0.5 mg/mL Inulin from dahlia tubers, 25 M HEPES (pH 7.5, BioPerformance Certified, ≥99.5% (titration), cell culture tested), 60% Leibovitz's L-15 Media with Galactose, Phenol Red, L-Glutamine and Sodium Pyruvate; without Glucose, HEPES and Sodium Bicarbonate, 20% Fetal Bovine Serum (FBS, Heat Inactivated). Aliquot 1 mL of 100% FBS in 1.5 mL tubes. Store at −20 °C indefinitely. Combine all components in a sterile environment within a tissue culture hood. Do not filter. Prepare 1 mL stock aliquots in 1.5 mL sterile tubes and store at −20 °C for up to one year. Thaw and thoroughly mix a 1 mL stock aliquot, when needed, to make single-use 50 μL working aliquots. Store the working aliquots at −20 °C. Do not refreeze thawed 50 μL working aliquots.

### 2.3 Imaging Setups

#### 2.3.1 Dissection Scope

Worm dissection is carried out using a stereomicroscope featuring a diascopic transmitted light illumination stand. An adjustable zoom capability with high magnification proves advantageous for the recognition of dissected embryo stages. However, opting for higher magnification objectives entails a trade-off, resulting in a reduced working distance and a more confined field of view, potentially posing challenges during dissection. To achieve this, we utilize a Nikon SMZ1500 stereomicroscope equipped with a Plan 1x objective, C-W10X eyepieces, and a C-DSS diascopic transmitted light illumination stand. Dissection scope images were acquired using a fixed focal length CMOS camera (Jeulin Industrial Digital Camera) equipped with a microscope adapter ∅23.2–∅30 mm, and using the ToupView software (https://www.touptek.com).

#### 2.3.2 Microscope (See Note 3)

The precise microscopy setup can vary depending on applications and budget. The use of a confocal microscope is recommended to improve the signal-to-noise ratio, and a fast-scanning system is essential to film rapid processes and avoid excessive light-induced toxicity. We use a Nikon Ti inverted microscope equipped with a Yokogawa CSU-X1 spinning disk head and emission filter wheel. Acquisitions are done through a Nikon CFI APO λS 60×/NA 1.4 oil immersion objective, equipped with a CoolSNAP HQ2 CCD camera (Photometrics Scientific) and controlled by the MetaMorph 7 software (Molecular Devices). Illumination is performed with a fiber-coupled laser bench engine with a 150 mW–488 nm and a 110 mW–561 nm laser (Roper Scientific). Cargille Type A immersion oil (Cat. #16482) is used.

#### 2.3.3 Temperature Control System (Optional) (See Note 4)

Employing a temperature control system is imperative when working with fast-acting temperature-sensitive (fast ts) mutants for filming at the permissive temperature (16 °C). It ensures the maintenance of this stable temperature and also facilitates precise timing for switching to the restrictive temperature (typically 26 °C). While using a temperature controller may be considered optional for routine live imaging, we have made it a standard practice across all experimental conditions. This decision is aimed at maintaining precise and stable temperature control during imaging, regardless of external factors, thus enhancing the reproducibility of our results across various experiments. To execute our experiments with utmost precision, we rely on the CherryTemp controller temperature system from Cherry Biotech (https://www.cherrybiotech.com/). This system includes a universal insert designed to accommodate a 24 × 60 mm coverslip for sample dissection. Additionally, a silicon spacer is provided to avoid evaporation of the medium during filming and to ensure that the temperature-controlling chip (thermalizing chip hereafter) does not cause any compressive damage to the dissected embryos (Fig. 1a). The thermalizing chip must be as close to the heat exchangers as possible to limit temperature variations in the connective tubing. The provided holder for the heat exchangers must thus be installed as close as possible to the microscope stage.

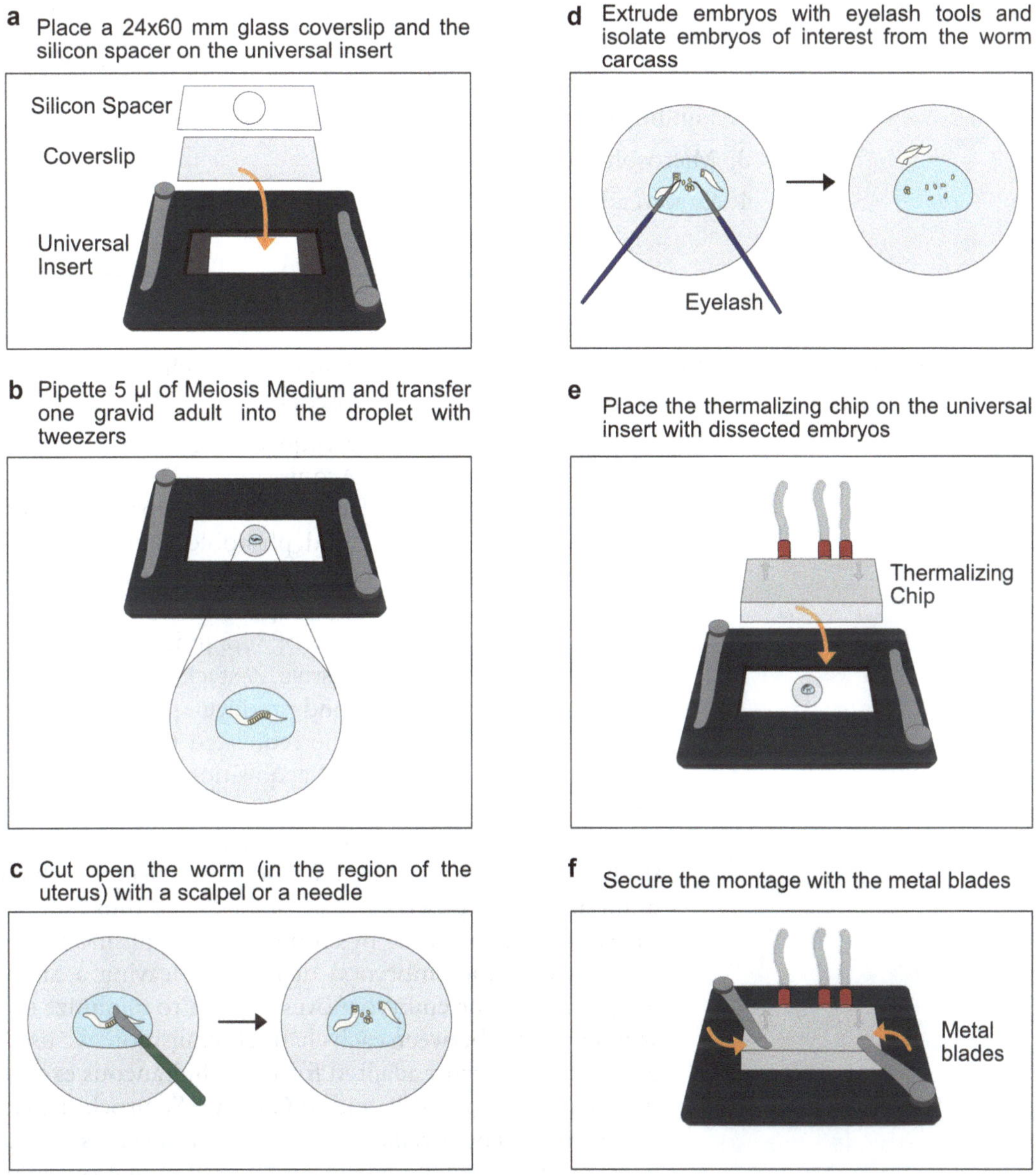


**Fig. 1** Workflow for dissection and mounting. (**a**) The 24 $\times$ 60 mm coverslip is placed onto the CherryTemp universal insert in the notch designed for this purpose and covered with the silicon spacer. (**b**) A 5 μL drop of meiosis medium is placed on the coverslip at the center of the silicon spacer opening, and one gravid adult hermaphrodite is transferred into the droplet. (**c**) The worm is cut open as close to the vulva as possible using a scalpel. (**d**) Oocytes and embryos are extruded from the gonad using eyelash tools, and embryos at the desired developmental stage are isolated. The worm carcass is removed from the drop of meiosis medium using eyelash tools. (**e**) The thermalizing chip is placed on the coverslip containing the dissected embryos, and secured in place with the two metal blades of the universal insert (**f**). The entire montage is placed on the inverted microscope stage for imaging

### 2.4 Software

1. Imaris: All our analyses were performed using Imaris 10.1.1. and Imaris File Converter (https://imaris.oxinst.com/big-data).
2. Fiji: https://imagej.net/software/fiji/downloads [18].
3. Microsoft Excel for Mac 2021.
4. GraphPad Prism 10.2.3.

## 3 Methods

### 3.1 Preparation of the Acquisition System

1. Once the microscope and computer are switched on, set the parameters controlling the microscope.
   (a) In the Metamorph software, acquisition parameters are set in the Multi-Dimensional Acquisition (MDA) window. It is essential to find a balance between depth sampling (Z-axis), time resolution, and photobleaching to minimize phototoxicity. We typically acquire 70 Z-planes at each time point, spaced 0.5 μm apart, resulting in a stack of images with a total thickness of 34.5 μm. We maintain a 20-s interval between successive Z-stack acquisitions to allow embryo recovery and reduce photobleaching effects. To minimize the time requested to acquire each Z-stack, we use the Stream Z acquisition mode in MetaMorph, with a 40 ms exposure time for each excitation channel (488 and 561 nm for GFP and mCherry, respectively). We also reduce the XY field of acquisition, which minimizes the exposed CCD sensor area and thus reduces the sensor readout time by cropping the acquisition window close to the embryo(s) of interest, leaving a safety margin in case the embryo moves. Finally, to minimize the switching time between each channel acquisition, we use a dual dichroic mirror adapted for the simultaneous excitation/emission of GFP and mCherry (Semrock Di01-T488/568). The specific channel of acquisition is selected by successively switching between the two excitation lasers and the two corresponding specific emission filters (Chroma ET525/50m for GFP and ET630/75m for mCherry).
   (b) All acquisitions are performed with 2 × 2 binning. Although 2 × 2 binning on a CCD sensor reduces the resolution by a factor of 2, it improves the signal-to-noise ratio by a factor of 4, thus improving image quality and facilitating object tracking.
   (c) Control the laser intensities at the objective level using a power meter (Ophir VEGA Meter, Photonics) and adjust the percentage of laser power used for imaging through

software control. Maintaining consistent laser intensities when filming embryos across multiple sessions enhances reproducibility between acquisitions and is essential for accurate image quantifications (*see* **Note 5**).

2. Switch on the CherryTemp and CherryLoop controllers. Start the CherryTemp user interface and run the Initialization step (this takes 5′30″). Select your temperature configuration (Dry or Immersion) depending on the objective used for acquisition. Set the temperature for Heat Engine and Fluid Engine. For routine live imaging, we typically set both temperatures at 23 °C. The graph in the interface window shows the actual temperature imposed by the thermalizing chip.

### 3.2 Dissection of Embryos

1. Place a 24 × 60 mm coverslip in the universal CherryTemp insert. Cover it with the provided silicon spacer (Fig. 1a).
2. Dispense a 5 μL drop of meiosis medium at the center of the spacer opening. Make sure that the medium doesn't touch the spacer. Place the CherryTemp insert under the dissection scope. Carefully transfer one or two healthy gravid worms (*see* **Note 6**) from an NGM plate to the droplet of meiosis medium using tweezers or an eyelash tool (*see* **Note 7**). Be cautious to prevent any harm to the worm, as this could result in the loss of embryos (Fig. 1b).
3. Using a scalpel or the beveled tip of a needle, cut open the worm as close to the vulva as possible (Fig. 1c). Many embryos will spontaneously emerge from the worm carcass. These are usually the oldest multicellular embryos. To harvest younger embryos stuck in the uterus, employ two eyelash tools: use one to maintain the worm cuticle in place, while applying pressure with the other to expel the embryos from the uterus (Fig. 1d) (*see* **Note 8**). Our preferred magnification for this step is typically set at 20× (Fig. 2a).
4. Stage and identify embryos of interest by increasing the zoom factor to 90× (Fig. 2b) (*see* **Note 9**). If necessary, use our provided embryo staging guide for this (Fig. 2b). If several embryos are at the desired stage, use the eyelash tool to arrange them side by side, ideally ensuring they do not physically contact each other (*see* **Note 10**). Locate the embryos of interest within the meiosis medium droplet to facilitate their finding under the microscope (*see* **Note 11**).
5. Place the thermalizing chip above the universal insert that contains the dissected embryos so that the chip is directly in contact with the meiosis medium droplet (Fig. 1e). Secure the thermalizing chip in place with the two metal blades attached to the universal insert (Fig. 1f).

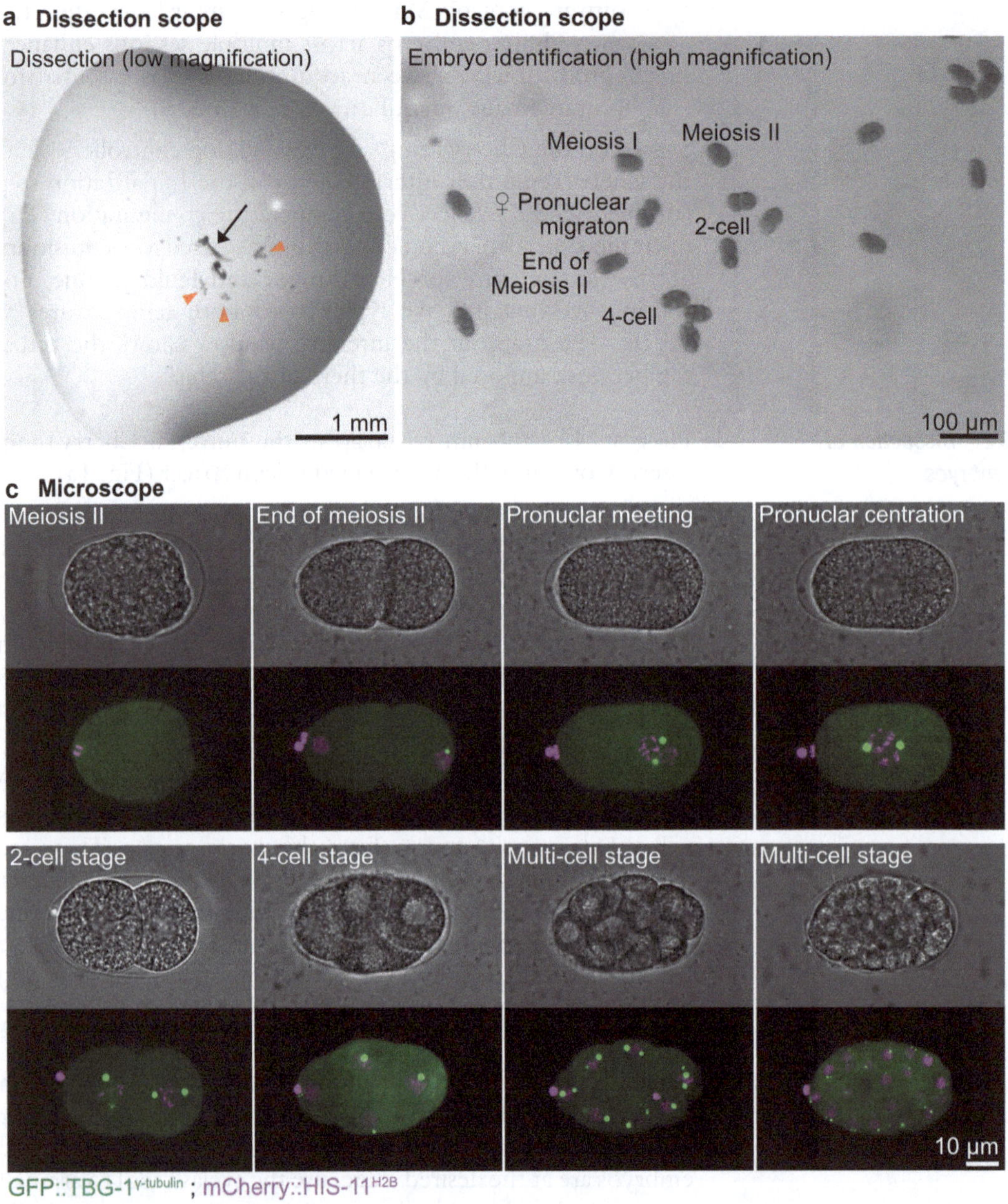


**Fig. 2** An illustrated guide for identifying distinct embryonic stages. (**a**) Still image taken after the worm dissection procedure in the meiosis medium droplet. The image was captured at a low magnification (20×) using a dissection scope. The black arrow indicates a fragment of worm carcass. The orange arrowheads point to embryos. Scale bar, 1 μm. (**b**) Still image of embryos at a higher magnification (90×). The developmental stage of embryos (indicated on the image) can be determined at this step according to their morphology. Scale bar, 100 μm. (**c**) Still images of embryos at the indicated developmental stages observed with brightfield (top) or spinning-disk confocal fluorescence microscopy (bottom). Centrosomes are in green (GFP::TBG-1), chromosomes are in magenta (mCherry::HIS-11). Scale bar, 5 μm

### 3.3 Live Imaging

1. Apply a tiny droplet of immersion oil onto the 60× objective. Place the montage (comprised of the universal insert and thermalizing chip, Fig. 1f) on the microscope stage using the provided microscope insert.
2. Use transmitted light and the 60× objective to locate the embryos of interest through the eyepieces of the microscope (Fig. 2c). If you find embryos at the desired stage, initiate acquisition immediately with the preset parameters. However, if embryos are younger than expected, wait until they reach the desired stage before initiating acquisition to minimize unnecessary light exposure.
3. Upon completion of acquisition, separate the temperature-controller chip from the montage, retrieve the silicon spacer and discard the coverslip containing the dissected embryos. Start a new dissection if needed.

### 3.4 Semi-automated Analysis Using Imaris

1. Download the free Imaris File Converter. Drag and drop the ND file obtained from Metamorph for each acquisition, and start their conversion into the imaris file format (.ims), which can be opened with the Imaris Software. Alternatively, the image sequence can be opened with the Fiji software and sent directly to Imaris (Pluggin > Imaris > Image to Imaris).
2. To track objects in XYZ and over time (t), utilize the Surface tool in Imaris. Create a new Surface for each acquired channel. The surface module requires several setting steps. Navigate between steps by clicking on the left and right blue arrows.
   - *Step 1: Algorithm.* Select "Segment only a region of interest" and "Track surface over time." Unselect "Classify Surfaces."
   - *Step 2: Region of Interest.* Define a box encompassing all the objects you wish to track within the desired time frame.
   - *Step 3: Segmentation Set Up.* Specify the channel for your objects. We use the Smooth segmentation parameter, with the "Absolute Intensity Thresholding" selected.
   - *Step 4: Threshold.* Imaris will suggest an initial threshold: a signal above this threshold is considered a potential object, while a signal below the threshold is deemed a background. You can manually adjust this threshold to better fit the shape of your objects (*see* **Notes 12** and **13**).
   - *Step 5: Filter Surfaces.* This step allows to filter surfaces according to their size. We use default parameters.
   - *Step 6: Edit Surfaces.* At this step, Imaris shows all detected objects within the Region of Interest for each timepoint. Any detected object that does not correspond to actual objects of interest can be removed by selecting them and

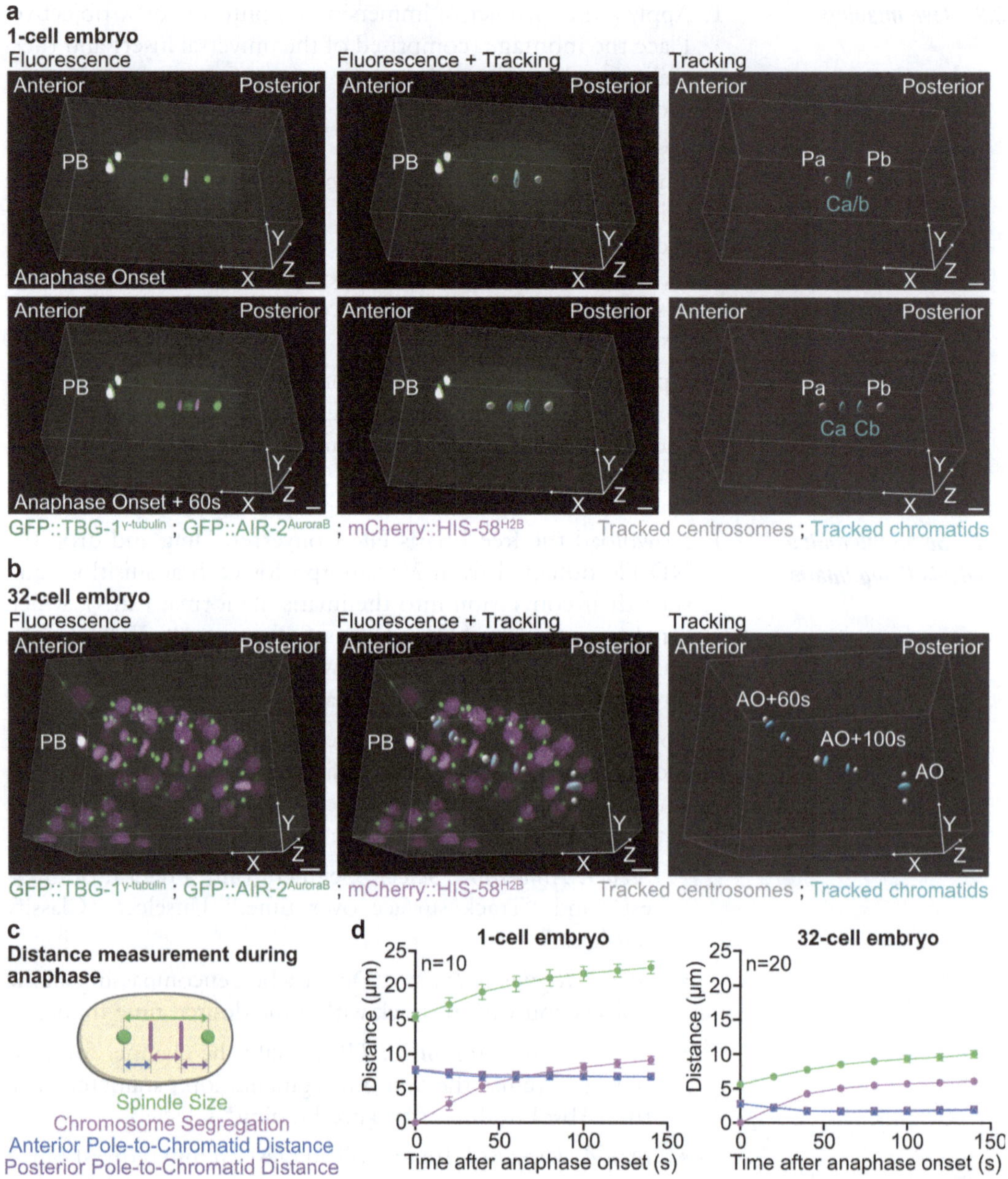


**Fig. 3** 3D tracking and distance measurement. (**a**, **b**) Still images of fluorescent signals and object tracking after 3D reconstitution on Imaris for one-cell (**a**) and 32-cell (**b**) embryos, captured at different times relative to Anaphase Onset (AO). Embryos are oriented with the anterior pole on the left and the posterior pole on the right. Centrosomes are in green (GFP::TBG-1), and their corresponding tracked objects are highlighted in grey (Pa: anterior pole, Pb: posterior pole). Chromosomes are shown in magenta (mCherry::HIS-58), and their corresponding tracked objects are highlighted in blue (Ca: anterior set of chromatids, Cb: posterior set of chromatids, Ca/b: chromosomes before anaphase onset). This strain also expresses GFP::AIR-2$^{AuroraB}$, which is not used for tracking. PB, Polar Bodies. Scale bar, 5 μm. (**c**) The schematic illustrates the different measured distances: spindle size (green double-headed arrow) is determined by measuring the distance between the two centrosomes, while sister chromatid segregation (magenta double-headed arrow) is determined by measuring the distance between the two sets of segregating chromatids. The pole-to-chromatid distance is determined by measuring the distance between the anterior pole and chromatids (blue double-headed arrow) and between the posterior pole and chromatids (purple double-headed arrow).

clicking on the "Delete" button. Multiple objects selection can be performed by clicking on them while holding down the Control key. If Imaris mistakenly identifies two objects that are close together as a single entity, they can be manually separated. To do this, position the object such that the intended cutting plane is vertical. While pressing the Shift key, choose the cutting location and then click on the "Cut Surface" button.

- *Step 7: Tracking.* Imaris provides a range of algorithms for tracking surfaces. The "Connected Components" algorithm is our preferred choice.
- *Step 8: Filter Tracks.* By default, the filter is set over the track duration feature, which is not a relevant parameter here. This filter thus needs to be turned off. We do not set any other filter.

Once every step is achieved, click on the green arrow to finish the Surface creation.

3. Imaris-generated tracks over time are displayed in the Edit Tracks menu. If tracks are misconnected, they can be manually adjusted. Individual objects within a track can be selected by pressing the "o" key prior to selection. To select an entire track, press the "t" key before making a selection. If an object is wrongly connected to a track, it can be isolated from the track. To do this, navigate to the Track menu, select the individual object, and click on "Disconnect the selected object." Conversely, an isolated object can be assigned to an existing track by selecting the isolated object and the adjacent object included in a track (while holding down the Control key) and clicking on "Connect selected objects."
4. After ensuring that the tracks align accurately with your biological objects (as shown in Fig. 3a, b), proceed to label them appropriately. Navigate to the Edit Label menu and add as many classes as needed to label all your tracks (one class per track). Typically, we label spindle poles (identified by GFP::TBG-1$^{\gamma\text{-tubulin}}$) as Pa (Pole a or anterior pole) and Pb (Pole b or posterior pole) and chromatids (identified by mCherry::HIS-58$^{H2B}$) as Ca (Chromatids a or anterior set of chromatids) and Cb (Chromatids b or posterior set of chromatids) (*see* **Note**

**Fig. 3** (continued) Following XYZ coordinate extraction for each labeled object during anaphase, the Euclidean distance between two objects can be calculated as $\text{distance}_{AB} = \sqrt{(X_A - X_B)^2 + (Y_A - Y_B)^2 + (Z_A - Z_B)^2}$. (**d**) Graph representing the mean distances measured from anaphase onset for one-cell (left, $n = 10$) and 32-cell (right, $n = 20$) embryos. The colors correspond to the distances depicted in (**c**). Error bars represent the 95% confidence interval

14). Select a track and click "Assign" to apply the appropriate label (*see* **Note 15**).

5. Once all the tracks are correctly labeled and assigned, navigate to the "Statistics" menu. Click on "Configure list of visible statistics values" to ensure that the "Surfaces" parameters are selected as they contain the XYZ coordinates of every object tracked by Imaris. Export all the statistics to an Excel file.
6. The Excel file comprises various features, each stored in distinct tabs. The "Position" tab contains XYZ coordinates for every labeled object at each time point (*see* **Note 16**). To streamline data analysis, extract the XYZ position for each object (Pa, Pb, Ca and Cb) and compile them into a single tab of a new Excel file. Create a pivot table with time points as rows and XYZ coordinates for each labeled object as columns. Utilize the Euclidean distance formula to compute the necessary distance measurements. We measure the spindle size by calculating the distance between Pa and Pb, sister chromatid segregation by calculating the distance between Ca and Cb, and Pole-to-Chromatid distance at the anterior and posterior side of the spindle by calculating the distance between Pa and Ca, and Pb and Cb, respectively (Fig. 3c). Distance measurements can be performed for any object of interest.
7. Compile these distances from all experimental conditions into a single tab of a new Excel file. Create another pivot table to categorize each type of distance measurement according to the respective experimental conditions. Export this pivot table into Prism software or any other suitable software for comprehensive graphing and statistical analysis (*see* **Note 17**).

## 4 Notes

1. Eyelash tools can be easily crafted by affixing the root end of a clean eyelash onto the tip of a Pasteur pipette (Sigma Cat. #S6018) using a small amount of nail polish. It is essential to note that the stiffness of the eyelash significantly impacts the dissection process. Stiff eyelashes aid embryo extrusion but risk damaging delicate oocytes or embryos. Balancing stiffness and flexibility is essential when choosing or crafting eyelash tools for dissection.
2. After anaphase II, fertilized oocytes and embryos are enclosed within an eggshell, effectively shielding them from the external environment. When imaging embryonic divisions, dissection can be performed in various media such as M9 (22 mM $KH_2PO_4$, 42 mM $Na_2HPO_4$, 85,5 mM NaCl, 1 mM $MgSO_4$), Egg salt (118 mM NaCl, 48 mM KCl, 2 mM

$CaCl_2$, 2 mM $MgCl_2$, 5 mM HEPES, pH 7.5), and even water. However, opting for dissection in the meiosis medium, which is meticulously controlled for osmotic balance, allows preserving fertilized oocytes, which are permeable and sensitive to osmotic pressure until anaphase II of meiosis [19].

3. Embryonic divisions occur rapidly, underscoring the necessity of positioning the dissection scope in close proximity to the microscope, ideally within the same room. This ensures that the desired developmental stages are not missed.
4. The method outlined here is tailored for use with the CherryTemp system. However, for those without access to the CherryTemp system, detailed instructions for setting up a custom filming chamber can be found in [13].
5. The stability of the laser intensities can fluctuate over time due to laser and optical component misalignment, caused by thermal fluctuation and vibrations. If small intensity variations are measured, laser intensities are adjusted through software control. However, for more significant deviations that cannot be rectified through software adjustments alone, manual realignment of the various optical components within the laser engine becomes necessary.
6. When selecting adult worms for filming one-cell embryos, it is advisable to choose young individuals with only one row of embryos inside the uterus. Older worms tend to contain more embryos, but these are typically older multicellular embryos.
7. To minimize fluorescent background and prevent interference during imaging, it is important to avoid excessive bacterial contamination when dissecting worms. Avoiding the inclusion of small larvae with adult worms is also important, as their vigorous movement can disrupt embryo imaging by physically shifting them. It is thus recommended to pick adults outside of the bacterial lawn, where bacterial contamination is lower and larvae are less abundant, ensuring cleaner and more stable imaging conditions.
8. During this step, it is crucial to remove any large moving segments of worm carcasses as well as any swimming larvae from the meiosis medium droplet. This precautionary measure helps prevent unexpected embryo movement during acquisition, which may occur due to the liquid flux induced by the movement of worm carcasses and larvae. Furthermore, this practice reduces the risk of meiosis medium contamination with bacteria released from the worm gut, ensuring cleaner imaging conditions.
9. One-cell embryos at the pronuclear migration stage form a pseudo-cytokinetic furrow [20], which gives them the appearance of 2-cell stage embryos. To distinguish between the

pronuclear migration stage and the actual 2-cell stage, observe the position of the pronuclei relative to the cytoplasm. If one nucleus is located very close to the posterior cortex, the embryo is likely at the pronuclear migration stage, and the nucleus close to the cortex is the male pronucleus.

10. Embryos tend to stick together and no longer lie flat on the glass coverslip.
11. Preferably, embryos selected for filming the first embryonic division should be in the pronuclear migration stage. This allows sufficient time for the seamless transfer of the entire montage to the microscope and for finding embryos under the 60× objective.
12. By default, Imaris gives the XYZ coordinates of the centroid for each selected object. Thus, adjusting the threshold impacts the volume of the object shape but has very little effect on its centroid coordinates.
13. If the intensity of an object fluctuates over time, using a single threshold for the entire acquisition may not be suitable. It can be beneficial to create an object track with an appropriate threshold for a series of time points with similar intensities. Subsequently, generate another track for the same object using a different threshold for subsequent time points with varying intensities. This approach helps ensure accurate object detection and tracking throughout the imaging process.
14. Ensure consistent labeling of all objects across different experimental conditions to streamline subsequent analysis.
15. At this step, objects and tracks are still editable. The "Connect" and "Disconnect" functions are located in the Edit Tracks menu, while the "Cut Surface" and "Delete" functions are located in the Edit menu.
16. At metaphase, which is the last time point before anaphase onset (AO) or $t = 0$ relative to AO, only one chromosome mass is visible. This chromosome mass corresponds to both the anterior and posterior sets of sister chromatids. Therefore, we attribute the same XYZ coordinates to both the anterior and posterior sets of sister chromatids at this time point.
17. In this chapter, we have outlined a method for filming early embryonic divisions of *C. elegans* in 3 dimensions over time. Furthermore, by using a transgenic *C. elegans* strain expressing GFP-tagged TBG-$1^{\gamma\text{-tubulin}}$ and mCherry-tagged HIS-$58^{H2B}$, we demonstrated how to successfully segment and track the spindle poles and chromosomes during mitosis at various early embryonic stages. The simplicity of genome-editing techniques in *C. elegans*, such as MosSCI transgenesis or CRISPR/Cas9 genome editing, allows for straightforward tracking of

the subcellular dynamics of any fluorescently tagged protein of interest using light microscopy approaches [21, 22]. By combining fluorescent labeling of key proteins with the method described in this review, any aspects of mitotic cell divisions throughout early embryonic divisions can be explored.

## Acknowledgments

We thank all members of the Dumont lab for their support and advice, especially Nicolas Macaisne for his comments on the manuscript. We are grateful to Patricia Moussounda, Clarisse Picard, Téo Bitaille, and Vincent Maupu-Massamba for providing technical support. We acknowledge the ImagoSeine core facility of Institut Jacques Monod, member of France-BioImaging (ANR-10-INBS-04) and IBiSA, with the support of Labex "Who Am I," Inserm Plan Cancer, Region Ile-de-France and Fondation Bettencourt Schueller. This work was supported by CNRS and University Paris Cité.

### References

1. Charalambous C, Webster A, Schuh M (2023) Aneuploidy in mammalian oocytes and the impact of maternal ageing. Nat Rev Mol Cell Biol 24(1):27–44. https://doi.org/10.1038/s41580-022-00517-3
2. Maiato H, Silva S (2023) Double-checking chromosome segregation. J Cell Biol 222(5): e202301106. https://doi.org/10.1083/jcb.202301106
3. Valdez VA, Neahring L, Petry S et al (2023) Mechanisms underlying spindle assembly and robustness. Nat Rev Mol Cell Biol 24(8): 523–542. https://doi.org/10.1038/s41580-023-00584-0
4. Renda F, Khodjakov A (2021) Role of spatial patterns and kinetochore architecture in spindle morphogenesis. Semin Cell Dev Biol 117: 75–85. https://doi.org/10.1016/j.semcdb.2021.03.016
5. Hoffmann I (2021) Centrosomes in mitotic spindle assembly and orientation. Curr Opin Struct Biol 66:193–198. https://doi.org/10.1016/j.sbi.2020.11.003
6. Gönczy P, Hatzopoulos GN (2019) Centriole assembly at a glance. J Cell Sci 132(4): jcs228833. https://doi.org/10.1242/jcs.228833
7. Conduit PT, Wainman A, Raff JW (2015) Centrosome function and assembly in animal cells. Nat Rev Mol Cell Biol 16(10):611–624. https://doi.org/10.1038/nrm4062
8. Kraus J, Alfaro-Aco R, Gouveia B et al (2023) Microtubule nucleation for spindle assembly: one molecule at a time. Trends Biochem Sci 48(9):761–775. https://doi.org/10.1016/j.tibs.2023.06.004
9. Brenner S (1974) The genetics of Caenorhabditis elegans. Genetics 77(1):71–94. https://doi.org/10.1093/genetics/77.1.71
10. Sulston JE, Horvitz HR (1977) Post-embryonic cell lineages of the nematode, Caenorhabditis elegans. Dev Biol 56(1):110–156. https://doi.org/10.1016/0012-1606(77)90158-0
11. Corsi AK, Wightman B, Chalfie M (2015) A transparent window into biology: a primer on Caenorhabditis elegans. WormBook 18:1–31. https://doi.org/10.1895/wormbook.1.177.1
12. Hattersley N, Lara-Gonzalez P, Cheerambathur D et al (2018) Employing the one-cell C. elegans embryo to study cell division processes. Methods Cell Biol 144:185–231. https://doi.org/10.1016/bs.mcb.2018.03.008
13. Laband K, Lacroix B, Edwards F et al (2018) Live imaging of C. elegans oocytes and early embryos. Methods Cell Biol 145:217–236. https://doi.org/10.1016/bs.mcb.2018.03.025
14. Maton G, Edwards F, Lacroix B et al (2015) Kinetochore components are required for

central spindle assembly. Nat Cell Biol 17(7): 953. https://doi.org/10.1038/ncb3199

15. Edwards F, Maton G, Gareil N et al (2018) BUB-1 promotes amphitelic chromosome biorientation via multiple activities at the kinetochore. elife 7:e40690. https://doi.org/10.7554/eLife.40690
16. Rose L, Gönczy P (2014) Polarity establishment, asymmetric division and segregation of fate determinants in early C. elegans embryos. WormBook 30:1–43. https://doi.org/10.1895/wormbook.1.30.2
17. Pitayu-Nugroho L, Aubry M, Laband K et al (2023) Kinetochore component function in C. elegans oocytes revealed by 4D tracking of holocentric chromosomes. Nat Commun 14(1):4032. https://doi.org/10.1038/s41467-023-39702-z
18. Schindelin J, Arganda-Carreras I, Frise E et al (2012) Fiji: an open-source platform for biological-image analysis. Nat Methods 9(7): 676–682. https://doi.org/10.1038/nmeth.2019
19. Stein KK, Golden A (2018) The C. elegans eggshell. WormBook 2018:1–36. https://doi.org/10.1895/wormbook.1.179.1
20. Oegema K, Hyman AA (2006) Cell division. WormBook 19:1–40. https://doi.org/10.1895/wormbook.1.72.1
21. El Mouridi S, Frøkjær-Jensen C (2022) Targeted and random transposon-assisted single-copy transgene insertion in C. elegans. Methods Mol Biol 2468:239–256. https://doi.org/10.1007/978-1-0716-2181-3_12
22. Paix A, Folkmann A, Rasoloson D et al (2015) High efficiency, homology-directed genome editing in Caenorhabditis elegans using CRISPR-Cas9 ribonucleoprotein complexes. Genetics 201(1):47–54. https://doi.org/10.1534/genetics.115.179382

Chapter 11

# Assessment of Chromosome Oscillations in Mammalian Cells by Live Cell Imaging

**Kenji Iemura and Kozo Tanaka**


## Abstract

In metaphase, chromosomes undergo a back-and-forth movement between two spindle poles called chromosome oscillation. This dynamic is necessary to maintain the robustness of chromosome segregation. This chapter describes the materials and methods required to observe chromosome oscillation in mammalian cell lines and calculate the deviation of amplitude (DAP), which is used to quantify the degree of chromosome oscillation.


**Key words** Chromosome dynamics, Kinetochore, Mitotic spindle, Metaphase, Microscope, Time-lapse imaging

## 1 Introduction

During metaphase, aligned chromosomes at the spindle equator move back and forth between the spindle poles [1–4]. These dynamics, known as chromosome oscillation, is a common phenomenon in mitosis and meiosis, from yeast to human cells [2, 5]. The pulling force of the microtubules attached to the kinetochores is the primary driver of chromosome oscillation [6–10]. However, it is regulated by the balance of various other forces, including the polar ejection force (PEF), which pushes on the chromosome arms to achieve repetitive movements, and the force of sister chromatid cohesion [6, 10–13]. Previous studies have shown that chromosome oscillation promotes Hec1 tail phosphorylation in non-transformed mammalian cells, ensuring faithful chromosome segregation [1, 14–19]. Mathematical modeling of Hec1 tail phosphorylation and kinetochore-microtubule attachment has also confirmed that chromosome oscillation promotes the correction of erroneous kinetochore-microtubule attachment [17, 20–22]. Chromosome oscillation is attenuated in cancer cells, which may cause chromosomal instability through inefficient

Tomomi Kiyomitsu (ed.), *The Mitotic Spindle: Methods and Protocols*, Methods in Molecular Biology, vol. 2872,
https://doi.org/10.1007/978-1-0716-4224-5_11,

correction of erroneous kinetochore-microtubule attachment [1, 23–28].

The accurate quantification of chromosome oscillations can measure the robustness of chromosome segregation. However, even the comparative analysis of chromosome oscillations between cells of the same origin is often inconsistent due to the multiple parameters that affect the measurement, including rotation and translocation of the spindle, as well as the phototoxicity during live cell imaging using fluorescence microscopy and individual differences in identifying cells that are suitable for observation. This paper describes the criteria for choosing cells for observation, preparation of cells for observation, quantification of chromosome oscillation, and calculation of the deviation from average position (DAP, [29]) as an indicator of chromosome oscillation.

## 2 Materials

1. $CO_2$ incubator for cell culture.
2. Water bath to keep the cell culture medium warm.
3. Glass bottom dish or glass bottom chamber.
4. Adherent mammalian cell lines in which kinetochores and microtubules are fluorescently labeled (*see* **Note 1**).
5. Culture medium containing fetal bovine serum (FBS) suitable for the cell line used.
6. Proteasome inhibitor (MG132).
7. The microscope is equipped with a temperature-controlled chamber and motorized stage for time-lapse observation (*see* **Note 2**).
8. ImageJ [30] or Fiji [31].
9. The ManualTracking plug-in in ImageJ.
10. The MultiStackReg plug-in in ImageJ [32].
11. Spreadsheet and graphical software (e.g. Microsoft Excel).

## 3 Methods

### 3.1 Preparation of Metaphase Cells

1. Cells are seeded to 70% confluency in a glass-bottom dish or glass-bottom chamber and cultured for 1 h at 37 °C in a 5% $CO_2$ atmosphere.
2. After cell adhesion (about 10 h after seeding in the case of HeLa cells), MG132 is added to a final concentration of 20 μM and incubated for 1 h to synchronize mitotic cells in metaphase (*see* **Note 3**).

***3.2 Observation of Chromosome Oscillation***

1. Cell culture dish or chamber is put on a fluorescence microscope warmed to 37 °C in a 5% $CO_2$ atmosphere, and observed with a ×100 oil objective lens to find metaphase cells in which all chromosomes are aligned at the spindle equator based on the CENP-A fluorescence signals. The cells with a bipolar spindle, where two spindle poles are in the same focal plane, are selected based on the fluorescent signal of α-tubulin (Fig. 1a and b, *see* **Note 4**).
2. Z-series of five sections in 2.5 μm increments are acquired at 2 s intervals for 5 min (*see* **Note 5**).
3. Repeat the selection and imaging of other metaphase cells as quickly as possible; it is recommended that images of several cells are captured to evaluate the DAP.

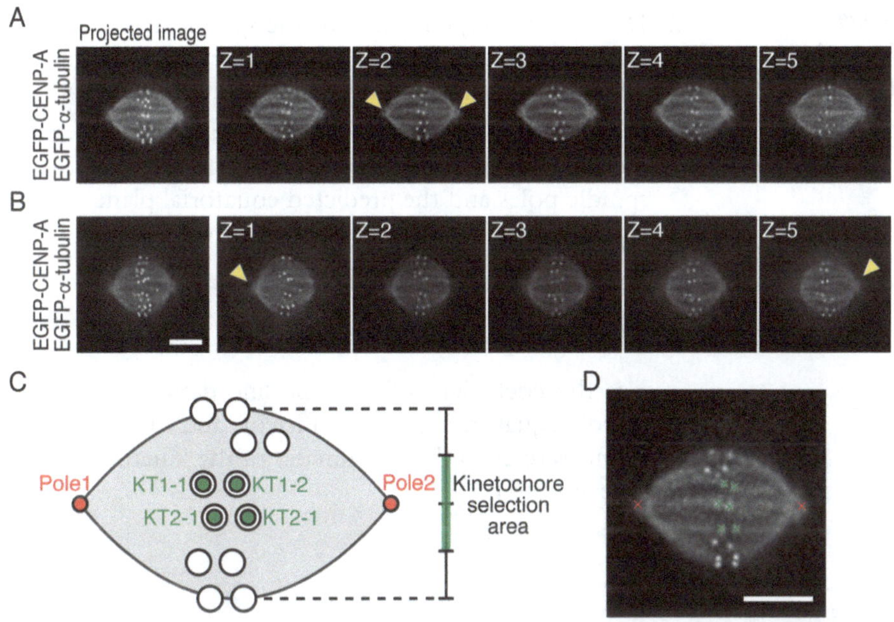


**Fig. 1** Selection of suitable cells for tracking chromosome oscillation. (**a**) A metaphase human epithelial cell expressing EGFP-CENP-A and EGFP-α-tubulin, in which both spindle poles are located on the same Z-slice. Cells were arrested in metaphase by MG132 treatment for 2 h and observed by live-cell imaging. The series of Z-sections are shown in the right panels. Yellow arrowheads indicate spindle poles. Scale bar: 5 μm. (**b**) A metaphase human epithelial cell expressing EGFP-CENP-A and EGFP-α-tubulin in which spindle poles are located on the different Z-slices. Cells were arrested in metaphase by MG132 treatment for 2 h and observed by live-cell imaging. The series of Z-sections are shown in the right panels. Yellow arrowheads indicate spindle poles. Scale bar: 5 μm. (**c**) Schematic diagram of the kinetochores selected for tracking chromosome oscillation. Tracking of both spindle poles (red dots) and kinetochores (green dots) localized at the center of the metaphase plate (green area) in a metaphase cell. (**d**) Spindle poles and kinetochores selected for tracking in A. Z = 1 to Z = 3 are projected, and spindle poles and kinetochores are marked with red and green crosses, respectively. Scale bar: 5 μm

4. Deconvolution is performed on the acquired images for cells whose spindle poles were in the same Z-slice during the 5 min, and the maximum intensity projection is performed on a total of three sections, the Z-slice containing both spindle poles and the Z-slices one before and one after.
5. Open the time-lapse imaging data in Fiji and use the Multi-StackReg plug-in to adjust the spindle orientation using the α-tubulin signal.
6. Track the spindle poles (edges of the long axis of the spindle) and the sister kinetochores (CENP-A dots) using the Manual-Tracking plug-in to obtain the X and Y coordinates (*see* **Note 6**).

### 3.3 Drawing Trajectories of Chromosome Oscillation and Calculation of DAP

1. Calculate the midpoint between the two spindle poles for each time frame.
2. Draw a line perpendicular to the spindle axis passing through the midpoint between the spindle poles for each time frame and define it as the predicted spindle equator (Fig. 2a).
3. Calculate the distance between the kinetochores and the predicted equatorial plane (ΔK) and the distance between the spindle poles and the predicted equatorial plane (ΔP) for each time frame using the following equation (Fig. 2b, *see* **Note 7**):

$$\Delta = \frac{ax_t + by_t + c}{\sqrt{a^2 + b^2}}$$

*a* is the coefficient of *x* in the line defined as the predicted spindle equator, *b* is its *y* coefficient, and *c* is its constant term. $x_t$ and $y_t$ refer to the coordinates of the kinetochore or spindle pole at time *t*.

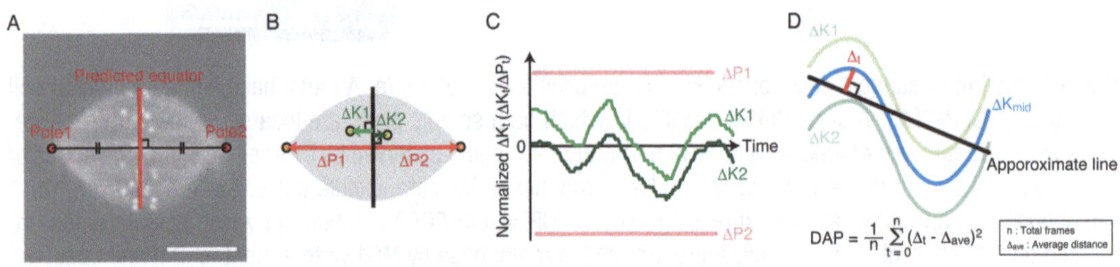


**Fig. 2** Calculation of chromosome oscillation. (**a**) Define the spindle equator. Draw a line between Pole1 and Pole2 and define the predicted spindle equator as a perpendicular line passing through the midpoint of the line between Pole1 and Pole2 (red line). (**b**) Calculate the distance from each kinetochore (ΔK1 or ΔK2, green arrow) or each spindle pole (ΔP1 or ΔP2, red arrow) to the predicted equatorial plane defined in **a**. (**c**) Plot the trajectory of ΔK1 and ΔK2 over time, which are normalized by ΔP1 and ΔP2. (**d**) Plot the trajectory of the midpoint of the normalized ΔK1 and ΔK2 (ΔKmid, blue line). Calculate the DAP, which is the standard deviation of the distance of ΔKmid from its regression line

4. Normalize ΔK by ΔP (*see* **Note 8**).
5. Draw line graphs for the normalized ΔK using graphing software (e.g., Microsoft Excel). This graph is the trajectory of the chromosome oscillation (Fig. 2c).
6. Calculate the coordinates of the midpoint of the sister kinetochore at each time frame from the normalized ΔK and draw the trajectory ($\Delta K_{mid}$). Then, draw a regression line for the $\Delta K_{mid}$ trajectory.
7. Calculate the distance between the $\Delta K_{mid}$ and the regression line ($\Delta_t$) using the following equation:

$$\Delta t = \frac{| ax_t + by_t + c |}{\sqrt{a^2 + b^2}}$$

$a$ is the coefficient of $x$ in the regression line, $b$ is its $y$ coefficient, and $c$ is its constant term. $x_t$ and $y_t$ refer to the coordinates of the $\Delta K_{mid}$ at time $t$.

8. Calculate the DAP as the standard deviations of $\Delta t$ (Fig. 2d).

$$\mathrm{DAP} = \frac{1}{n}\sum\nolimits_{t=1}^{n} (\Delta_t - \Delta_{ave})^2$$

$n$ is the number of total time frame, $\Delta_{ave}$ is an average of $\Delta_t$ for the whole-time frame.

## 4 Notes

1. For the observation of chromosome oscillation, kinetochores and microtubules should be visualized with fluorescent substances excited at the same wavelength in order to shorten the imaging interval and to minimize damage to the cells by the excitation light. We routinely use cell lines in which EGFP-CENP-A and α-Tubulin-EGFP are stably expressed. Different fluorophores can be used to visualize kinetochores and microtubules, but the imaging conditions must be optimized. Compounds such as SiR-tubulin can also be used to visualize microtubules.
2. Thermostatic equipment and motorized Z-stages in microscopes are essential for observing chromosome oscillation. For wide-field fluorescence microscopy, an LED light source and a piezo stage should be available to accelerate image acquisition speed. If confocal microscopy is used, high-speed scanning, e.g., a spinning disc system, is required. The deconvolution process can also improve the measurement accuracy of chromosome oscillation.

3. If the microscope is not equipped with a $CO_2$ regulator, replace the medium with the one that supports cell growth without $CO_2$ equilibration (e.g., Leibovitz's L-15 medium) warmed to 37 °C when treating cells with MG132.
4. Cells in which the two spindle poles are in the same Z-section during the imaging must be carefully selected, as a subtle shift in the focal plane of the spindle poles will significantly affect the quantification of chromosome oscillation.
5. In the presence of MG132, the mitotic cells are arrested in metaphase. All observations should be completed within 3 hours when cells are treated with MG132, as prolonged metaphase has been reported to cause premature separation of sister chromatids, known as cohesion fatigue.
6. It has been reported that chromosome oscillation is suppressed for the kinetochores localized at the edges of the metaphase plate [33]. Therefore, the kinetochores located at the center of the metaphase plate must be quantified for comparison between different conditions (Fig. 1c and d).
7. In oscillation trajectories, DK and DP take both positive and negative values. This can be achieved by removing the absolute value symbol from the numerator in the equation.
8. When examining chromosome oscillation in different cell types or under different conditions, such as drug treatment, spindle length may change with cell type or condition. As the relative amplitude of chromosome oscillation changes as a function of spindle length, DK needs to be normalized by DP when performing these comparative analyses.

## References

1. Iemura K, Yoshizaki Y, Kuniyasu K et al (2021) Attenuated chromosome oscillation as a cause of chromosomal instability in cancer cells. Cancers (Basel) 13:4531. https://doi.org/10.3390/cancers13184531
2. Nicklas RB (1997) How cells get the right chromosomes. Science (1979) 275:632–637. https://doi.org/10.1126/science.275.5300.632
3. Maiato H (2004) The dynamic kinetochore-microtubule interface. J Cell Sci 117:5461–5477. https://doi.org/10.1242/jcs.01536
4. Skibbens RV, Skeen VP, Salmon ED (1993) Directional instability of kinetochore motility during chromosome congression and segregation in mitotic newt lung cells: a push-pull mechanism. J Cell Biol 122:859–875. https://doi.org/10.1083/jcb.122.4.859
5. JKitajima TS, Ohsugi M, Ellenberg J (2011) Complete kinetochore tracking reveals error-prone homologous chromosome biorientation in mammalian oocytes. Cell 146:568–581. https://doi.org/10.1016/j.cell.2011.07.031
6. Dumont S, Salmon ED, Mitchison TJ (2012) Deformations within moving kinetochores reveal different sites of active and passive force generation. Science (1979) 337:355–358. https://doi.org/10.1126/science.1221886
7. Tanaka K, Kitamura E, Kitamura Y et al (2007) Molecular mechanisms of microtubule-dependent kinetochore transport toward spindle poles. J Cell Biol 178:269–281. https://doi.org/10.1083/jcb.200702141
8. Civelekoglu-Scholey G, Sharp DJ, Mogilner A et al (2006) Model of chromosome motility in Drosophila embryos: adaptation of a general

mechanism for rapid mitosis. Biophys J 90: 3966–3982. https://doi.org/10.1529/biophysj.105.078691

9. Wan X, O'Quinn RP, Pierce HL et al (2009) Protein architecture of the human kinetochore microtubule attachment site. Cell 137:672–684. https://doi.org/10.1016/j.cell.2009.03.035
10. Jaqaman K, King EM, Amaro AC et al (2010) Kinetochore alignment within the metaphase plate is regulated by centromere stiffness and microtubule depolymerases. J Cell Biol 188: 665–679. https://doi.org/10.1083/jcb.200909005
11. Ke K, Cheng J, Hunt AJ (2009) The distribution of polar ejection forces determines the amplitude of chromosome directional instability. Curr Biol 19:807–815. https://doi.org/10.1016/j.cub.2009.04.036
12. Stumpff J, Wagenbach M, Franck A et al (2012) Kif18A and chromokinesins confine centromere movements via microtubule growth suppression and spatial control of kinetochore tension. Dev Cell 22:1017–1029. https://doi.org/10.1016/j.devcel.2012.02.013
13. Almeida AC, Maiato H (2018) Chromokinesins. Curr Biol 28:R1131–R1135. https://doi.org/10.1016/j.cub.2018.07.017
14. DeLuca KF, Lens SMA, DeLuca JG (2011) Temporal changes in Hec1 phosphorylation control kinetochore–microtubule attachment stability during mitosis. J Cell Sci 124:622–634. https://doi.org/10.1242/jcs.072629
15. DeLuca KF, Meppelink A, Broad AJ et al (2018) Aurora A kinase phosphorylates Hec1 to regulate metaphase kinetochore–microtubule dynamics. J Cell Biol 217:163–177. https://doi.org/10.1083/jcb.201707160
16. Iemura K, Natsume T, Maehara K et al (2021) Chromosome oscillation promotes Aurora A–dependent Hec1 phosphorylation and mitotic fidelity. J Cell Biol 220. https://doi.org/10.1083/jcb.202006116
17. Campos Medina MA, Iemura K, Kimura A et al (2021) A mathematical model of kinetochore-microtubule attachment regulated by Aurora A activity gradient describes chromosome oscillation and correction of erroneous attachments. Biomed Res 42:203–219. https://doi.org/10.2220/biomedres.42.203
18. DeLuca JG, Gall WE, Ciferri C et al (2006) Kinetochore microtubule dynamics and attachment stability are regulated by Hec1. Cell 127: 969–982. https://doi.org/10.1016/j.cell.2006.09.047
19. Guimaraes GJ, Dong Y, McEwen BF et al (2008) Kinetochore-microtubule attachment relies on the disordered N-terminal tail domain of Hec1. Curr Biol 18:1778–1784. https://doi.org/10.1016/j.cub.2008.08.012
20. Cheeseman IM, Anderson S, Jwa M et al (2002) Phospho-regulation of kinetochore-microtubule attachments by the aurora kinase Ipl1p. Cell 111:163–172. https://doi.org/10.1016/S0092-8674(02)00973-X
21. Welburn JPI, Vleugel M, Liu D et al (2010) Aurora B phosphorylates spatially distinct targets to differentially regulate the kinetochore-microtubule interface. Mol Cell 38:383–392. https://doi.org/10.1016/j.molcel.2010.02.034
22. Auckland P, Clarke NI, Royle SJ et al (2017) Congressing kinetochores progressively load Ska complexes to prevent force-dependent detachment. J Cell Biol jcb.201607096. https://doi.org/10.1083/jcb.201607096
23. Cimini D, Howell B, Maddox P et al (2001) Merotelic kinetochore orientation is a major mechanism of aneuploidy in mitotic mammalian tissue cells. J Cell Biol 153:517–528. https://doi.org/10.1083/jcb.153.3.517
24. Janssen A, van der Burg M, Szuhai K et al (2011) Chromosome segregation errors as a cause of DNA damage and structural chromosome aberrations. Science (1979) 333:1895–1898. https://doi.org/10.1126/science.1210214
25. Burrell RA, McGranahan N, Bartek J et al (2013) The causes and consequences of genetic heterogeneity in cancer evolution. Nature 501: 338–345. https://doi.org/10.1038/nature12625
26. Bakhoum SF, Thompson SL, Manning AL et al (2009) Genome stability is ensured by temporal control of kinetochore–microtubule dynamics. Nat Cell Biol 11:27–35. https://doi.org/10.1038/ncb1809
27. Funk LC, Zasadil LM, Weaver BA (2016) Living in CIN: mitotic infidelity and its consequences for tumor promotion and suppression. Dev Cell 39:638–652. https://doi.org/10.1016/j.devcel.2016.10.023
28. Ricke RM, van Deursen JM (2013) Aneuploidy in health, disease, and aging. J Cell Biol 201: 11–21. https://doi.org/10.1083/jcb.201301061
29. Stumpff J, von Dassow G, Wagenbach M et al (2008) The kinesin-8 motor Kif18A suppresses kinetochore movements to control mitotic chromosome alignment. Dev Cell 14:252–262. https://doi.org/10.1016/j.devcel.2007.11.014

30. Schneider CA, Rasband WS, Eliceiri KW (2012) NIH Image to ImageJ: 25 years of image analysis. Nat Methods 9:671–675
31. Schindelin J, Arganda-Carreras I, Frise E et al (2012) Fiji: An open-source platform for biological-image analysis. Nat Methods 9: 676–682. https://doi.org/10.1038/nmeth.2019
32. Thévenaz P, Ruttimann UE, Unser M et al (1998) A pyramid approach to subpixel registration based on intensity. IEEE Trans Image Process 7:27–41. https://doi.org/10.1109/83.650848
33. Civelekoglu-Scholey G, He B, Shen M et al (2013) Dynamic bonds and polar ejection force distribution explain kinetochore oscillations in PtK1 cells. J Cell Biol 201:577–593. https://doi.org/10.1083/jcb.201301022

# Part IV

## Screening

# Chapter 12

# Genetic Engineering and Screening Using Base Editing and Inducible Gene Knockout

**Esther Feng Ying Ng and Franz Meitinger**

## Abstract

Genetic engineering and screening in human cells are powerful techniques for the precise and comprehensive identification and analysis of gene and protein domain functions. Genome-wide knockout screens have been extensively utilized to discover essential genes, tumor suppressors, and genes that regulate responses to various chemicals, including antimitotic and therapeutic drugs. The advent of base editors, which facilitate the targeted mutation of single amino acids, has advanced the identification of critical and functional domains or motifs. In this context, we outline methods for creating efficient base editor and inducible knockout cell lines for targeted gene manipulation and conducting genetic screens to elucidate the roles of genes and their domains within a specific cell biological context.

**Key words** Base editing, Gene editing, Gene knockout, Genetic screening, Gene engineering, CRISPR, Cas9

## 1 Introduction

The development of the CRISPR-Cas9 system has revolutionized gene engineering in human cells [1–3]. This technology enables the knockout, overexpression, and repression of genes and the manipulation of specific gene coding sequences to determine the consequences of a mutation in a cell type-specific manner. In this chapter, we describe methods for generating inducible gene knockout and base editing cell lines and conducting genetic screens (Fig. 1).

The CRISPR-Cas9 system consists of two components: a guide RNA (gRNA) and the endonuclease Cas9 [4]. The programmable gRNA directs Cas9 to a specific site in the genome, where the endonuclease introduces a double-strand break (DSB) (Fig. 1a). Following DNA cleavage, the error-prone DNA repair mechanism known as non-homologous end joining (NHEJ) repairs the DSB but often introduces *in*sertions or *del*etions (indels) at the cut site. When a DSB occurs in the protein-coding region of a gene, these

Tomomi Kiyomitsu (ed.), *The Mitotic Spindle: Methods and Protocols*, Methods in Molecular Biology, vol. 2872, https://doi.org/10.1007/978-1-0716-4224-5_12, 

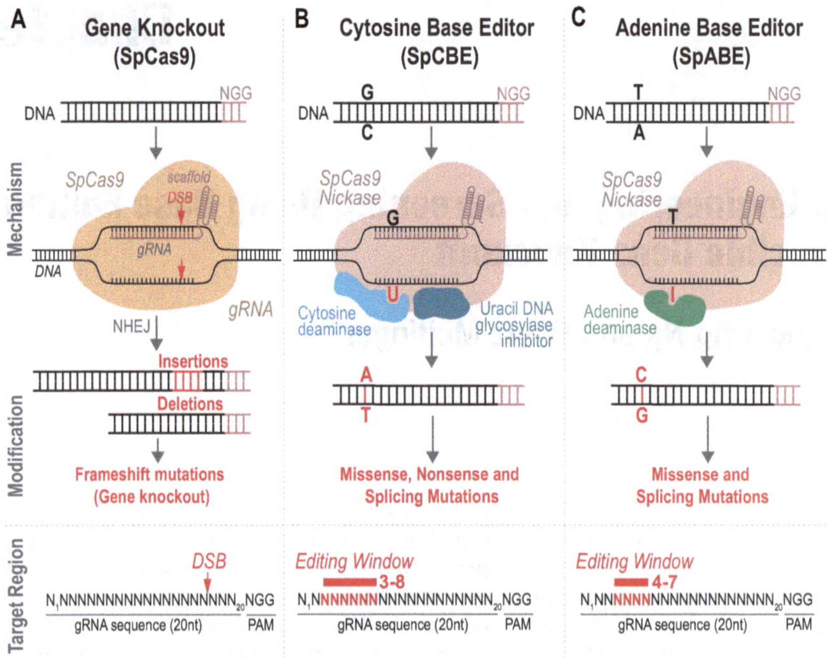


**Fig. 1** Mechanism of action of SpCas9 and base editors. (**a**) The endonuclease Streptococcus pyogenes Cas9 (SpCas9) is directed to the genomic DNA in a gRNA-dependent manner and introduces a double-strand break (DSB; red arrows). Non-homologous end joining (NHEJ) repairs the DSB and introduces base insertions or deletions that cause a frameshift mutation, which impairs protein expression and function. (**b**) Cytosine base editors mutate cytosine to uracil by deamination within the editing window. Uracil is then replaced by thymine. The Uracil DNA glycosylase inhibitor domain (UGI) prevents the correction of the base editing event. (**c**) Adenine base editors deaminate adenine to inosine, which is then replaced by guanine. The editing window for SpABE (N4-N7 within 20 nt upstream of PAM sequence) is narrower than for SpCBE (N3–N8)

indels can lead to deleterious frameshift mutations. This system is widely used to investigate gene function or to screen for genes with specific roles.

To investigate the function of genes in a timely manner, inducible systems are required. A frequently used strategy is the expression of Cas9 with a doxycycline-dependent promoter. This system is particularly useful for examining the function of essential genes [5, 6].

Recently, the development of base editors has enabled targeted mutations of cytosine to thymine (C-to-T) or adenine to guanine (A-to-G) without creating a DNA DSB [7, 8] (Fig. 1b, c). To avoid DSBs, a nickase mutation is introduced into Cas9 (nCas9 or Cas9 (D10A)). These single base modifications can alter coding sequences, resulting in missense or nonsense mutations. They can also modify splicing sites if the guide RNA (gRNA) targets the boundary between an exon and intron. Consequently, base editors are well-suited for investigating the functions of specific amino acids within protein domains. Cytosine base editors are fusion proteins composed of nCas9, a cytosine deaminase (e.g., APOBEC-1) and a uracil DNA glycosylase inhibitor (UGI) to prevent DNA repair and thus increase the mutation rate (Fig. 1b) [8]. Adenine base editors (ABE) consist of nCas9 and an adenine deaminase (Fig. 1c) [7]. In recent years, significant improvements have been made to base editors to optimize on-target activity and minimize off-target effects [9–17].

Genome-wide knockout screens have been extensively used to identify essential genes, tumor suppressors, and genes that regulate responses to various chemicals, including antimitotic and therapeutic drugs [18, 19]. In contrast, base editor screens are still relatively novel, with only a few reported to date [20, 21].

Given the potential of genetic screens in cell biology, this chapter outlines protocols for generating inducible knockout and base-editor cell lines (Figs. 2 and 3) as well as their application in genetic screens (Fig. 4). We focus on CRISPR-Cas9 systems based on the Cas9 variant derived from Streptococcus pyogenes (SpCas9, SpCBE, SpABE), which are highly efficient for studying cell

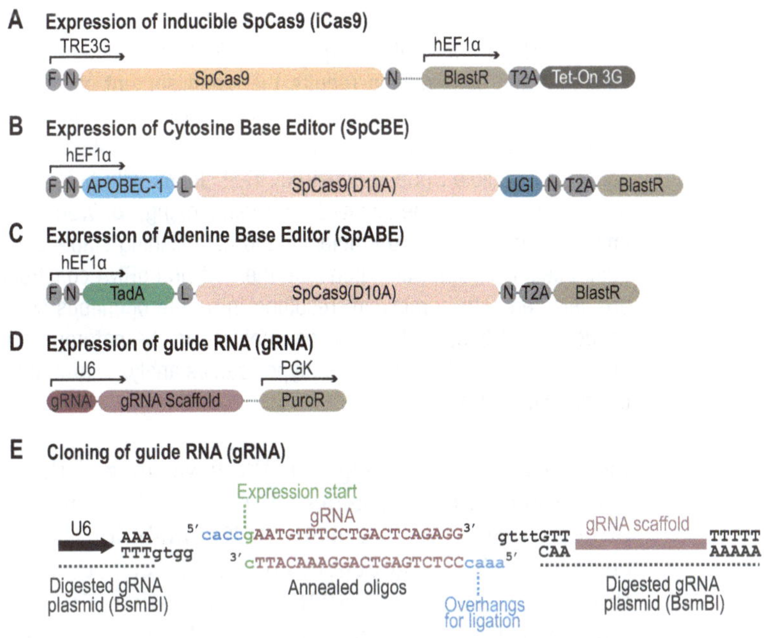


**Fig. 2** Lentiviral constructs used to generate gene knockout or base editing. (**a**) SpCas9 is expressed from a TRE3G promoter. Constitutively expressed Tet-On 3G activates SpCas9 expression upon binding to doxycycline. (**b**) The SpCas9(D10A) nickase mutant is fused to APOBEC-1 and UGI to enable efficient C-to-T editing. (**c**) The SpCas9(D10A) nickase mutant is fused to TadA to enable efficient A-to-G editing. (**d**) The gRNA is fused to the scaffold sequence and expressed from a U6 promoter. This construct is suitable for all gene knockout and base editing purposes using Cas9 derived from Streptococcus pyogenes. (**e**) Complementary oligos containing the gRNA sequence are annealed and ligated into the digested gRNA plasmid. The oligos also contain a 5′ guanosine, which is required for the U6 dependent expression and overhangs for the ligation. *F* 3xFLAG, *N* NLS, *L* linker

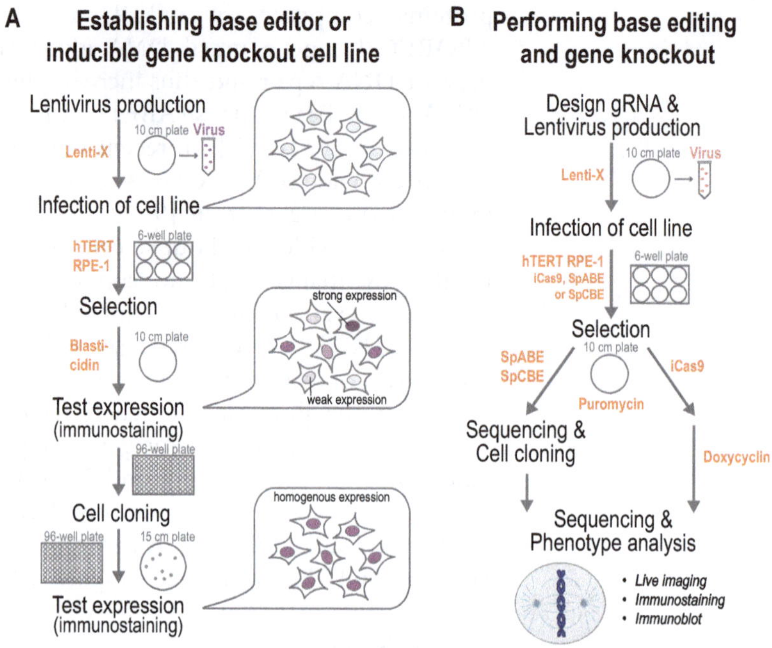


**Fig. 3** Workflow for establishing cell lines for base and gene editing. (**a**) Efficient knockout and base editing require the establishment of cell lines that homogenously express SpCas9 (upon induction with doxycycline) or base editors SpABE/SpCBE. Following infection, cells express heterogeneously the base editor or inducible SpCas9 construct. Cell clones with homogenous expression of the construct can be isolated after cell cloning. (**b**) Workflow for testing the knockout or base editing efficiency and establishing stable and homogenously edited cell lines. After infection with the lentiviral gRNA construct, positive cells are selected with puromycin, resulting in a heterogeneous cell pool. The consequences of base editing can be analyzed in the heterogeneous cell pool or individual clones. Knockout phenotypes can be analyzed after three or more days of doxycycline treatment

biological questions (Fig. 2). We have utilized these methods to study essential mitotic genes (PLK4, p31$^{Comet}$) and to characterize proteins that sensitize cells to PLK4 inhibition [5, 6, 19]. In this chapter, we show how genetic engineering can be used to manipulate three genes—53BP1, USP28 and p53—that play a crucial role in arresting cells after defective, prolonged mitosis (Fig. 5) [19, 22, 23].

## 2 Materials

We assume that basic laboratory equipment and consumables for molecular biology (plasmid cloning and purification, genomic DNA purification, DNA gel extraction) and cell culture are available.

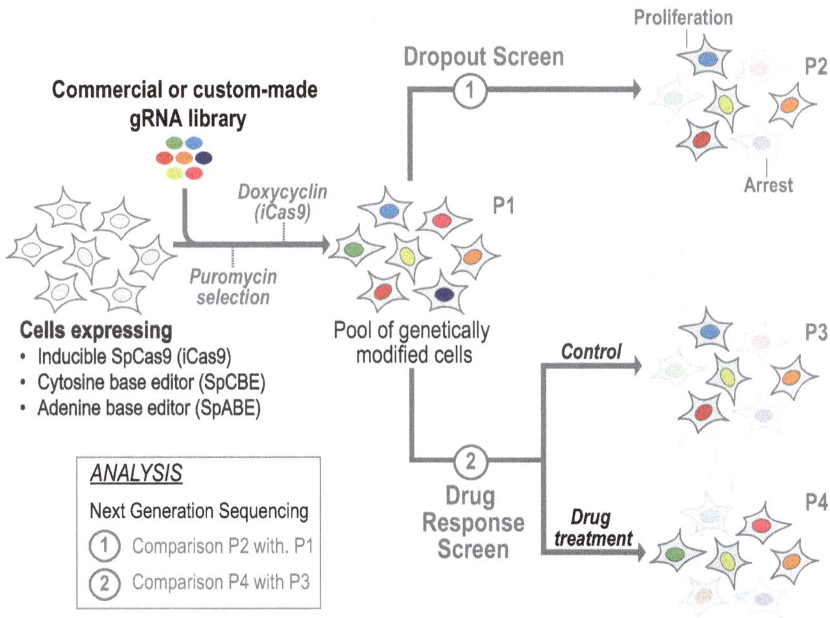


**Fig. 4** Workflow for performing a base editor or gene knockout screen. Inducible knockout and base editing cell lines can be used to perform dropout and drug response screens. The analysis of genetic screens is performed by next-generation sequencing to determine the gRNA pool in the surviving cell population. See text for description

### *2.1 Establishing Cell Line with Stable Expression of iCas9, SpABE, or SpCBE*

1. Plasmids: iCas9 (Edit-R Inducible Lentiviral Cas9; Horizon Discovery); SpABE (pMO1170: pLenti-EF1a-FLAG-ABE8e); SpCBE (pMO819: pLenti-FNLS-P2A-BlaR) (Fig. 2a–c)
2. Lenti-X 293-T (Takara; 632180) (*see* **Note 1**)
3. hTERT RPE-1 (ATCC; CRL-4000)
4. Dulbecco's Phosphate-Buffered Saline (PBS)
5. TrypLE Express Enzyme (Trypsin)
6. Lenti-X Packaging Single Shots (Takara)
7. Growth medium for Lenti-X 293-T: DMEM (1×) with 2 mM L-alanyl-L-glutamine containing 1× Pen/Strep and 10% Tet system approved FBS
8. Growth medium for hTERT RPE-1: DMEM/F12 containing 1× Pen/Strep and 10% Tet system approved FBS
9. 10 mg/mL Polybrene
10. Nuclease-free water
11. 10 mg/mL Blasticidin S HCl
12. Dimethyl Sulfoxide (DMSO)
13. Fixing reagent: 16% Paraformaldehyde aqueous solution. Prepare 4% solution in PBS fresh for each experiment.

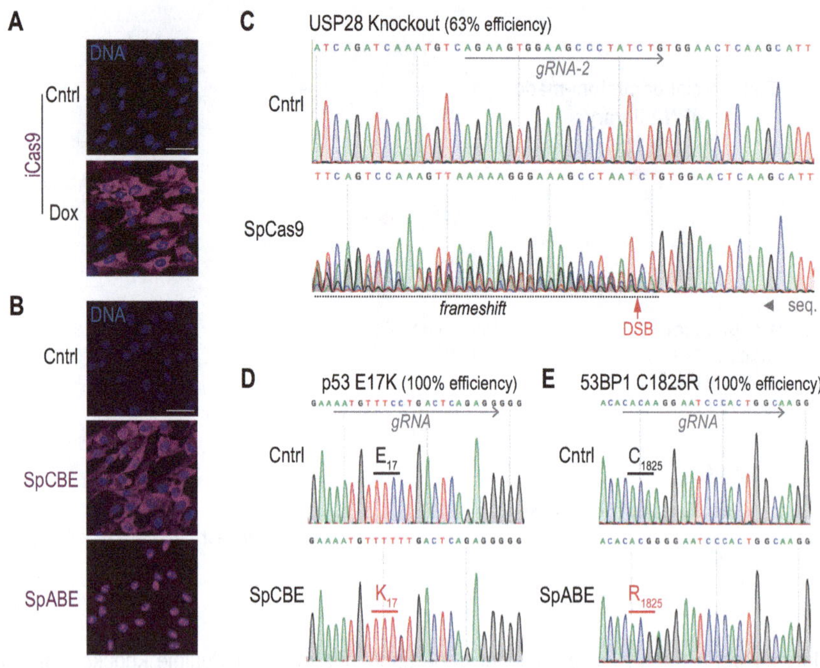

**Fig. 5** Gene expression and editing efficiency. (**a**, **b**) Expression of SpCas9 after induction (**a**), SpCBE and SpABE (**b**) after cell cloning are shown. Immunostaining was performed using antibodies for FLAG which is fused to the n-terminus of the respective genes. (**c**) Example of gene knockout efficiency using gRNA-2 from the Brunello library for the gene USP28. The efficiency was determined with the TIDE online tool. Note that the sequencing direction is from 3′ to 5′. (**d**, **e**) Base editing efficiency of cytosine and adenine base editors. The examples show guide RNAs targeting amino acid 17 of p53 (**d**) and 1825 of 53BP1 (**e**). For the example shown, efficient base editing was achieved without single-cell cloning. Scale bars, 50 μm; *seq.* sequencing direction, *Dox* doxycycline

14. Washing buffer: PBS with 0.1% Triton-X100
15. Blocking buffer: PBS with 2% Bovine Serum Albumin (BSA), 0.1% Triton-X100
16. 96-well imaging plate
17. Monoclonal ANTI-FLAG M2 antibody (Sigma-Aldrich; F1804-200UG; mouse; working concentration 1:1000 in blocking buffer)
18. 10 mg/mL Hoechst 33342 Trihydrochloride Trihydrate
19. Cy™5 AffiniPure Donkey Anti-Mouse IgG (Jackson Immunoresearch; 715-175-150)
20. Suitable microscope for multiwell plate imaging (e.g., Cell-Voyager CQ1 Benchtop High-Content Analysis System, Yokogawa)
21. Reservoir for multichannel pipettes

### 2.2 Base Editing and Gene Knockout

1. Plasmids for gRNA expression: LentiGuide-Puro (Addgene; 52963) (Fig. 2d)
2. Oligos for gRNA cloning:
   - p53-forward (5′ caccgAATGTTTCCTGACTCAGAGG 3′)
   - p53-reverse (5′ aaacCCTCTGAGTCAGGAAACATTc 3′)
   - 53BP1-forward (5′ caccgCACAAGGAATCCCACTGGCA 3′)
   - 53BP1-reverse (5′ aaacTGCCAGTGGGATTCCTTGTGc 3′)
   - USP28-forward (5′ caccgAGAAGTGGAAGCCCTATCTG 3′)
   - USP28-reverse (5′ aaacCAGATAGGGCTTCCACTTCTc 3′)
3. Oligo for sequencing gRNA plasmid: seq-gRNA (5′ GGCCT ATTTCCCATGATTCC 3′)
4. Oligos for sequencing gene and base editing efficiency:
   - p53-seq1 (5′ CCAACCCTTGTCCTTACCAGAACG 3′)
   - p53-seq2 (5′ GCTTGGGTTGTGGTGAAACATTGG 3′)
   - 53BP1-seq1 (5′ CTGCCTATACAATTCTGGATGGGCT TC 3′)
   - 53BP1-seq2 (5′ GGTGAGGCAGAGAGAACATCTATGG 3′)
   - USP28-seq1 (5′ CACAGTGAAGGTCACAGGTAGT TAGTC 3′)
   - USP28-seq2 (5′ TGGAAGGGAATGCTCATTGGAGC 3′)
5. BsmBI-v2 restriction enzyme
6. QIAquick Gel Extraction Kit
7. T4 Ligase and buffer
8. Stellar competent cells (Takara)
9. 10mg/ml Puromycin
10. Quick-DNA MiniPrep Kit
11. Plasmid Plus Midi Kit (Qiagen)
12. Plasmid Mini Kit

### 2.3 Base Editor or Gene Knockout Screen

Make sure to obtain sufficient cell culture reagents and culture plates before starting the screen.

1. Cell line (e.g., hTERT RPE-1) with stable expression of iCas9, SpABE, or SpCBE
2. gRNA library (commercial or custom-made)
3. 10 mg/mL Polybrene
4. 10 mg/mL Puromycin
5. QIAamp Blood Midi Kit (Qiagen)
6. Sterile flasks (500 or 1000 mL)

7. Growth medium for hTERT RPE-1: DMEM/F12 containing 1× Pen/Strep and 10% Tet system approved FBS
8. Dulbecco's Phosphate-Buffered Saline (PBS)
9. TrypLE Express Enzyme (Trypsin)

## 3 Methods

Here, we describe methods for generating stable cell lines for base and gene editing using lentiviral constructs to deliver and express iCas9, SpABE, SpCBE, and gRNAs (Fig. 2). While the protocols are optimized for the hTERT RPE-1 cell line, they can be easily adapted for use with other cell lines. Maintain cells in the recommended culture medium and ensure they are in the proliferation phase (below 80% confluency) at 37 °C with 5% $CO_2$. The number of cells to plate on a fresh plate depends on the replating efficiency, cell size, and doubling time of the respective cell line. Always follow your institute's safety guidelines when handling lentiviruses.

### 3.1 Establishing Cell Line with Stable Expression of iCas9, SpABE or SpCBE

The workflow for establishing a stable cell line for the expression of iCas9, SpABE and SpCBE is shown in Fig. 3a.

#### 3.1.1 Lentivirus Production

1. The first step is to package the transgenes into lentiviral particles. Plate $3 \times 10^6$ Lenti-X 293T cells in 10 cm cell culture plates. Incubate for 24 h at 37 °C, 5% $CO_2$. Prepare for each lentiviral construct one plate (iCas9, SpABE, SpCBE) (*see* **Note 2**).
2. Add 600 μL nuclease-free water and 7 μg each lentiviral construct in a 1.5 mL tube. Mix well by vortexing for 5 s, then spin down in a mini centrifuge (*see* **Note 3**).
3. Transfer the 600 μL DNA solution into a Lenti-X Single Shot tube containing transfection mix. Vortex for at least 20 s to completely dissolve the pellet and spin the liquid briefly down. Incubate the transfection mixture for 10 min at room temperature.
4. Transfer the 600 μL transfection mixture onto the Lenti-X 293T cells (*see* **Note 4**).
5. Incubate cells for 2–3 days at 37 °C, 5% $CO_2$. The lentivirus is produced in the Lenti-X 293T cells and can be collected from the medium.
6. Carefully transfer the lentivirus-containing medium into a 15 mL conical tube and centrifuge for 15 min at 500× *g* to separate cell debris from the supernatant (*see* **Note 5**).

7. After centrifugation, transfer the supernatant into a new 15 mL conical tube or screw-cap tubes for storage. The supernatant contains lentiviruses, which can be directly used to infect the target cell line (*see* **Notes 6** and 7).

#### 3.1.2 Infection of Cell Lines and Selection

1. Thaw and recover the cell line of interest (e.g., hTERT RPE-1) at least 5 days before the lentiviral infection, using the appropriate culture medium.
2. Plate $1.0 \times 10^5$ cells per well into 6-well plates (*see* **Note 8**). For each lentiviral construct, one well is required for the infection. Three wells are required for controls (Control 1: no treatment; Control 2: polybrene; Control 3: polybrene and later selection compound).
3. Add 8 μg/mL of the infection reagent polybrene to each well, except Control 1 (*see* **Note 9**).
4. Infect cells with 100–500 μL lentivirus-containing supernatant and incubate cells for 2 days at 37 °C, 5% $CO_2$ (*see* **Notes 10** and **11**).
5. After 2 days, remove the lentivirus-containing medium carefully, wash with 1 mL PBS and add 150 μL trypsin per well to detach cells. Incubate for 5 min at 37 °C, 5% $CO_2$ (*see* **Note 12**).
6. Resuspend cells with 1 mL fresh medium and transfer cells to a 10 cm cell culture plate containing 9 mL medium. Add 8 μg/mL blasticidin S to the cells infected with lentivirus and to Control 3 for selection.
7. Incubate cells with blasticidin S until all cells in the Control 3 plate are dead and detached from the plate (*see* **Note 13**). Control 1 shows unperturbed cell proliferation and Control 2 indicates the sensitivity to polybrene treatment. After blasticidin S selection, grow the surviving cells in a regular cell culture medium. Freeze cells in a growth medium with 7.5% DMSO or proceed to the next step.

#### 3.1.3 Test Expression by Immunostaining

1. To test the expression of the transgenes, plate 8000 cells per well in 100 μL growth medium in a 96-well imaging plate for immunostaining. Include a negative control, which does not express a FLAG tag. Avoid shaking the plate to ensure uniform distribution of cells across the well surface. Add 1 μg/mL doxycycline to iCas9 expressing cells (not required for SpABE and SpCBE). Incubate plate for 24 h, 5% $CO_2$ (*see* **Notes 14** and **15**).
2. After 24 h briefly wash each well with 100 μL PBS and fix cells with 100 μL 4% paraformaldehyde for 10 min at room temperature.

3. Wash three times for 5 min each with 100 μL washing buffer.
4. Remove the washing buffer and add 100 μL blocking buffer to each well. Incubate the plate overnight at 4 °C or for 2 h at room temperature.
5. Remove the blocking buffer and add 100 μL of primary antibody solution (ANTI-FLAG M2; 1:1000 in blocking buffer) to each well. Incubate for 2 h at room temperature.
6. Wash wells three times for 5 min each with 100 μL washing buffer and add 100 μL blocking buffer solution containing secondary antibody (Cy™5 AffiniPure Donkey Anti-Mouse IgG; 1:1000) and DNA dye Hoechst (1:2000). Incubate for 1 h at room temperature in the dark (*see* **Note 16**).
7. Cover the plate with aluminum foil to protect it from light and detect FLAG and DNA staining using a microscope with multiwell plate imaging capability. The immunostaining will show the heterogeneous expression of the transgene (iCas9, SpABE or SpCBE), which cannot be seen in the control cells (*see* **Notes 17** and **18**).

#### 3.1.4 Cell Cloning

1. To obtain cells with homogenous expression of the transgenes, single-cell cloning is recommended. Two different methods can be used: 96-well plate (**steps 2–4**) or 15 cm plate (**steps 5–7**) (*see* **Notes 19** and **20**).
2. For the 96-well plate method, prepare a cell suspension containing 70 polyclonal transgene-expressing cells in 40 mL medium in a 50 mL conical tube (*see* **Note 21**). Mix well and transfer the cell suspension into a pipette reservoir suitable for multi-channel pipettes. Use multi-channel pipettes and aspirators when working with 96-well plates. Add 200 μL of the cell suspension into each well of two 96-well plates so that each well contains a maximum of one cell. Incubate 96-well plates for 10–14 days at 37 °C, 5% $CO_2$. Inspect colony growth using an inverted microscope. Mark wells that contain a single colony (*see* **Note 22**).
3. In the next step, the colonies will be transferred into 6-well plates. Prepare enough 6-well plates to isolate all colonies. Add 2 mL of fresh medium into each well and prewarm the plates at 37 °C, 5% $CO_2$.
4. Remove medium from 96-well plates. Wash wells with 100 μL PBS. Add 20 μL trypsin to every well and incubate for 5 min at 37 °C, 5% $CO_2$. Resuspend cells in fresh medium and transfer the single clones into the prewarmed 6-well plates. Incubate plates at 37 °C, 5% $CO_2$ until cells reach 60–80% confluency, after which they are plated to be tested for transgene expression while the remaining are frozen for storage.

5. For the 15 cm plate method, plate 1000 transgene-expressing cells in a 15 cm plate. Gently shake the plate to distribute the cells evenly across the plate. Incubate the plate for 10–14 days at 37 °C, 5% $CO_2$.
6. Start isolating colonies when they contain around 100–300 cells. Place an inverted microscope inside the cell culture hood to pick single clones in a sterile environment. With the help of the microscope, pick single clones using a 200 μL pipette. To do so, carefully push the pipette tip underneath the cell colony to detach it from the plate surface, while at the same time sucking the detached cells into the pipette tip.
7. Transfer each colony to a separate well of a 12-well plate containing 1 mL prewarmed medium. Expand cells until they cover around 60–80% of a 6-well plate, after which they are plated to be tested for transgene expression while the remaining are frozen for storage (*see* **Note 23**).

#### 3.1.5 Test Expression by Immunostaining

Test expression of the transgenes in the single clones as previously described (*see* Subheading 3.1.3). Examples of homogenous expression of iCas9, SpABE and SpCBE are shown in Fig. 5a, b. Make sure to freeze aliquots of each clone (*see* **Note 24**).

### 3.2 Base Editing and Gene Knockout

The workflow for gene knockout or base editing is depicted in Fig. 3b.

#### 3.2.1 Design gRNA Plasmid and Lentivirus Production

1. Design or retrieve gRNA sequences for efficient base and gene editing. For gene knockout we recommend testing gRNAs from the Brunello library [24]. For base editing, identify sites in your gene of interest where the PAM sequence (NGG) is located using software such as Snapgene or Benchling (www.benchling.com). Ensure the correct bases (Adenine for ABE, Cytosine for CBE) are located within the editing window (Fig. 1b, c). The 20 bp sequence immediately prior to the PAM sequence is the gRNA sequence (e.g., Target sequence for p53 (5′ AATGTTTCCTGACTCAGAGG 3′), 53BP1 (5′ CACAAGGAATCCCACTGGCA 3′) and USP28 (5′ AGAAGTGGAAGCCCTATCTG 3′). See illustration of the gRNA cloning process in Fig. 2e.
2. The gRNA promoter U6 requires a Guanosine nucleotide at the 5′ site of the sequence to initiate gRNA expression. Add "g" to the 5′ end of the gRNA sequence and create a reverse complement sequence (*see* gRNA sequences in Subheading 2.2).
3. Add overhangs "*CACC*" to the 5′ end of the forward oligo and "*AAAC*" to the 5′ end of the reverse oligo to facilitate the cloning into the BsmBI digested plasmid (*see* gRNA sequences in Subheading 2.2). Obtain oligos from vendors.

4. Clone gRNAs into the lentiGuide-Puro backbone. To do so, digest the plasmid with BsmBI-v2 for 2 h at 55 °C and isolate digested construct by gel purification.
5. Anneal the gRNA oligos to generate a gRNA duplex. To do so, mix 1 μL of forward oligo (100 μM) and 1 μL of reverse oligo (100 μM) with 8 μL sterile water. Incubate solution at 95 °C for 5 min. Let samples slowly cool down at room temperature for 30 min. Dilute the annealed oligo duplex with sterile water (1:100; final concentration 0.1 μM) (*see* **Note 25**).
6. Ligate the oligo duplex into the digested lentiGuide-Puro construct. To do so, mix the digested lentiGuide-Puro (50 ng), gRNA duplex (1 μL; 0.1 μM), T4 ligase buffer (1 μL of 10× buffer) and fill up the mixture with sterile water to 9.5 μL. Mix well and add T4 ligase (0.5 μL). Mix well and incubate the ligase mix for 2 h or overnight at room temperature (*see* **Note 26**).
7. Transform 1 μL of the ligation mix into 10 μL competent stellar cells, plate transformed cells on LB-agar plates containing 100 μg/mL ampicillin and incubate overnight at 37 °C. Isolate plasmid from 2 to 8 single clones and confirm successful ligation by sequencing using the oligo for gRNA plasmid sequencing (seq-gRNA; *see* Subheading 2.2). Perform a plasmid midi prep to obtain a pure toxin-free plasmid solution.
8. Produce lentiviral particles with the cloned lentiGuide-Puro constructs that express the selected gRNAs as previously described (*see* Subheading 3.1.1).

#### 3.2.2 Infection of Cell Line and Selection

1. Infect hTERT RPE-1 cell line expressing iCas9, SpABE or SpCBE (*see* Subheading 3.1) with the corresponding lentivirus for expression of the selected gRNAs (*see* Subheading 3.1.2).
2. Select infected cells with 8 μg/mL puromycin (*see* **Note 27**).
3. Test the infected cell lines for the efficiency of gene knockout or base editing around 8–10 days after the infection. For cell lines carrying iCas9, induce the expression of Cas9 with 1 μg/mL doxycycline for 3–4 days before harvesting the cells.

#### 3.2.3 Sequencing and Cell Cloning

1. To test the gene knockout and base editing efficiency, harvest 50,000–1,000,000 cells and extract the genomic DNA using Quick-DNA MiniPrep Kit. Extract the genomic DNA in 15 μL sterile water. Also extract genomic DNA from unmodified cells for comparison.
2. Design oligos to amplify the region around the edited DNA by PCR. The oligos should bind around 300–500 bp up- and downstream of the edited site (e.g., p53-seq1 and p53-seq2; 53BP1-seq1 and 53BP1-seq2; USP28-seq1 and USP28-seq2).

Use 0.5 μL (5–20 ng) of the purified genomic DNA for the PCR reaction. Purify PCR product by gel extraction. Sequence PCR product using one of the oligos that was used for the PCR amplification (seq1 or seq2).

3. The gene knockout efficiency can be determined by the online tool TIDE (http://shinyapps.datacurators.nl/tide/). Base editing efficiency can be determined with Snapgene or similar programs. See example for gene knockout efficiency in Fig. 5c and for base editing efficiency Fig. 5e, d.
4. To study the consequences of gene knockout of non-essential genes or base editing, we recommend cloning single cells, which carry the modified region (*see* Subheading 3.1.4).

#### 3.2.4 Phenotype Analysis

To study the phenotype of cells with essential genes knocked out, we recommend analyzing cells 3–4 days after starting doxycycline treatment [5, 6]. Mitotic phenotypes in living cells can be analyzed with specific dyes that label DNA, tubulin, or actin (SiR-DNA, SiR-Tubulin, SiR-Actin). For long-term experiments, we recommend the usage of endogenously expressed markers such as H2B-RFP (DNA) or Ruby-MAP4-MBD (Tubulin) [5]. Alternatively, cells can be subjected to immunoblotting and immunostaining to confirm gene knockout and visualize the phenotype.

### 3.3 Base Editor or Gene Knockout Screen

The workflow for a base editor or gene knockout screen is depicted in Fig. 4. Two types of screens can be performed. The dropout screen detects mutations that alter the cell proliferation rate. Knockout of essential genes or mutations of essential amino acids leads to the dropout of the corresponding gRNAs. A drug response screen identifies mutations that alter the response to chemical inhibitors such as antimitotic drugs.

1. Commercial gRNA libraries are available for whole genome knockout screens in humans and mice (e.g., Brunello library) [21]. For base editor screens, a gene-specific gRNA library must be designed and established. The library can be synthesized and packed into lentiviral particles by external services (e.g., vectorbuilder). Protocols to produce lentiviral libraries have been described [20].
2. We recommend the free online software Benchling to create custom-made gRNA libraries. For base editor screens, include all possible gRNAs for each gene of interest. In addition, include non-targeting gRNAs as controls (5–10% of the total number of gRNAs). The gRNAs must be designed according to the PAM sequence specific to the Cas9 used. In our example, we used SpCas9, which has the PAM sequence NGG.
3. The calculations in this example are made for a library with 10,000 gRNAs. Adjust the numbers according to the size of the gRNA library used.

4. The genetic screen should be performed with a 300–1000× coverage and a multiplicity of infection (MOI) of 0.4–0.8 (*see* **Note 28**). The coverage means that each gRNA should be 300–1000× present throughout the screen. For smaller libraries, we recommend using 1000× coverage. For large libraries 300× is sufficient (*see* **Note 29**).
5. In our example we use a coverage of 1000× and a MOI of 0.6. The screen-specific cell number can be calculated with the following equation: gRNAs × Coverage/MOI = 10,000 × 1000/0.6 = 16,666,667 cells (17 × $10^6$).
6. Double the number of cells to account for a control sample. Thus, the required number of cells for the screen is 34 × $10^6$ (*see* **Note 30**).
7. Thaw and recover frozen hTERT RPE-1 cells expressing iCas9, SpABE or SpCBE for at least 5 days before infection. The required cell number for the infection is 34 × $10^6$.
8. On Day 1 of the screen, prepare the cell culture hood and wear appropriate PPE (personal protective equipment) for working with lentiviruses. Treat all consumables and liquids that were exposed to lentiviruses according to safety regulations.
9. Prepare the cell suspension for lentiviral infection in a sterile 500 mL flask.
   - 14 × 20 mL = 280 mL growth medium pre-warmed to 37 °C
   - 34 × $10^6$ hTERT RPE-1 (iCas9, SpCBE or SpABE) cells
   - 224 μL Polybrene (10 mg/mL) to make up a final concentration of 8 μg/mL
   - Lentiviral gRNA library
10. Calculation of the required amount of lentivirus.
   - Calculation for lentiviral gRNA library with a titer of 5.0 × $10^6$ TU/mL (transduction units/mL) (*see* **Note 31**).
   - For a 1000× coverage, 1000 × 10,000 transduction units are required (10,000,000 TU).
   - This number is further doubled since we double the number of cells for infection to include a control sample (20,000,000 TU).
   - Required volume: 20 × $10^6$/5.0 × $10^6$ = 4 mL of the lentiviral gRNA library.
11. Gently mix the suspension to create a homogenous cell/lentivirus mixture.
12. Pipette 20 mL of cell/lentivirus mix into each 15 cm plate, a total of 14 plates.

13. Incubate cells for 24 h at 37 °C, 5% $CO_2$. During the 24-h incubation, cells attach to the plate surface and are infected by the lentivirus.
14. On Day 2, harvest and count cells after 24 h of incubation. To do so, briefly transfer the medium from the cell culture plates using a 25 ml pipette into a waste container containing bleach and wash each plate with 10 ml PBS. Discard the PBS into the waste container.
15. Incubate each plate with 2 mL trypsin for 5 min at 37 °C, 5% $CO_2$ and resuspend detached cells in 8 mL fresh medium per plate. Collect cell suspension in 50 mL conical centrifuge tubes and centrifuge at 250× *g* for 5 min at room temperature to pellet cells. Remove the supernatant carefully without disturbing the cell pellet at the bottom of the tube. Resuspend all cells in 20 mL fresh medium and count the cell number/ml of the suspension.
16. Replate a total of $17 \times 10^6$ cells on 34 × 15 cm plates ($0.50 \times 10^6$ cells per plate). To do so, prepare a sterile 1000 mL flask and mix cells with 680 mL medium. Distribute 20 mL of cell suspension to each plate. Incubate for 24 h.
17. Freeze remaining cells (infection control; $>17 \times 10^6$ cells). To do so, centrifuge at 250× *g* for 5 min at room temperature to pellet cells. Remove the supernatant carefully without disturbing the cell pellet at the bottom of the tube. Resuspend the cells in 1 mL PBS and transfer the cell suspension into a 1.5 mL microcentrifuge tube. Pellet the cells in a microcentrifuge at 500× *g* for 5 min at room temperature. Remove the supernatant carefully without disturbing the cell pellet at the bottom of the tube. Freeze cells immediately in liquid nitrogen or dry ice. Store cell pellet at −80 °C.
18. Base editor screen (**steps 18–21**): 24 h after replating cells, add puromycin (8 μg/mL for hTERT RPE-1 cells) to select for infected cells (*see* **Note 27**).
19. Continue incubation of cells at 37 °C, 5% $CO_2$ for another 2 days.
20. Replate $17 \times 10^6$ cells in 34 × 15 cm plates in fresh medium with puromycin as previously described. Incubate cells for another 3 days (*see* **Note 32**).
21. Collect the remaining cells for DNA extraction ($>17 \times 10^6$ cells; Sample 1).
22. Knockout screen (**steps 22–25**): 24 h after replating cells, add puromycin (8 μg/mL for hTERT RPE-1 cells) to select for infected cells (*see* **Note 27**).
23. Continue incubation of cells at 37 °C with 5% $CO_2$ for another 2 days.

24. Replate $17 \times 10^6$ cells in $34 \times 15$ cm plates in fresh medium. Induce Cas9 expression with 1 μg/mL Doxycycline. Incubate cells for 3 days (*see* **Note 33**).
25. Collect the remaining cells for DNA extraction (>$17 \times 10^6$ cells; Sample 1).
26. If performing a drug response screen, divide the infected and selected cells into two pools. One part will be treated with the drug of choice and the other pool will serve as control. If performing a dropout screen, continue with one sample (Fig. 4).
27. For a dropout screen, prepare one 1000 mL sterile flask with $17 \times 10^6$ cells in 680 mL growth medium. For a drug response screen, prepare 2 flasks. Add the drug of choice to one flask and the same amount of drug vehicle (e.g., DMSO) to the other flask.
28. Plate 20 mL cell suspension in 15 cm plates. Use 34 plates for each condition.
29. Maintain cells and replate cells every three days. Collect at each step cell samples for DNA extraction.
30. Continue the procedure for up to 24 days after infection.
31. At the end of the screen, you will have one infection control sample and several samples for analysis.
    - Infection control: Reference for infection efficiency and gRNA representation
    - Sample 1: 3 days after infection
    - Sample 2 (+/− drug): 6 days after infection (3 days after Cas9 induction)
    - Sample 3 (+/− drug): 9 days after infection (3 days of drug treatment)
    - Sample 4 (+/− drug): 12 days after infection (6 days of drug treatment)
    - Sample 5 (+/− drug): 15 days after infection (9 days of drug treatment)
    - Sample 6 (+/− drug): 18 days after infection (12 days of drug treatment)
    - Sample 7 (+/− drug): 21 days after infection (15 days of drug treatment)
    - Sample 10 (+/− drug): 24 days after infection (18 days of drug treatment)
32. The duration of the screen and the frequency of sample collection can be modified according to the needs.

33. Use a DNA extraction kit, which is appropriate for the cell number and suitable for next generation sequencing (NGS; QIAamp Blood Midi Kit). Refer to the manufacturer's handbook and protocol for detailed instructions on the DNA purification procedure.
34. Measure the concentration of sample DNA (*see* **Note 34**).
35. Analyze gRNA representation of each sample by NGS.
36. NGS will reveal which gRNAs are depleted or enriched in the specific samples.

## 4 Notes

1. Alternatively, HEK-293T cells are suitable for virus production.
2. Cells should be around 80% confluent on the day of transfection.
3. Plasmids for lentivirus production should be free of endotoxins and have a concentration of higher than 0.5 μg/μL. We recommend purifying plasmids with the Plasmid Plus Midi Kit.
4. Disperse the transfection mixture equally across the 10 cm plate using a 1000 μL pipette. Carefully shake the plate to distribute the transfection mix without spilling the liquid.
5. Use seals for the centrifuge to prevent contamination in case a tube breaks.
6. We recommend aliquoting the lentivirus-containing medium into freezer-compatible screw-cap tubes for storage. The lentivirus can be stored at -80 °C and reused for at least 5 years.
7. If required, the lentivirus can be concentrated with a Lenti-X$^{TM}$ concentrator (Takara).
8. The cell number depends on the cell line. Differences in cell size and cell proliferation rate should be considered.
9. hTERT RPE-1 cells require 8 μg/mL Polybrene for efficient infection. A similar concentration is toxic for many other cell lines. Test sensitivity to Polybrene of your cell line of interest in a separate experiment. Most cell lines require 3 μg/mL Polybrene for an optimal outcome.
10. In most cases, 500 μL of lentiviral supernatant works well. However, the amount of supernatant to be used depends on the titer of the lentivirus. Higher concentrations of lentiviruses may be toxic. We recommend initially testing different amounts of lentivirus-containing supernatant (e.g., 100 μL; 250 μL; 500 μL). The concentration of the lentivirus depends

on the size of the plasmid packed into the lentivirus. The upper limit of DNA that can be efficiently packed into lentiviral particles is around 10,000 bp.

11. Some cell lines are sensitive to 48 h of exposure to polybrene. For these cell lines we recommend starting the infection with a higher cell number and only incubating the cells for 24 h with the lentiviral particle.
12. The duration of incubation with trypsin varies between cell lines. Some cell lines may require longer to efficiently detach.
13. The amount of blasticidin S might vary between cell lines. If all cells die during selection, we recommend optimizing blastidicin S concentration and selection duration. Allow the cells to recover for two days after infection before adding the selection compound to improve the resistance to blastidicin S. The selection takes between 5 and 7 days for hTERT RPE-1 cells.
14. To observe Cas9 expression, 24 h induction is sufficient. To observe gene editing, we recommend inducing Cas9 for 3–4 days.
15. The cell number to be plated depends on the cell line. Cells should cover 50–80% of the surface 24 h after plating. Alternatively, cells can be plated and investigated by immunostaining on glass coverslips.
16. The 96-well plate can be placed in a closed drawer or alternatively wrapped with aluminum foil during incubation to prevent bleaching of the fluorophore and DNA dye.
17. Plates can be stored at 4 °C for a few days. However, since the fluorescence signals will fade over time, it is best to image the cells as soon as possible.
18. Ideally, at least 10% of the cell population expresses the transgene.
19. The 15 cm plate method works better for cell lines that have slower proliferation rate or do not grow well when isolated to a single cell.
20. The number of plates required depends on the percentage of positive cells in the cell population. Around 10–20 clones can be isolated from a 96-well plate and 20–30 clones from a 15 cm plate. If 10% of the population expresses the transgene, at least 50 clones should be screened.
21. To obtain 70 cells per 40 mL medium, prepare serial dilutions of the cell suspension. For example, if the cell suspension has 1,000,000 cells/mL, transfer 20 μL into a 50 mL conical tube with 20 mL growth medium to obtain a concentration of one cell per μL. Mix suspension well and transfer 70 μL into a 50 mL conical tube with 40 mL growth medium.

22. Avoid wells that have more than one colony as they might originate from different clones. Colonies should cover at least 15% of the well before continuing with the next step.
23. Refresh the growth medium during colony growth if necessary.
24. At this point, any frozen cell colonies that did not express the transgene can be discarded. Freeze a larger stock of each clone, which shows a uniform and high expression of the transgene.
25. The oligo duplex is stable for several days at 4 °C and can be stored for longer at −20 °C.
26. Prepare a control ligation of the digested plasmid without gRNA. The differences in colony number growing on the LB-AMP plates indicate the efficiency of the cloning process.
27. The required concentration of puromycin depends on the cell line. For most cell lines, the optimal concentration is 0.5–1.0 μg/mL. hTERT RPE-1 cells require a higher concentration because they already carry one copy of a puromycin-resistant gene. The selection in hTERT-RPE-1 cells can take up to one week, while in other cell lines it takes only 1–2 days. Passage cells during the selection process if necessary.
28. An MOI of 0.6 means that theoretically out of 10 cells only 6 cells are infected with a gRNA-containing virus. With an MOI of 1.0 each cell receives, on average, one gRNA. However, the infection of cells follows a Gaussian distribution, which means that at an MOI of 1.0 some cells receive 2 or more gRNAs, and some cells don't receive any gRNA. We recommend an MOI smaller than one to limit the percentage of cells that are infected with two lentiviral particles and thus express two independent gRNAs. Double infections can increase the noise and limit the sensitivity of the screen.
29. When using a large library, a high coverage can easily lead to a large number of plates, which cannot be handled. Choose sufficient coverage that keeps the scope of the task manageable.
30. Throughout the screen it is important to maintain cells in the proliferating phase. Plates should never reach 100% confluency. However, it is also important to maintain a high density of cells for optimal infection efficiency and to minimize the number of plates that must be maintained throughout the screen. For hTERT RPE-1 cells we use $2.5 \times 10^6$ cells per 15 cm plate for the lentiviral infection. During a 24-h infection the cell number doubles once. The maximum capacity of a 15 cm plate is around $5 \times 10^6$ hTERT RPE-1 cells. Thus, the infection of $34 \times 10^6$ cells requires $14 \times 15$ cm plates.
31. The titer of lentiviral gRNA library is provided by the company from which the library was obtained. The titer of infectious particles can also be obtained by infecting cells with different

volumes of the library followed by the selection with puromycin. Only the infected cells can form colonies. The titer can be calculated by considering the number of cells that were infected and the volume of the library that was used to infect the cells. For example, if 1 μL of library leads to the formation of 5000 colonies then the titer of library is 5,000,000 or $5 \times 10^6$ TU/mL

32. Efficient base editing may take up to eight days.
33. Avoid unwanted toxic effects as a consequence of simultaneous Cas9 induction and puromycin selection.
34. Make sure that the purity, amount, and concentration of the purified DNA solution meets the requirements for Next Generation Sequencing.

## Acknowledgments

We thank Midori Ohta for feedback on the manuscript and members of the Cell Proliferation and Gene Editing Unit for discussion. This work was supported by Okinawa Institute of Science and Technology and the JSPS KAKENHI grant to F.M. (23K05773).

### References

1. Jinek M, Chylinski K, Fonfara I et al (2012) A programmable dual-RNA-guided DNA endonuclease in adaptive bacterial immunity. Science 337(6096):816–821. https://doi.org/10.1126/science.1225829
2. Cong L, Ran FA, Cox D et al (2013) Multiplex genome engineering using CRISPR/Cas systems. Science 339(6121):819–823. https://doi.org/10.1126/science.1231143
3. Mali P, Yang L, Esvelt KM et al (2013) RNA-guided human genome engineering via Cas9. Science 339(6121):823–826. https://doi.org/10.1126/science.1232033
4. Ran FA, Hsu PD, Wright J et al (2013) Genome engineering using the CRISPR-Cas9 system. Nat Protoc 8(11):2281–2308. https://doi.org/10.1038/nprot.2013.143
5. Meitinger F, Ohta M, Lee KY et al (2020) TRIM37 controls cancer-specific vulnerability to PLK4 inhibition. Nature 585(7825):440–446. https://doi.org/10.1038/s41586-020-2710-1
6. Meitinger F, Belal H, Davis RL et al (2024) Control of cell proliferation by memories of mitosis. Science 383(6690):1441–1448. https://doi.org/10.1126/science.add9528
7. Gaudelli NM, Komor AC, Rees HA et al (2017) Programmable base editing of A*T to G*C in genomic DNA without DNA cleavage. Nature 551(7681):464–471. https://doi.org/10.1038/nature24644
8. Komor AC, Kim YB, Packer MS et al (2016) Programmable editing of a target base in genomic DNA without double-stranded DNA cleavage. Nature 533(7603):420–424. https://doi.org/10.1038/nature17946
9. Zafra MP, Schatoff EM, Katti A et al (2018) Optimized base editors enable efficient editing in cells, organoids and mice. Nat Biotechnol 36(9):888–893. https://doi.org/10.1038/nbt.4194
10. Koblan LW, Doman JL, Wilson C et al (2018) Improving cytidine and adenine base editors by expression optimization and ancestral reconstruction. Nat Biotechnol 36(9):843–846. https://doi.org/10.1038/nbt.4172
11. Hu JH, Miller SM, Geurts MH et al (2018) Evolved Cas9 variants with broad PAM compatibility and high DNA specificity. Nature 556(7699):57–63. https://doi.org/10.1038/nature26155

12. Richter MF, Zhao KT, Eton E et al (2020) Phage-assisted evolution of an adenine base editor with improved Cas domain compatibility and activity. Nat Biotechnol 38(7): 883–891. https://doi.org/10.1038/s41587-020-0453-z
13. Sakata RC, Ishiguro S, Mori H et al (2020) Base editors for simultaneous introduction of C-to-T and A-to-G mutations. Nat Biotechnol 38(7):865–869. https://doi.org/10.1038/s41587-020-0509-0
14. Zhang X, Zhu B, Chen L et al (2020) Dual base editor catalyzes both cytosine and adenine base conversions in human cells. Nat Biotechnol 38(7):856–860. https://doi.org/10.1038/s41587-020-0527-y
15. Grunewald J, Zhou R, Lareau CA et al (2020) A dual-deaminase CRISPR base editor enables concurrent adenine and cytosine editing. Nat Biotechnol 38(7):861–864. https://doi.org/10.1038/s41587-020-0535-y
16. Neugebauer ME, Hsu A, Arbab M et al (2023) Evolution of an adenine base editor into a small, efficient cytosine base editor with low off-target activity. Nat Biotechnol 41(5): 673–685. https://doi.org/10.1038/s41587-022-01533-6
17. Chen L, Zhu B, Ru G et al (2023) Re-engineering the adenine deaminase TadA-8e for efficient and specific CRISPR-based cytosine base editing. Nat Biotechnol 41(5): 663–672. https://doi.org/10.1038/s41587-022-01532-7
18. Hart T, Chandrashekhar M, Aregger M et al (2015) High-resolution CRISPR screens reveal fitness genes and genotype-specific cancer liabilities. Cell 163(6):1515–1526. https://doi.org/10.1016/j.cell.2015.11.015
19. Meitinger F, Anzola JV, Kaulich M et al (2016) 53BP1 and USP28 mediate p53 activation and G1 arrest after centrosome loss or extended mitotic duration. J Cell Biol 214(2):155–166. https://doi.org/10.1083/jcb.201604081
20. Cuella-Martin R, Hayward SB, Fan X et al (2021) Functional interrogation of DNA damage response variants with base editing screens. Cell 184(4):1081–1097. e1019. https://doi.org/10.1016/j.cell.2021.01.041
21. Hanna RE, Hegde M, Fagre CR et al (2021) Massively parallel assessment of human variants with base editor screens. Cell 184(4): 1064–1080 e1020. https://doi.org/10.1016/j.cell.2021.01.012
22. Fong CS, Mazo G, Das T et al (2016) 53BP1 and USP28 mediate p53-dependent cell cycle arrest in response to centrosome loss and prolonged mitosis. elife 5. https://doi.org/10.7554/eLife.16270
23. Lambrus BG, Daggubati V, Uetake Y et al (2016) A USP28-53BP1-p53-p21 signaling axis arrests growth after centrosome loss or prolonged mitosis. J Cell Biol 214(2): 143–153. https://doi.org/10.1083/jcb.201604054
24. Sanson KR, Hanna RE, Hegde M et al (2018) Optimized libraries for CRISPR-Cas9 genetic screens with multiple modalities. Nat Commun 9(1):5416. https://doi.org/10.1038/s41467-018-07901-8

# Chapter 13

# Systematic Identification of Microtubule Posttranslational Modification "Readers" by Quantitative Proteomics

Takashi Hotta and Ryoma Ohi

## Abstract

Microtubules, dynamic polymers assembled from α, β-tubulin dimers, contribute to myriad cellular processes. This is largely attributed to microtubule-associated proteins (MAPs). How MAPs selectively bind microtubules to carry out various functions is not known. The "*Tubulin Code*" theory proposes that posttranslational modifications (PTMs) of microtubules serve as signs that can be read by specific MAPs, thereby conferring specific functional properties to the microtubules. In support of this hypothesis, "reader" MAPs have been identified for various tubulin PTMs, but, until recently, no systematic screening had been performed to identify readers in an unbiased manner. We addressed this by developing a reader identification pipeline that uses quantitative mass spectrometry to interrogate the microtubule proteome of cells programmed to express specific PTMs. This pipeline can be used to identify readers for any tubulin PTM from various cell types as long as the writer enzymes are known. We also provide an alternative, complementary approach to obtain modified microtubules using a generic writer enzyme in vitro.

**Key words** Microtubule, Posttranslational modifications, Microtubule-associated proteins, Quantitative proteomics

## 1 Introduction

Microtubules (MTs) play an essential role in a variety of cellular processes, including intracellular trafficking, cell motility, and cell division. To accomplish these tasks, MTs bind to various MT-associated proteins (MAPs), including stabilizers, crosslinkers, nucleators, severing factors, and motor proteins. Cellular MTs are not thought to be equal, and mechanisms that allow cells to differentiate between MT sub-populations within the cell have been sought. One idea, referred to as the "*Tubulin Code*," proposes that specific posttranslational modifications (PTMs) on MTs cause them to recruit specific MAPs. Therefore, cataloging MAPs that "read" specific tubulin PTMs (code "readers") is crucial to fully decipher the Tubulin Code. A handful of reader MAPs is known, but prior to our work, systematic screens for readers of any tubulin

Tomomi Kiyomitsu (ed.), *The Mitotic Spindle: Methods and Protocols*, Methods in Molecular Biology, vol. 2872,
https://doi.org/10.1007/978-1-0716-4224-5_13,

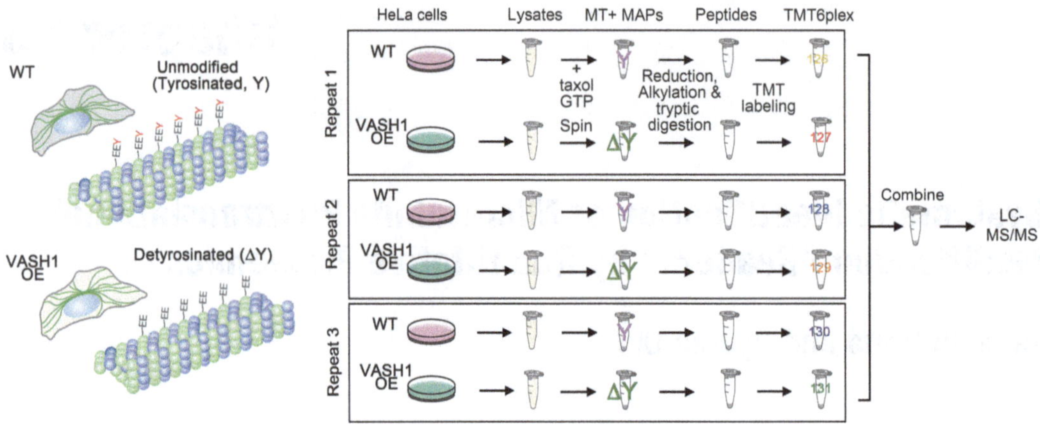


**Fig. 1** Experimental workflow of the reader screening pipeline optimized for α-tubulin Y/ΔY cycle. A total of six independent MT pellet fractions are labeled with six distinct isobaric mass tags. These tags are color-coded in the figure and are denoted by their respective mass-to-charge (m/z) ratios: 126, 127, 128, 129, 130, and 131, which correspond to the different isotopic compositions allowing for multiplexed quantitation in MS analysis

PTM had not been performed. In our studies, we focused on the tyrosination/detyrosination (Y/ΔY) cycle, the oldest and most studied tubulin PTM. During detyrosination, the C-terminal tyrosine residue of α-tubulin is enzymatically removed by the code writers (*e.g.*, vasohibin (VASH) or MATCAP) and added back by the code eraser (tubulin tyrosine ligase or TTL) [1–5]. To identify readers of the Y/ΔY system, we established an unbiased proteomic screening pipeline to identify code readers by quantitative mass spectrometry [6].

The workflow of the reader identification pipeline is summarized in Fig. 1. This pipeline employs the widely used MT-MAPs co-pelleting assay, in which MAPs are co-pelleted with MTs assembled in cell lysates supplemented with Taxol and GTP. By applying this assay to both wild-type HeLa cells and HeLa cells overexpressing VASH1 detyrosinase, along with its obligate binding partner *S*mall *V*asohibin *B*inding *P*rotein (SVBP), we generate two distinct MT samples—unmodified (tyrosinated) and detyrosinated. Theoretically, these two MT pellet fractions should preferentially contain Y- and ΔY-readers, respectively. After repeating this process three times, we obtain six MT-MAPs fractions (three each for Y and ΔY). These six samples are combined and analyzed by tandem mass tag (TMT) quantitative mass spectrometry (MS) [7]. Since a TMT-sixplex analysis includes three replicates of the MT-MAPs co-pelleting assay, the relative abundance of MAPs in either Y- or ΔY-conditions can be statistically evaluated.

The difference in the affinity of MAPs for unmodified *versus* modified MTs is often subtle. In our screen, even though our samples were nearly completely unmodified or modified, the difference in MT binding was minimal for many MAPs, including known Y-readers. This is thought to be because many MAPs exhibit multivalent binding to MTs, *i.e.*, the C-terminal tails are not the only binding sites on MTs for these MAPs. Therefore, it is critical to maximize the difference in modification levels between unmodified and modified samples, ideally in binary conditions. In our Y/ΔY reader screen, we use the HeLa Kyoto cell line as the material to isolate MTs and MAPs. Since tubulin in HeLa cells is known to be largely unmodified [8], wild-type HeLa cells serve as an excellent source of unmodified tubulin/microtubules (*see* **Note 1**). To obtain a highly modified tubulin/microtubule sample, several methods should be carefully considered and optimized. In this chapter, we will first address this optimization step, then describe the co-pelleting assay, quantitative proteomics, and finally hit validation.

## 2 Materials

### 2.1 Generation of Fully Modified MTs

1. pmEGFP-N1-VASH1-IRES-SVBP plasmid [9]
2. Wild-type HeLa Kyoto cells
3. Cell culture medium: DMEM supplemented with 10% fetal bovine serum (FBS) and penicillin-streptomycin.
4. Lipofectamine 2000 transfection reagent
5. Knock-in HeLa cell line expressing VASH1-SVBP [9]
6. 2 μg/mL doxycycline
7. Tetracycline-negative FBS
8. 1 μg/mL puromycin
9. *TTLΔ* HeLa cell line [10]
10. Tissue culture $CO_2$ incubator.
11. Benchtop ultracentrifuge, e.g. Optima TM MAX-TL (Beckman Coulter).
12. Rotors TLA45 and TLA110 (Beckman Coulter).
13. Ultracentrifuge tubes.
14. Water bath (37 °C).
15. NP-40 lysis buffer: 6 mM $Na_2HPO_4$, 4 mM $NaH_2PO_4$, 2 mM EDTA, 150 mM NaCl, 1% NP40, 1 mM benzamidine, 1 mM PMSF, and 10 μg/mL each of leupeptin, pepstatin, and chymostatin.
16. Sonicator
17. Anti-Y-, ΔY-, Δ2- and a total α-tubulin antibodies
18. Carboxypeptidase A (CPA)

#### 2.2 MT-MAPs Co-pelleting Assay

1. BRB80 buffer: 80 mM PIPES, 1 mM EGTA, 1 mM $MgCl_2$ pH 6.8.
2. 10× cOmplete mini EDTA free: Dissolve 1 tablet in 1 mL of BRB80 (freshly prepared).
3. 1 M DTT: Dissolve in water and store at −20 °C.
4. 10 mM Taxol: Dissolve in DMSO and store at −20 °C.
5. 100 mM MgGTP: Dissolve in water with equimolar $MgSO_4$, neutralize and store at −80 °C.
6. Sucrose cushion: 5% sucrose in BRB80 with 10 μM Taxol and 1 mM GTP. Warm up at 37 °C.
7. MT pellet wash buffer: BRB80 with 10 μM Taxol and 1 mM GTP. Warm up at 37 °C.
8. 8 M urea solution: Dissolve 480 mg of urea in 500 μL BRB80 and adjust the volume to 1 mL. Freshly prepared.

#### 2.3 Quantitative Proteomics (TMT)

1. TMTsixplex Isobaric Label Reagent Set, 1 × 0.8 mg (Thermo Fisher).
2. Sep-Pak C18 cartridges (Waters).
3. 15 mM 2-chloracetamide
4. Sequencing grade modified trypsin
5. Vacufuge.
6. 0.1 M TEAB
7. Anhydrous acetonitrile
8. 5% hydroxylamine
9. 0.1% formic acid/2% acetonitrile
10. LC-MS/MS. We use a RSLC Ultimate 3000 nano-UPLC (Dionex) with a 50 cm, 75 μm i.d. C18 column (Thermo Fisher) and an Orbitrap Fusion (Thermo Fisher).
11. Proteome Discoverer (v2.3; Thermo Fisher) or equivalent software

#### 2.4 Hit Validation

1. GFP-tagged mammalian expression plasmid for each hit.

## 3 Methods

#### 3.1 Strategies to Generate Fully Modified MTs

For successful reader identification, it is important to generate MTs that are post-translationally modified to a high stoichiometry. Various approaches can be used to prepare unmodified and fully modified MTs to use as bait to isolate specific reader MAPs. For example, tubulin PTMs can be increased either by overexpressing the PTM writer or by knocking out the eraser in cells (*see* **Note 2**).

Alternatively, tubulin PTMs can be promoted by incubating cell lysates with purified writer enzyme prior to (or during) MT polymerization. These approaches can be combined if one approach is not sufficient to generate high levels of tubulin PTM of interest. For the Y/ΔY reader screen, we have effectively used the writer overexpression and in vitro methods. Depending on the PTM of interest, this step requires optimization. Below, we detail our optimization step in the Y/ΔY reader screening.

### 3.2 PTM Upregulation in Cells

We compared the overexpression of the detyrosination writer (VASH1-SVBP) and the knockout of the eraser (TTL). For writer overexpression, we first transiently expressed a bi-cistronic VASH1-SVBP construct [9] using transfection. We confirmed the expression of VASH1-SVBP proteins and their activity in HeLa cells by detecting ΔY-tubulin in the cell lysate using Western blot analysis. Subsequently, a knock-in HeLa cell line that expresses VASH1-SVBP in a doxycycline-dependent manner was generated [9]. Western blotting of lysates prepared from cells cultured in the presence of doxycycline for 1–4 days confirmed the presence of ΔY-tubulin, similar to transient expression. Next, to test the eraser knockout approach, we knocked out TTL using the CRISPR-Cas9 system, generating a *TTLΔ* HeLa cell line [10]. When comparing the reduction of Y-tubulin and the increase of ΔY-tubulin, both methods of overexpressing the writer and knocking out the eraser proved to be equally effective, with an approximate 50% reduction of Y-tubulin (Fig. 2a). However, improved levels of detyrosination were observed after in vitro MT assembly, which was induced by the addition of Taxol and GTP to the lysates. Although Y-tubulin was still present in the MT fraction obtained from the TTL knockout lysate, it was undetectable in the lysate prepared from VASH1-SVBP overexpression cells (Fig. 2b), suggesting that detyrosination of Taxol-stabilized MTs occurred in the presence of VASH1-SVBP

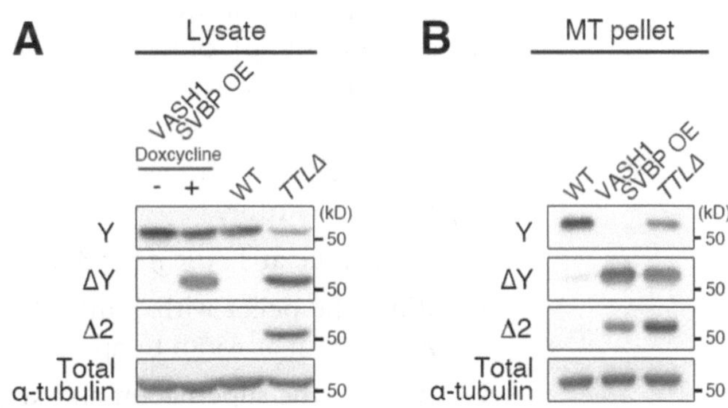


**Fig. 2** Comparison in Y-, ΔY- and Δ2-α-tubulin levels between wild-type HeLa cells (WT), VASH1-SVBP overexpressing (OE) cells cultured with or without 2 μg/mL doxycycline for 3 days, or TTL knockout (*TTLΔ*) cells. Lysates (**a**) and the MT pellet fractions (**b**)

in cell lysate. The combination of in-cells and in vitro detyrosination successfully generated a fully detyrosinated MT sample. Therefore, we chose the writer overexpression method rather than the eraser knockout condition for further reader screening (*see* **Notes 3–5**).

#### *3.2.1 Writer (VASH1-SVBP) Overexpression Method*

For the transfection approach, refer to **steps 1** through **3**. For the use of doxycycline-inducible knock-in cells, refer to **steps 4** through 7.

1. Seed wild-type HeLa Kyoto cells in a well of a 6-well plate and culture in DMEM supplemented with 10% fetal bovine serum (FBS) and penicillin-streptomycin.
2. The next day, transfect cells with a bicistronic construct pmEGFP-N1-VASH1-IRES-SVBP that harbors human VASH1-mEGFP and SVBP-FLAG using Lipofectamine 2000 transfection reagent.
3. Cells should be nearly confluent and ready for lysate preparation in the next 1–2 days.
4. Generate a knock-in HeLa cell line expressing VASH1-SVBP [9]. We use a recombination-mediated cassette exchange method to generate a doxycycline-inducible cell line [11, 12] (*see* **Note 2**).
5. Seed VASH1-SVBP knock-in HeLa cells in a well of a 6-well plate and culture in DMEM supplemented with 10% tetracycline-negative FBS, penicillin-streptomycin, 1 μg/mL puromycin (*see* **Note 6**).
6. Initiate VASH1-SVBP expression by adding 2 μg/mL doxycycline in the culture medium.
7. Cells should be ready for lysate preparation 1–4 days after the addition of doxycycline (*see* **Note** 7).
8. Aspirate the culture medium, rinse cells with 1 mL of DPBS and add 300 μL of trypsin to detach cells (volume is given per well). Incubate at 37 °C for a few minutes.
9. Once cells are detached, add 1 mL of culture media to resuspend cells and transfer the whole content to a 1.5 mL Eppendorf tube.
10. Centrifuge at 170× *g* for 3 min at room temperature.
11. Rinse the pellet with 1 mL of DPBS and centrifuge at 500× *g* for 1 min at room temperature.
12. Remove the supernatant and keep the tube on ice.
13. Add 100 μL of ice-cold NP-40 lysis buffer, resuspend cells and incubate on ice for 3 min.
14. Sonicate the cell mixture.

15. Centrifuge at 21,000× *g* for 15 min at 4 °C.
16. Collect the supernatant as cleared whole cell lysate.
17. Measure the protein concentration using the Bradford protein assay.
18. Load 15 μg of lysate proteins per well of a 15-well mini protein gel and run an SDS-PAGE. Along with lysate proteins, load 5 μg of MT-MAPs fraction per lane. For the preparation of MT-MAPs fractions, refer to Subheading 3.3 but omit CPA in **step 10**.
19. Transfer to a PVDF membrane and perform western blotting using anti-Y-, ΔY-, Δ2- and total α-tubulin antibodies as well as anti-GAPDH antibody as a loading control.

#### 3.2.2 Eraser (TTL) Knockout Method

1. Establish a TTL knockout HeLa cell line using the CRISPR-Cas9 system [10, 13]. We detected three distinct INDELs (note that HeLa cells are hypertriploid) and also confirmed the complete loss of TTL protein in cell lysate by western blotting [10].
2. Seed cells on a 6-well plate and follow **steps 8–19** shown above.

### 3.3 Enzyme Addition In Vitro

An alternative approach to generating highly modified MTs is adding a modification enzyme to the lysate prepared from wild-type cells. Notably, tubulin modification only occurs post-lysate preparation. The advantage of this in vitro method over in-cell modification methods is that this method requires only wild-type cells, negating any requirement for genetically modified cell lines. Crucially, the preparation of two MT samples, both unmodified and modified, from the same cell lysate with an identical proteomic profile eliminates the need to check the expression level of each MAP in lysates prepared from independent cell lines.

To induce the modification of interest in vitro, it is ideal to use recombinantly prepared tubulin-specific writers. However, if applicable, generic enzymes can also promote tubulin PTMs in vitro. When generating ΔY-tubulin/MTs, we utilize Carboxypeptidase A (CPA), a potent enzyme commonly used in generating ΔY-tubulin in vitro [8]. CPA is capable of removing tubulin's C-terminal tyrosine even on ice for brief periods, making it advantageous as it does not require warm incubation and, subsequently, minimizes undesired MT polymerization prior to full detyrosination. Since the modification process should continue after MT polymerization, we skipped monitoring the modification status in lysate and instead proceeded to the MT assembly to check the Y/ΔY state in the MT-MAPs pellet fractions. In the case of CPA-treated lysate, we detect nearly fully detyrosinated MTs in the MT-MAPs fraction (*see*

**Note 8**). Again, depending on the PTMs of interest, this process needs careful optimization.

1. Culture wild-type HeLa Kyoto cells on a 2 × 150 mm dish.
2. For each dish, aspirate media, rinse cells with 4 mL DPBS, and add 2 mL trypsin to detach cells. Once cells are detached, add 8 mL DMEM to collect cells. Do the same for all dishes, combine them all, and transfer them to a 50 mL conical tube.
3. Centrifuge, 170× *g* for 3 min at room temperature.
4. Rinse cell pellet with 35 mL DPBS and centrifuge.
5. Resuspend the pellet in 15 mL DPBS and transfer it to a 15 mL tube.
6. Resuspend the pellet with 1 mL DPBS and transfer it to a 1.5 mL Eppendorf tube.
7. Centrifuge 500× *g* for 1 min at room temperature.
8. Remove supernatant and approximately 150 μL of cell pellet is recovered. Keep the tubes on ice.
9. Add 1.5× volume of ice-cold 1xBRB80 containing 1.67× cOmplete mini EDTA free protease inhibitor cocktail (final 1×) and 1.67 mM DTT (final 1 mM). For example, if the cell pellet is 150 μL, add 187.5 μL of 1xBRB80, 37.5 μL of 1× cOmplete mini EDTA free protease inhibitor cocktail and 0.375 μL of 1 M DTT.
10. Add 1/100 volume of CPA (*see* **Note 9**), Taxol (final 10 μM) and GTP (final 1 mM) to lysate on ice.
11. After 3 min on ice, incubate at 37 °C for 25 min.
12. Put the reaction mix on 100 μL of pre-warmed sucrose cushion in a 1.5 mL tube.
13. Centrifuge with TLA45 rotor at 80,000× *g* for 30 min at 37 °C.
14. Wash the pellet with 100 μL of warm MT wash buffer twice and resuspend the pellet in 40 μL of cold 8 M urea solution.
15. Resuspend the solution well, incubate on ice for 10 min and transfer to a new tube.
16. Measure protein concentration with Bradford protein assay.
17. Load 250 ng of protein per lane and run an SDS-PAGE.
18. Evaluate the Y/ΔY state by western blotting with Y-, ΔY-, Δ2- and total α-tubulin antibodies.

### 3.4 MT-MAPs Co-pelleting Assay for MS Analysis

The MT-MAPs co-pelleting assay facilitates the isolation of MTs and MAPs *via* centrifugation following the in vitro assembly of MTs in the lysate, based on the affinity of MAPs to MTs. Initially, whole protein lysates are prepared from cells, and subsequent incubation of the lysates in the presence of Taxol and GTP promotes MT polymerization. To minimize the contamination of unbound proteins to the MT-MAPs pellet, the assembled MT mix is placed on a sucrose cushion for ultracentrifugation. Eventually, MAPs that co-pellet with MTs are subjected to identification *via* MS analysis. Below, we provide the procedure for the isolation of MT-MAPs fractions using VASH1-SVBP overexpressing method (Subheading 3.2.1). If the in vitro modification method is used, scale up the procedure described above (Subheading 3.3).

1. Culture cells on 5 × 150 mm dish per condition. Use wild-type HeLa cells and HeLa cells overexpressing VASH1-SVBP (in the presence of 2 mg/mL doxycycline for 3–4 days).
2. For each dish, aspirate media, rinse cells with 4 mL DPBS, and add 2 mL trypsin to detach cells. Once cells are detached, add 8 mL DMEM to collect cells. Do the same to all dishes, combine all and transfer it to a 50 mL conical tube.
3. Centrifuge, 1000× *g* for 3 min at room temperature.
4. Rinse cell pellets with 35–50 mL DPBS and centrifuge.
5. Rinse with DPBS one more time.
6. Resuspend the pellet with 1 mL DPBS and transfer it to a 1.5 mL Eppendorf tube.
7. Centrifuge 500× *g* for 1 min at room temperature.
8. Remove supernatant and approximately 500–600 μL of cell pellet is recovered (for each condition). Keep the tubes on ice.
9. Add 1.5× volume of ice-cold BRB80 containing 1.67× cOmplete mini EDTA free protease inhibitor cocktail (final 1×) and 1.67 mM DTT (final 1 mM). For example, if the cell pellet is 600 μL, add 750 μL of BRB80, 150 μL of 1× cOmplete mini EDTA free protease inhibitor cocktail and 1.5 μL of 1 M DTT.
10. Sonicate cells for 10 s, repeat 4 times.
11. Clear lysates by centrifugation at 100,000× *g* for 1 h at 4 °C using the TLA110 rotor.
12. Transfer the supernatant to a new 1.5 mL tube. The volume is typically ~750 μL. Save ~50 μL for protein quantification and SDS-PAGE analysis.
13. Transfer 650 μL of lysate in a tube and add 0.65 μL of 10 mM Taxol (final 10 μM) and 6.5 μL of 100 mM GTP (final 1 mM) to initiate MT assembly.
14. Incubate at 37 °C for 25 min on a water bath (*see* **Note 10**).
15. Layer assembly mix on 500 μL of pre-warmed sucrose cushion. Centrifuge at 80,000× *g* for 30 min at 37°C.
16. Rinse MT pellets twice with 100 μL of warm MT wash buffer.

17. Resuspend pellets in BRB80 supplemented with 85 μL of cold 8 M urea and incubate on ice for 10 min.
18. Transfer disassembled tubulin/MAP mix to new tubes.
19. Measure protein concentration using Bradford Protein Assay.
20. Transfer MT-MAPs solution that contains a total of 25 μg of protein and adjust the sample volume to 50 μL by adding BRB80 (final protein concentration is 0.5 μg/μL).
21. Snap-freeze the samples with liquid nitrogen and store them at −80 °C until MS analysis.
22. Snap freeze and store the rest of proteins at −80 °C for the future validation step (see Subheading 3.4).
23. Repeat this assay two more times. A total of 6 samples (each 25 μg of tubulin + MAPs) should be subjected to a TMT analysis.

### 3.5 TMT Quantitative Proteomics

A critical part of the reader identification pipeline is the use of quantitative proteomics for identifying reader MAPs. In a label-free MS analysis, in which we simply compare the peptide spectra counts between different samples analyzed by independent MS runs, it should be noted that comparisons of protein abundance are not quantitative due to the lack of internal control. Since peptide detection can be significantly affected by various factors unique to each sample, it is crucial to combine multiple samples and analyze them in a single run.

In quantitative MS methods such as TMT or Stable Isotope Labeling by Amino acids in Cell culture (SILAC), proteins/peptides are labeled with mass tags or stable isotopes, respectively, to distinguish the peptide spectra belonging to each sample. This enables the evaluation of the relative abundance of specific peptides across different samples. Since TMT-multiplex allows the simultaneous analysis of multiple samples, implementing this approach with multiple biological replicates provides statistical interpretation of the results. The obtained hits should be evaluated based on their enrichment in either unmodified or modified samples, which can be visualized using a volcano plot (Fig. 3). To facilitate the evaluation of the results, it is important to first identify positive controls. In the case of the Y/ΔY reader screen, we first evaluated the MS result by checking known Y-readers, CAP-Gly proteins such as CLIP-170, CLIP-115 and p150$^{glued}$ (Fig. 3, blue dots). MS analysis shows that these proteins pellet more efficiently with wild-type MTs than with those from VASH-SVBP overexpressing cells. This serves as a proof of principle and suggests that this pipeline may reflect the actual difference in the abundance of readers depending on the PTM states of MTs.

For the TMT procedure, we simply follow the instructions provided by the manufacturer. After results are obtained, it is

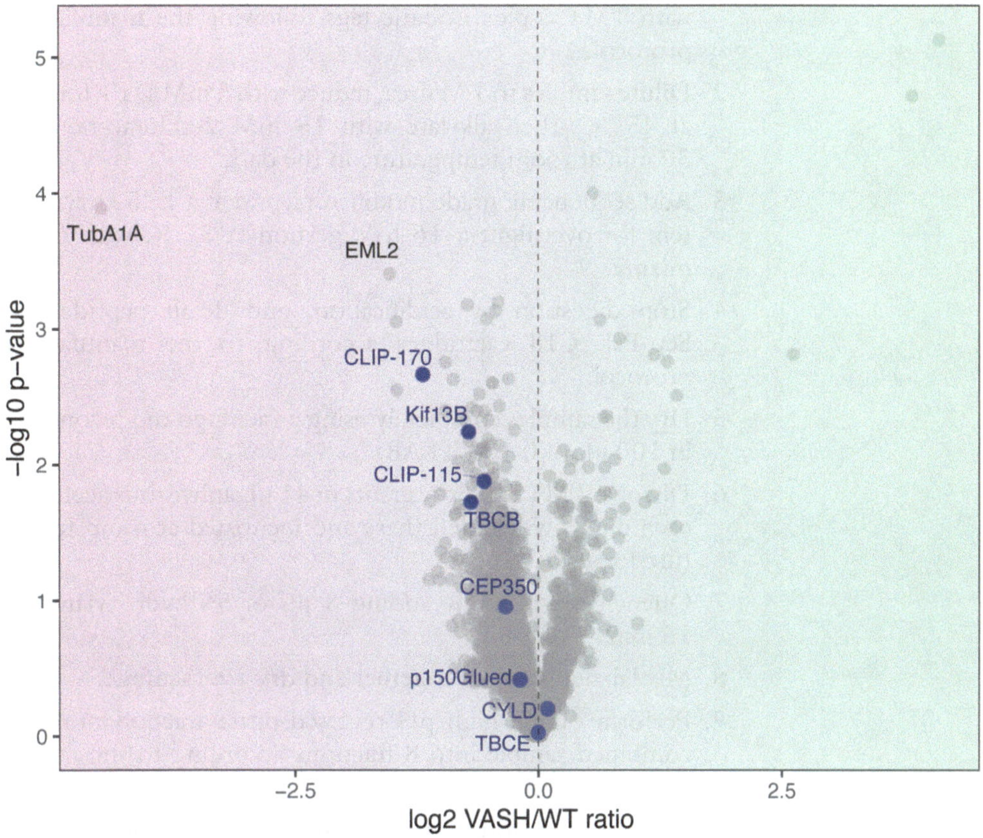


**Fig. 3** An example of a volcano plot showing proteins co-sedimented with Y- or ΔY-MTs in magenta or green background, respectively. X-axis represents $\log_2$ relative abundance of each protein in VASH1 OE over wild-type samples (VASH/WT). Y-axis shows $\log_{10}$ p value. Blue dots highlight CAP-Gly proteins

essential to properly normalize the MS result to adjust the slightly uneven amount of total proteins in each sample. Alternatively, each score can be normalized against the total amount of tubulin in each sample. We have compared normalization methods against several factors, such as total β-tubulin (average of all β-tubulin isotypes), median β-tubulin isotype, or total protein. However, we saw the same overall trends in the MAP distribution between the unmodified and modified samples. The optimal normalization method should be determined based on specific needs. Note that we did not use the total α-tubulin abundance for normalization because the identification of α-tubulin isoforms relies on the C-terminal peptides, but due to the loss of terminal tyrosine in the modified sample, it appears as if the α-tubulin abundance has reduced significantly in the ΔY condition (VASH1-SVBP overexpression) (Fig. 3, the leftmost dot).

1. For the 6 samples generated in the MT-MAPs co-sedimentation assay, each 25 μg protein sample was digested and labeled

with TMT 6-plex isobaric tags following the manufacturer's protocol.

2. Dilute samples to 1 M urea, reduce with 5 mM DTT for 30 min at 45 °C, then alkylate with 15 mM 2-chloracetamide for 30 min at room temperature in the dark.
3. Add sequencing grade modified trypsin at a 1:25 enzyme:protein for overnight (~16 h) digestion at 37 °C with constant mixing.
4. Stop digestion by acidification, and desalt peptides using Sep-Pak C18 cartridges according to the manufacturer's protocol.
5. Dry the samples completely using a vacufuge and reconstituted in 100 μL of 0.1 M TEAB.
6. Dissolve TMT 6-plex reagents in 41 μL anhydrous acetonitrile; each digest was added there and incubated at room temperature for 1 h.
7. Quench reactions by adding 8 μL of 5% hydroxylamine for 15 min.
8. Mix labeled samples together and dry *via* vacufuge.
9. Perform offline high pH reversed-phase fractionation of the combined sample into 8 fractions according to the manufacturer's protocol.
10. Dry and reconstitute fractions in 9 μL of 0.1% formic acid/ 2% acetonitrile in preparation for LC-MS/MS analysis.
11. For each fraction, analyze a 2 μL aliquot by LC-MS/MS using a RSLC Ultimate 3000 nano-UPLC (Dionex) with a 50 cm, 75 μm i.d. C18 column (Thermo Fisher Cat # ES903) and an Orbitrap Fusion (Thermo Fisher). A 3-h gradient at 300 nL/min using 0.1% formic acid/acetonitrile (2–22% acetonitrile in 150 min; 22–32% acetonitrile in 40 min; 20 min wash at 90% followed by 50 min re-equilibration) was used.
12. Introduce eluent in the mass spectrometer *via* an EasySpray source (Thermo Fisher).
13. Set the mass spectrometer to collect MS1 scans (Orbitrap; 120 K resolution; AGC target 2×105; max IT 100 ms) followed by Top Speed MS2 scans (1 m/z isolation width, collision-induced dissociation; ion trap; NCE 35; AGC 5×103; max IT 100 ms).
14. For multinotch-MS3, the top 10 precursors from each MS2 were fragmented by HCD followed by Orbitrap analysis (NCE 55; 60 K resolution; AGC 5×104; max IT 120 ms, 100–500 m/z scan range).
15. For data analysis, use Proteome Discoverer (v2.3; Thermo Fisher) or equivalent software.

16. Search spectral files against the SwissProt human protein database using the following search parameters: Set MS1 and MS2 tolerance to 10 ppm and 0.6 Da, respectively; carbamidomethylation of cysteines and TMT labeling of lysine and N-termini of peptides were considered static modifications; oxidation of methionine and deamidation of asparagine and glutamine were considered variable.
17. Filter identified proteins and peptides to retain only those passing a 1% FDR threshold.
18. Perform quantification only using high-quality MS3 spectra (average signal-to-noise ratio >6 and <30% isolation interference).

### 3.6 Hit Validation

Hits are prioritized based on preferential enrichment in either Y- or ΔY- samples. Although the TMT-based quantitation reliably reflects the relative abundance of each MAP in two opposite PTM states (see Fig. 1 in [6]), a validation step using Western blot analysis is useful to check the differential binding of each candidate reader to either Y- or ΔY-MT fractions. This is particularly important when multiple isoforms are present for a reader and only specific isoform(s) show preferential binding to the PTM of MTs, as in the case of EML2 (Figs. 1 and 2 in [6]).

It is noteworthy that the expression level of some MAPs can be affected by the overexpression of VASH1-SVBP (or knockout of TTL) (*see* **Note 5**). Therefore, it is important to confirm that the protein abundance is comparable in both the wild-type and VASH1-SVBP overexpression lysates. Western blot analysis of the six lysates and MT fractions used for the MS analysis should provide a comprehensive assessment of the reader.

As a final validation, we make constructs of GFP-tagged readers for transient expression in wild-type HeLa cells, followed by the MT-MAPs co-pelleting assay using CPA to confirm each reader's preferential binding to either Y- or ΔY-MTs. As with any other experiments, it is important to check the functionality of the GFP-tagged reader by live cell imaging or immunofluorescence. If necessary, the position (N- or C-terminal) or choice of tags should be considered. GFP-tagging interferes with MT binding of some MAPs such as EML2-S (Figure S4A in [6]). If this is the case, small affinity tags such as FLAG, c-myc, HA or PA should be considered.

#### 3.6.1 Western Blotting Validation for Endogenous Reader Candidates (Lysate and MTs)

1. Use the 6 lysates and MT pellet fractions used for the TMT MS analysis.
2. Load 15 μg of lysate proteins and 5 μg (for MAP blots) or 200 ng (for tubulin blots) of MT pellet fractions per lane for both unmodified (WT) and detyrosinated (VASH1-SVBP) conditions and run SDS-PAGE.

3. Perform western blot and analyze the Y/ΔY state of tubulin and enrichment of reader candidates.

*3.6.2 Western Blotting Validation Using GFP-Tagged Reader Candidates*

1. Transfect wild-type HeLa cells with constructs of GFP-tagged reader candidates and perform MT-MAPs co-pelleting assay using CPA for the ΔY-condition.
2. Analyze the differential binding of GFP-tagged reader candidates in Y- and ΔY-MT fractions by SDS-PAGE and western blotting using anti-GFP antibody.

## 4 Notes

1. As tubulin PTM levels vary among different cell lines, the specific PTM profile in the cell line of choice should be assessed in each study.
2. Alternatively a stable cell line can be generated using lentiviral transduction.
3. Overexpression of writer or knockout of eraser may impact cellular activities and proliferation, or expression and PTMs of a wider variety of proteins, including tubulin. It is important to monitor cell proliferation and morphology and MT configurations (interphase and mitosis) by flow cytometry, live cell imaging or immunofluorescence.
4. It is critical to monitor tubulin PTMs in addition to the PTM of interest. In the case of ΔY upregulation, Δ2-α-tubulin was also generated (Fig. 2b), although it was negligible compared to brain tubulin, which contains abundant Δ2-α-tubulin. If other tubulin PTM levels are significantly altered, this must be taken into account when interpreting the MS result.
5. Overexpression of writer or knockout of eraser may also alter the expression level of other proteins, especially MAPs. Such a differential expression of MAPs between unmodified and modified samples will lead to biased identification of the MAPs in downstream MS analysis. Therefore, it is important to monitor the protein expression level in the lysate for each reader candidate (*see* Subheading 3.6).
6. Our knock-in system uses a tetracycline promoter. To prevent leaky protein expression, it is important to maintain the cell line using tetracycline-negative FBS.
7. When an inducible system is used for overexpression, it is important to optimize the induction time to maximize the tubulin PTMs. In the case of VASH1-SVBP, we compared 1–4 days after doxycycline addition and decided to use a 3 or 4-day induction period.

8. CPA also generates Δ2- and perhaps Δ3-α-tubulin. Also keep in mind that it may cause proteolytic digestion of other proteins.
9. Before use, mix CPA solution well. Do not centrifuge and remove white matter in the vial.
10. For optimum results regarding the efficient binding of specific MAP, the conditions of the MT assembly step, such as incubation time and salt concentration, may need to be customized. We compared the MT assembly incubation time ranging from 25 to 90 min and found that 25 min was sufficient for various MAPs. After longer incubation, binding efficacy decreased for some MAPs, likely due to protein degradation or gradual changes in tubulin PTM states over time.

## References

1. Aillaud C, Bosc C, Peris L et al (2017) Vasohibins/SVBP are tubulin carboxypeptidases (TCPs) that regulate neuron differentiation. Science 358:1448–1453
2. Nieuwenhuis J, Adamopoulos A, Bleijerveld OB et al (2017) Vasohibins encode tubulin detyrosinating activity. Science 358:1453–1456
3. Landskron L, Bak J, Adamopoulos A et al (2022) Posttranslational modification of microtubules by the MATCAP detyrosinase. Science 376:eabn6020
4. Nicot S, Gillard G, Impheng H et al (2023) A family of carboxypeptidases catalyzing alpha- and beta-tubulin tail processing and deglutamylation. Sci Adv 9:eadi7838
5. Ersfeld K, Wehland J, Plessmann U et al (1993) Characterization of the tubulin-tyrosine ligase. J Cell Biol 120:725–732
6. Hotta T, McAlear TS, Yue Y et al (2022) EML2-S constitutes a new class of proteins that recognizes and regulates the dynamics of tyrosinated microtubules. Curr Biol 32:3898–3910 e3814
7. Thompson A, Schafer J, Kuhn K et al (2003) Tandem mass tags: a novel quantification strategy for comparative analysis of complex protein mixtures by MS/MS. Anal Chem 75:1895–1904
8. Souphron J, Bodakuntla S, Jijumon AS et al (2019) Purification of tubulin with controlled posttranslational modifications by polymerization-depolymerization cycles. Nat Protoc 14:1634–1660
9. Hotta T, Haynes SE, Blasius TL et al (2021) Parthenolide destabilizes microtubules by covalently modifying tubulin. Curr Biol 31: 900–907 e906
10. Hotta T, Plemmons A, Gebbie M et al (2023) Mechanistic analysis of CCP1 in generating DeltaC2 alpha-tubulin in mammalian cells and photoreceptor neurons. Biomolecules 13
11. Khandelia P, Yap K, Makeyev EV (2011) Streamlined platform for short hairpin RNA interference and transgenesis in cultured mammalian cells. Proc Natl Acad Sci 108:12799–12804
12. Sturgill EG, Norris SR, Guo Y et al (2016) Kinesin-5 inhibitor resistance is driven by kinesin-12. J Cell Biol 213:213–227
13. McKinley KL, Cheeseman IM (2017) Large-scale analysis of CRISPR/Cas9 cell-cycle knockouts reveals the diversity of p53-dependent responses to cell-cycle defects. Dev Cell 40:405–420 e402

# Part V

## Centrosome Manipulation and Purification

# Chapter 14

# Comparative Pharmacological Analysis of Mitotic Inhibitors Using Isogenic Ploidy Series of HAP1 Cells

**Koya Yoshizawa and Ryota Uehara**

## Abstract

Drastic changes in chromosome number and cellular contents upon ploidy alterations profoundly affect the stability of mitotic regulation in different biological and pathological processes. Isogenic ploidy series of somatic cell lines are useful for studying the effects of ploidy differences on mitotic regulation at cellular and molecular levels. This chapter describes experimental procedures using isogenic human HAP1 cell lines that cover haploid, diploid, and tetraploid states. We first describe methods to establish and maintain these isogenic HAP1 ploidy series using a flow cytometer. We then describe a procedure of comparative pharmacological assay for analyzing ploidy-dependent changes in the functionality of the mitotic spindle components.

**Key words** Ploidy, Whole-genome duplication, Mitotic spindle, Size scaling, Centrosome, Chromosome segregation, Flow cytometry, HAP1 cells

## 1 Introduction

Ploidy alterations, including drastic gain or loss of chromosome copies, occur in broad biological processes such as development, tumorigenesis, and evolution [1]. Ploidy alterations are accompanied by drastic changes in cell size and cellular contents, profoundly affecting the mitotic spindle structure through the organelle size scaling principle [2–4]. However, the influences of ploidy differences on the properties and functionality of the mitotic spindle remain largely unknown.

Isogenic ploidy series, a series of cell lines derived through ploidy alterations in a common original line, are useful tools for studying the biological impacts of ploidy differences. Since the genetic background of cell lines is identical in the series, any differences among them can be attributed to their ploidy differences. Therefore, comparative analyses using isogenic ploidy series provide unique opportunities to identify ploidy-dependent factors

Tomomi Kiyomitsu (ed.), *The Mitotic Spindle: Methods and Protocols*, Methods in Molecular Biology, vol. 2872, https://doi.org/10.1007/978-1-0716-4224-5_14,

affecting cellular regulatory processes. For example, ploidy-dependent changes in functionality of mitotic spindle components can be addressed by comparing the effects of their inhibition or gene suppression among isogenic ploidy series [2, 5, 6]. In somatic cell culture models, ploidy alteration is induced by artificially blocking chromosome segregation or cytokinesis to cause whole-genome duplication (WGD), followed by clonal isolation of the post-WGD cells [7]. Near-diploid mammalian cell lines are often used as "ground state" models to establish their tetraploid derivatives. Further induction of WGD in tetraploid cells often compromises chromosome stability, precluding the establishment of stable octaploid cell lines [8]. This technical limitation has hindered comparative studies in a broader range of ploidy differences in mammalian cell lines.

However, the establishment of the rare human near-haploid somatic cell line HAP1 enabled the development of an isogenic cell line series covering the range from haploid to tetraploid states in human cellular background [4, 9, 10]. HAP1 is an adherent cell line derived through reprogramming near-haploid chronic myelogenous leukemia cell line KBM7 [11]. HAP1 has a single copy of all chromosomes except for the duplication of a portion of chromosome 15, making it an ideal research tool for human genetic studies. In mammals, somatic haploid cells are generally unstable, converting to diploid through spontaneous WGD in consecutive cultures for days to weeks [11]. The diploidized HAP1 cell population can be isolated from the original haploid population by cell sorting based on cell size or DNA content [4, 12]. Isogenic tetraploid HAP1 can be obtained by inducing WGD in the diploid HAP1 cells by blocking cell division and isolating the resulting cell clones. Because of the unstable nature of haploid HAP1 cells, periodic purification of the haploid population by size sorting is required for their maintenance. Since ploidy alteration makes substantial differences in various aspects of cellular properties and gene regulations [13], unnoticed progression of diploidization in haploid cell culture potentially compromises the reliability of the data obtained using HAP1 cells. Therefore, as detailed below, the regular check of ploidy states of cell lines is highly recommended.

In this chapter, we explain methods of establishing and maintaining isogenic ploidy series of HAP1 cells (Fig. 1a), and comparative pharmacological assay for analyzing ploidy-dependent changes in the functionality of the mitotic spindle components.

## 2 Materials

### 2.1 Compounds

1. Cell counting kit-8 (CCK-8; CK04, Dojindo, Kumamoto, Japan).
2. Cytochalasin B (an inhibitor of actin polymerization for cytokinesis blockage).

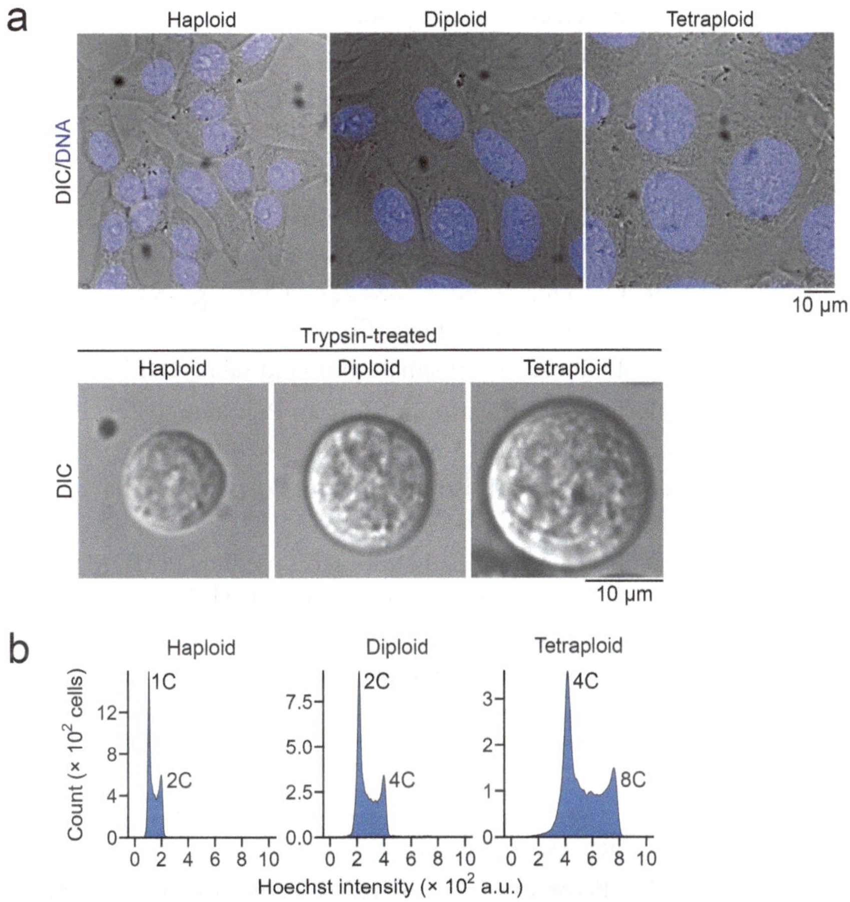


**Fig. 1** (**a**) Haploid, diploid, and tetraploid HAP1 cells cultured on substrates (top) or detached from substrates by trypsinization (bottom). DNA was stained using Hoechst 33342. (**b**) Histograms of Hoechst 33342 signal in haploid, diploid, or tetraploid HAP1 cells analyzed by a flow cytometer. The labels on the plot indicate the relative DNA amount (C-value). In haploid cell culture, 1C and 2C correspond to G1 and G2/M cells, respectively. In isogenic diploid and tetraploid cell cultures, G1 and G2/M peaks shift to 2C and 4C, and 4C and 8C, respectively. (Adapted from "Tetraploidy-linked sensitization to CENP-E inhibition in human cells." (Yoshizawa et al. [10]). Published under a Creative Commons Attribution License (CC BY))

3. Dimethyl sulfoxide (DMSO, cell culture grade).
4. Hoechst33342 (for DNA content analysis).
5. Nocodazole (a microtubule depolymerizing reagent for mitotic cell arrest).

### *2.2 Cell Culture, Cell Proliferation Assay, and Cell Sorting*

1. HAP1 cells (RRID: CVCL_Y019; Haplogen GmbH, Vienna, Austria)
2. 0.05 w/v% Trypsin-0.53 mmol/L EDTA·4Na Solution with Phenol Red
3. 10 cm cell culture dish

4. 6 well cell culture plate
5. 96 well cell culture plate
6. Clear 96 Well Semi-Skirted PCR Plate (4ti-0760, Azenta Life Sciences, MA, US)
7. 25 mL reagent reservoir
8. Bambanker (a serum-free freezing medium, 302-14681, GC LYMPHOTEC, Tokyo, Japan)
9. Collagen Type I (354236, Corning, NY, US, used as 10% coating solution in DPBS)
10. Dulbecco's phosphate-buffered saline (DPBS (−)).
11. Falcon 5 mL Round Bottom Polystyrene Test Tube, with Cell Strainer Snap Cap (352235, Corning)
12. Falcon 5 mL Round Bottom Polystyrene Test Tube, with Snap Cap (352058, Corning)
13. Iscove's modified Dulbecco's medium (IMDM) supplemented with 10% fetal bovine serum (FBS) and 1× antibiotic-antimycotic solution.
14. A multichannel pipette

### 2.3 Analytical Device

1. Absorbance microplate reader. We use Sunrise (Tecan, Männedorf, Switzerland).
2. Flow cytometer. We use JSAN desktop cell sorter (Bay Biosciences, Tokyo, Japan).
3. Microscope. We use Ts2 (Nikon, Tokyo, Japan) with ×4 0.13 NA Plan Fluor lens or LWD ×20 0.40 NA Ph1 ADL lens.

## 3 Methods

### 3.1 DNA Content Analysis Using a Flow Cytometer

For a comparative study of ploidy series, it is crucial to check the ploidy states of cell lines used for experiments routinely. This section describes a procedure of DNA content analysis using a flow cytometer.

1. Seed HAP1 cells on a collagen-coated 6-well plate (*see* **Note 1**) at 5–10% confluency. Incubate for 36–48 h at 37 °C with 5% $CO_2$ (*see* **Note 2**).
2. Start a flow cytometer and finish laser adjustment following the manufacturer's instructions.
3. Rinse the cells once with 0.5 mL DPBS, and apply 0.5 mL 0.05% trypsin. Incubate for 5 min at 37 °C with 5% $CO_2$.
4. Apply 0.5 mL DPBS to the 6-well plate, gently pipette, and transfer the cell suspension to a 1.5 mL microtube.

5. Add Hoechst33342 to the cell suspension at the final concentration of 10 μg/mL and incubate for 15–30 min at 37 °C with 5% $CO_2$ (*see* **Note 3**).
6. Filter cells through a cell strainer and apply them to a flow cytometer.
7. Conduct DNA content analysis by measuring Hoechst33342 fluorescence intensity. We typically flow cells at an event rate of 100–150 per second. DIC images and a typical result of DNA content analysis in haploid, diploid, and tetraploid HAP1 cells are shown in Fig. 1.

### 3.2 Maintenance of Haploid HAP1 Cells by Size-Based Cell Sorting

Haploid HAP1 cells undergo spontaneous auto-diploidization in consecutive passages for a few weeks, requiring periodic purification (*see* **Note 4**). This section describes a procedure for keeping a pure haploid population of HAP1 cells. Since diploidized cells are about 2 times larger than original haploids in volume (Fig. 1a), haploid cells can be purified through size sorting using a flow cytometer. Haploid cell population can also be sorted based on DNA content through Hoechst33342 staining. However, we usually purify haploid cells using the size-based sorting of unstained cells to avoid the artificial effect of Hoechst33342 on chromatin state and genome stability.

1. Seed HAP1 cells containing the haploid population (*see* **Note 5**) on a collagen-coated 10-cm dish at 5–10% confluency. Incubate for 36–48 h at 37 °C with 5% $CO_2$.
2. Start a flow cytometer and finish laser adjustment.
3. Rinse the cells once with 1 mL DPBS, and apply 1 mL 0.05% trypsin. Incubate for 5 min at 37 °C with 5% $CO_2$.
4. Gently pipette and collect the trypsin-treated cells into a 15 mL conical tube. Rinse the dish with 2–3 mL DPBS and collect all cells into the tube.
5. Filter cells through a cell strainer and apply them to a flow cytometer to start size-based cell sorting.
6. Set a sorting gate to collect only the viable cell population with the bottom 15% size in FSC. An example of gate design on the FSC-SSC plot is shown in Fig. 2a. Collect cells in 1 mL prewarmed medium in a 5 mL polystyrene round-bottom tube. Finish cell sorting within 60 min to avoid excess cellular stress. We usually flow cells at an event rate of 700–1000 per second and collect 450,000 haploid cells.
7. Transfer the sorted cells on a collagen-coated 6-well plate at a density of 150,000 cells per well (3 wells in total).
8. Exchange media every 1 h, 3 times. Remove and apply media gently so that cells are not detached from the substrate (note that cell attachment is still weak, especially 1 h after sorting). Incubate for 2–3 days at 37 °C with 5% $CO_2$.

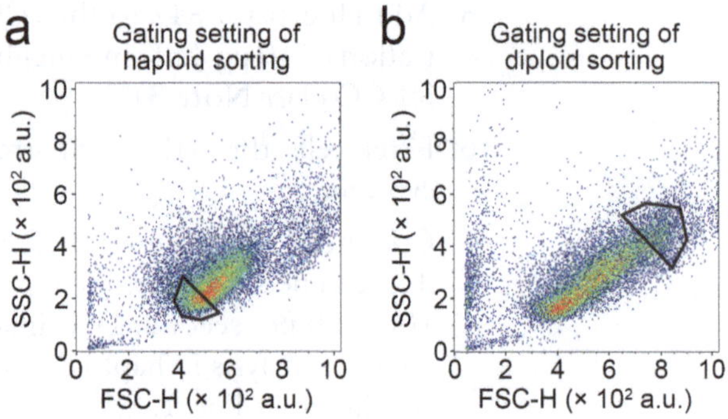


**Fig. 2** (**a**, **b**) Examples of gating settings for size-based sorting of haploid (**a**) and diploid cells (**b**). The gated regions on FSC-SSC plots are surrounded by black lines

9. Passage cells when the culture reaches 80–90% confluency and expand culture from each well to a 10-cm dish (3 dishes in total).
10. When the culture reaches 80–90% confluency, trypsinize the cells, gently pipette the suspension, and transfer it from all 3 dishes to a 15 mL conical tube. Rinse each dish twice with 1 mL medium to collect all cells into the tube.
11. Centrifuge the cell suspension at 1000 rpm for 3 min, discard the medium, and suspend cells into 9 mL pre-cooled Bambanker. Aliquot 1 mL cell suspension to each cell freezing vial (make nine vials in total). Vials can be kept at −80 °C for several months, but we recommend keeping them in liquid nitrogen for more extended storage.
12. To evaluate the sorting quality, check the DNA content of haploid cells as described in Subheading 3.1 right after recovering from the frozen stock. We routinely check the purity of haploid cells for each batch of frozen vials.

### 3.3 Establishment of Diploid HAP1 Cells by Size-Based Cell Sorting

This section describes a method to isolate a pure diploid HAP1 population. To obtain a diploidized population, one needs to culture haploid cells for a few weeks. Then, a diploidized population can be purified using size-based cell sorting.

1. Continue passages of haploid cell culture for a few weeks to achieve a sufficient degree of auto-diploidization (*see* **Note 6**).
2. Subject the haploid-diploid mixed culture to size-based cell sorting as described in Subheading 3.2, **step 2–6**, except that a sorting gate needs to be set to collect only the viable cell population with the top 10% size in FSC. An example of gate design on the FSC-SSC plot is shown in Fig. 2b.
3. Expand, freeze, and check the DNA content of the sorted diploid cells as described in Subheading 3.2, **step 7–12** (*see* **Note 7**).

### 3.4 Establishment of Tetraploid HAP1 Cells by Mitotic Shake-Off and Cytokinesis Blockage

This section describes how to establish tetraploid HAP1 subclones from diploid HAP1 cells obtained in Subheading 3.3. For this, whole-genome duplication is induced in diploid HAP1 by blocking cytokinesis using cytochalasin B. To minimize the time duration of cytochalasin B treatment, we first isolate mitotic cells by shake-off after short-term nocodazole arrest and then block subsequent cytokinesis. We have confirmed that this protocol is also applicable for establishing tetraploid subclones from HCT116, DLD-1, RKO, and HepG2 cells.

1. Seed diploid cells on a 10-cm dish at 5–10% confluency (*see* **Note 8**). Incubate for 36–48 h at 37 °C with 5% $CO_2$ until reaching 70–80% confluency.
2. Remove the culture medium from the dish and apply 5 mL prewarmed medium.
3. To remove the pre-existing mitotic cells (*see* **Note 9**), tap the dish on the side wall 5–10 times. We usually use a handle of scissors for tapping dishes.
4. Aspirate culture medium to remove floating mitotic cells. Rinse the dish with 3 mL prewarmed medium.
5. Add 10 mL prewarmed medium containing 20 ng/mL nocodazole. Incubate for 4 h at 37 °C with 5% $CO_2$ (*see* **Note 10**).
6. Aspirate culture medium gently to remove nocodazole. Rinse the dish gently with 5 mL prewarmed medium 3 times (*see* **Note 11**).
7. Apply 3 mL prewarmed medium containing 1 μg/mL cytochalasin B (*see* **Notes 11–13**).
8. Tap the dish on the side wall gently 1–3 times to detach the cells arrested in mitosis.
9. Transfer the detached mitotic cells to a collagen-coated 6-well plate. Incubate for 4 h at 37 °C with 5% $CO_2$.
10. Aspirate the culture medium to remove the cytochalasin B. Rinse the plate with 3 mL prewarmed medium 3 times.
11. Apply 2 mL prewarmed medium to the plate. Incubate for 2 h at 37 °C with 5% $CO_2$.
12. After most cells attach to the culture plate, roughly check the proportion of binucleated cells. Typically, more than 95% of cells are binucleated at this step.
13. Trypsinize the cells and subject them to serial dilution at a ratio of 1:5 in four 96-well culture plates. Incubate for 4–6 days at 37 °C with 5% $CO_2$. Exchange the medium every 3–4 days.
14. Check each well in 96-well plates under a microscope to find wells containing single colonies.

15. Expand, freeze, and check the DNA content of the established tetraploid cells as described in Subheading 3.2, **step 7–12** (*see* **Note 7**).

### 3.5 Comparative Pharmacological Analysis Using HAP1 Ploidy Series

This section explains a protocol for cell proliferation assay using the HAP1 ploidy series to identify inhibitors whose efficacy changes in ploidy-dependent manners. Using this assay, we recently found that CENP-E inhibitors preferentially suppress the proliferation of tetraploid cells over haploids and diploids through tetraploidy-linked aggravation of chromosome misalignment [9].

For comparing the cytotoxicity of various mitotic inhibitors among the HAP1 ploidy series, cells are treated with inhibitors at different concentrations in 96-well plates for 48 h. Dose-dependent effects of inhibitors on cell viability are then quantified by colorimetric assay using Cell Counting Kit-8. Figure 3a, b show examples of well design for the assay. For the side-by-side comparison, haploid, diploid, and tetraploid cells are seeded in parallel in the same dish. Wells containing only cell culture media should also be prepared for background subtraction. We do not use the outmost lines of wells of plates for the assay, and they are filled with water to avoid drying.

1. Thaw frozen stocks of haploid, diploid, and tetraploid HAP1 cells prepared in Subheadings 3.2, 3.3, and 3.4. Culture them on 10-cm dishes. Incubate for 40–48 h at 37 °C with 5% $CO_2$.
2. Trypsinize cells, count cell number, and adjust cell densities as follows (*see* **Note 14**):

   Haploid cells: 2.5 × $10^4$ cells/mL (4 mL in a 15 mL conical tube)

   Diploid cells: 1.25 × $10^4$ cells/mL (4 mL in a 15 mL conical tube)

   Tetraploid cells: 0.625 × $10^4$ cells/mL (4 mL in a 15 mL conical tube)
3. Use a multichannel pipette in the following steps. Apply 90 μL of cell suspension prepared in **step 2** to each well in the 96-well plates with the well design shown in Fig. 3a (*see* **Note 15**). Apply 100 μL water to the outmost lines of wells and 90 μL culture media (for background subtraction) to designated wells indicated in Fig. 3a (*see* **Note 16**). Incubate for 24 h at 37 °C with 5% $CO_2$.
4. Prepare intermediate dilution (10× inhibitor solution) of mitotic inhibitors in a culture medium in a 96-well PCR plate. We usually measure cell viability at 12 concentration points in a 1:2 dilution series (*see* **Note 17**).

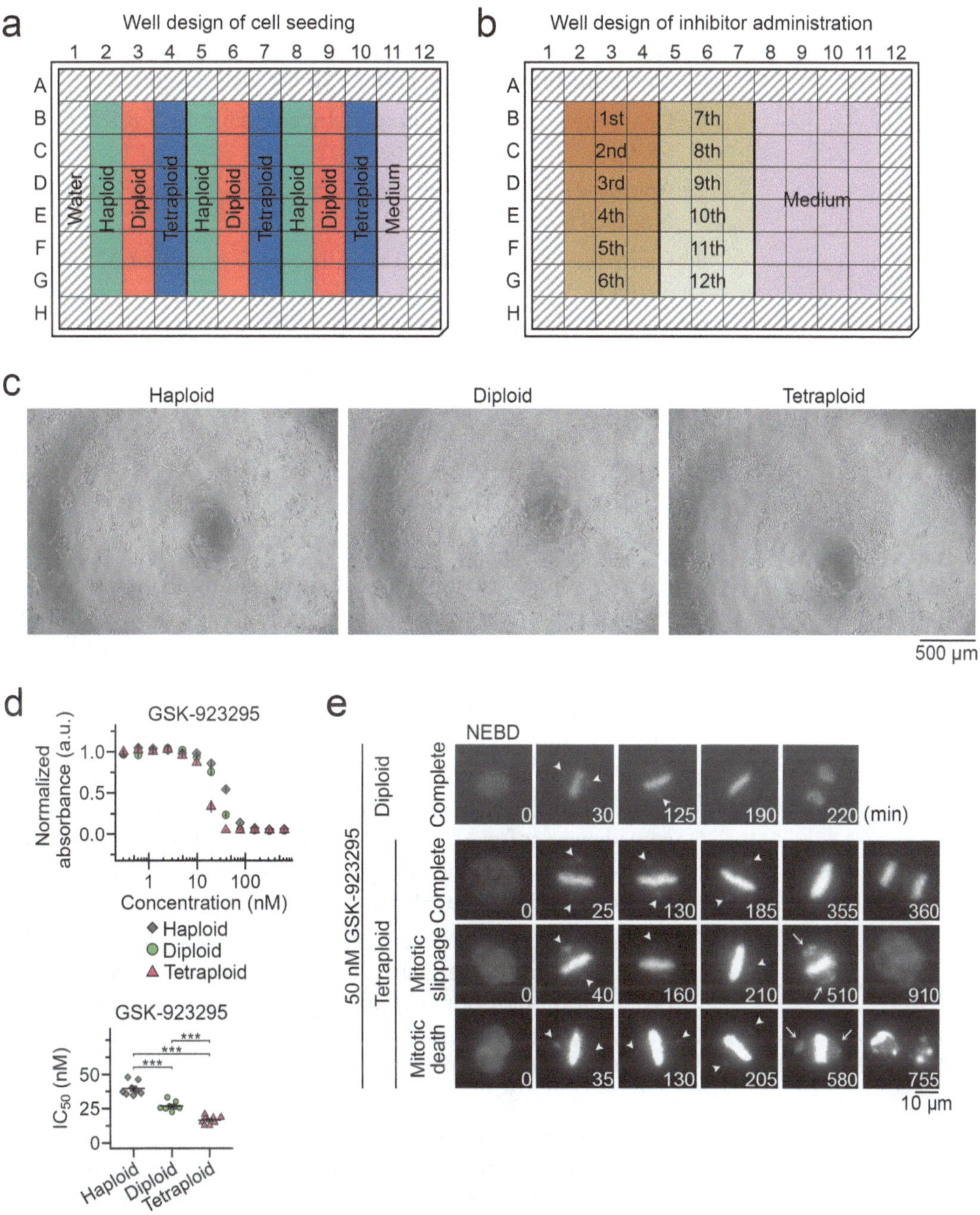


**Fig. 3** (**a**, **b**) A well design for cell seeding (**a**) and inhibitor administration (**b**) in a 96-well culture plate used in the experiment in Subheading 3.5. (**c**) Images of cell cultures immediately after CCK-8 absorbance measurement in Subheading 3.5, **step 9**. (**d**) An example of dose-response curve of normalized absorbance (top) or dot plot of $IC_{50}$ values (bottom) in a comparative colorimetric cell proliferation assay. Mean $\pm$ standard error (SE) of eight replicates from four independent experiments (***$P < 0.001$, the Steel-Dwass test). (**e**) Examples of live imaging of diploid and tetraploid cells treated with GSK-923295. Chromosomes were labeled by expression of histone H2B tagged with fluorescent proteins. Mitotic defects caused by CENP-E inhibition were aggravated in tetraploid cells compared to diploids (arrowheads for misaligned polar chromosomes and arrows for cohesion fatigue). (Figure **d** and **e** are adapted from "Tetraploidy-linked sensitization to CENP-E inhibition in human cells." (Yoshizawa et al. [10]). Published under a Creative Commons Attribution License (CC BY))

5. Apply 10 μL 10× inhibitor solution prepared in **step 4** to the corresponding wells (*see* **Note 18**). Apply 10 μL medium to the "medium-only" wells. See Fig. 3b for well design.
6. Centrifuge for 5 s at 1000 rpm. Incubate for 44 h at 37 °C with 5% $CO_2$.
7. Apply 5 μL CCK-8 to all wells except those filled with water (*see* **Note 18**).
8. Centrifuge for 5 s at 1000 rpm. Incubate for 4 h at 37 °C with 5% $CO_2$.
9. Measure the absorbances in 450 nm by a microplate spectrometer. Cell cultures should not reach confluency at the time of measurement (*see* Fig. 3c).
10. Calculate a normalized absorbance $Abs_{norm}$ in each sample as follows (*see* **Note 19**),

$$Abs_{norm} = \frac{Abs_{inh} - Abs_{medium}}{Abs_{cell} - Abs_{medium}},$$

where $Abs_{inh}$ is a sample absorbance reading from a well containing cells treated with the inhibitor at a concentration, $Abs_{cell}$ is an average non-treated cell absorbance reading in wells containing cells without the inhibitor, and $Abs_{medium}$ is an average background absorbance reading in wells containing medium only.

11. Draw a dose-response curve of cell proliferation (normalized absorbance readings calculated in **step 10**) using statistical or spreadsheet software. The R script with sample data has been uploaded to GitHub (https://github.com/koyayoshizawa/Curve_fitting_and_IC50_calculation_using_R).
12. Calculate the $IC_{50}$ of each inhibitor in each ploidy state by fitting a logistic regression curve to the dose-response cell proliferation plot obtained in **step 11** using an equation,

$$y = d + \frac{a - d}{1 + \left(\frac{x}{c}\right)^b},$$

where $y$ is the normalized absorbance, $x$ is inhibitor concentration, $a$ or $d$ is the absorbance at zero or infinite inhibitor concentration, and $b$ or $c$ is the slope factor or the inflection point, respectively.

13. Estimate $IC_{50}$ that is $x$ when 0.5 is substituted for $y$. An example of comparative $IC_{50}$ analysis of a CENP-E inhibitor, GSK-923295, is shown in Fig. 3d. A demonstration of tetraploidy-linked aggravation of chromosome alignment defects is shown in Fig. 3e.

## 4 Notes

1. Use collagen-coated dishes to cultivate haploid cells to support their substrate attachment.
2. The doubling time of HAP1 cells is 12–13 h [4]. We routinely passage HAP1 cells at a 1:12–1:16 split every 2 days.
3. Sufficient quality of Hoechst33342 staining is achieved by 15 min incubation in HAP1 cells. However, Hoechst staining quality improves by extending incubation time for tetraploid subclones of some other lines, such as HCT116 cells.
4. To avoid contamination of the diploidized population, we conduct all experiments using haploid HAP1 cells within 7 days after thawing the frozen stock of freshly sorted haploid cells. Freshly sorted haploids can be kept as frozen stocks without affecting the stability of the haploid state.
5. We usually conduct the size-based purification of haploid cells before an evident diploidized population appears in cell culture. However, using this protocol, one can purify haploid cells from a culture containing a substantial proportion (e.g., 60%) of diploidized cells. Freshly thawed cells should be passaged at least once before being subjected to size sorting to avoid excess cell damage.
6. When started from pure haploid culture, the proportion of the haploid population is reduced by half after 4 weeks in our experimental condition.
7. Unlike haploids, diploid and tetraploid cells are relatively stable in their ploidy state and can be kept without periodic purification.
8. Consider preparing multiple culture dishes in the case of failure to obtain a sufficient number of mitotic cells.
9. We usually remove pre-existing mitotic cells before treating nocodazole so that most cells freshly enter mitosis during 4-h nocodazole arrest.
10. We treat 40 ng/mL nocodazole to arrest other lines such as HCT116, DLD-1, RKO, and HepG2 cells.
11. This step should be conducted gently to avoid detaching mitotic cells from the dish.
12. We treat 10 μg/mL cytochalasin B for blocking cytokinesis in other lines such as HCT116, DLD-1, RKO, and HepG2 cells.
13. At this step, cells proceed to anaphase after nocodazole removal but fail to undergo cytokinesis in the presence of cytochalasin B, resulting in whole-genome duplication.

14. Cell volume is roughly proportional to the ploidy level in HAP1 cells. Therefore, cells are seeded at densities inverse to their ploidies to equalize the total cell mass in each sample.
15. To avoid cell aggregation, continuously pipette the cell suspension in a cell reservoir while dispensing it to 96-well plates.
16. To facilitate plate preparation, water and media can be aliquoted before starting trypsinization in **step 2**.
17. DMSO vehicle may affect cell growth when introduced at the final concentration exceeding 0.1%.
18. To avoid capillary action, we dispense the inhibitor or CCK-8 solution on the wall of each well (i.e., tips of microchip dispensers do not touch the surface of cell culture). The drops of the solution on the wall are then spun down into cell culture by a short centrifugation.
19. Normalize the absorbance readings of inhibitor-treated haploid, diploid, and tetraploid cells to those of non-treated cells with the corresponding ploidy states.

## Acknowledgments

Study in the Uehara laboratory is supported by JSPS KAKENHI (Grant Numbers JP19KK0181, JP19H05413, JP19H03219, JPJSBP120193801, and JP21K19244 to Ryota Uehara), the Princess Takamatsu Cancer Research Fund, the Orange Foundation, the Smoking Research Foundation, Daiichi Sankyo Foundation of Life Science, the Terumo Life Science Foundation, and the Akiyama Life Science Foundation. Koya Yoshizawa is supported by the Sasakawa Scientific Research Grant and the Hokkaido University-Hitachi Joint Cooperative Support Program for Education and Research. The authors declare no competing commercial interests. We thank Kan Yaguchi for providing unpublished data. The flow cytometer used in this work is managed by the Global Facility Center at Hokkaido University.

## References

1. Fox DT, Soltis DE, Soltis PS et al (2020) Polyploidy: a biological force from cells to ecosystems. Trends Cell Biol 30(9):688–694. https://doi.org/10.1016/j.tcb.2020.06.006
2. Viganó C, von Schubert C, Ahrné E et al (2018) Quantitative proteomic and phosphoproteomic comparison of human colon cancer DLD-1 cells differing in ploidy and chromosome stability. Mol Biol Cell 29(9): 1031–1047. https://doi.org/10.1091/mbc.E17-10-0577
3. Hara Y, Kimura A (2013) An allometric relationship between mitotic spindle width, spindle length, and ploidy in Caenorhabditis elegans embryos. Mol Biol Cell 24(9): 1411–1419. https://doi.org/10.1091/mbc.E12-07-0528
4. Yaguchi K, Yamamoto T, Matsui R et al (2018) Uncoordinated centrosome cycle underlies the instability of non-diploid somatic cells in mammals. J Cell Biol 217(7):2463–2483. https://doi.org/10.1083/jcb.201701151

5. Kuznetsova AY, Seget K, Moeller GK et al (2015) Chromosomal instability, tolerance of mitotic errors and multidrug resistance are promoted by tetraploidization in human cells. Cell Cycle 14(17):2810–2820. https://doi.org/10.1080/15384101.2015.1068482
6. Jemaà M, Manic G, Lledo G et al (2016) Whole-genome duplication increases tumor cell sensitivity to MPS1 inhibition. Oncotarget 7(1):885–901. https://doi.org/10.18632/oncotarget.6432
7. Bernhard SV, Gemble S, Basto R et al (2023) Experimental approaches to generate and isolate human tetraploid cells. Methods Mol Biol 2545:391–399. https://doi.org/10.1007/978-1-0716-2561-3_20
8. Graves JA, McMillan J (1984) Control of DNA synthesis in polyploid mammalian cells. J Cell Physiol 121(2):409–414. https://doi.org/10.1002/jcp.1041210219
9. Olbrich T, Vega-Sendino M, Murga M et al (2019) A chemical screen identifies compounds capable of selecting for haploidy in mammalian cells. Cell Rep 28(3):597–604.e594. https://doi.org/10.1016/j.celrep.2019.06.060
10. Yoshizawa K, Matsura A, Shimada M et al (2023) Tetraploidy-linked sensitization to CENP-E inhibition in human cells. Mol Oncol 17(6):1148–1166. https://doi.org/10.1002/1878-0261.13379
11. Carette JE, Raaben M, Wong AC et al (2011) Ebola virus entry requires the cholesterol transporter Niemann–Pick C1. Nature 477(7364):340–343. https://doi.org/10.1038/nature10348
12. Beigl TB, Kjosås I, Seljeseth E et al (2020) Efficient and crucial quality control of HAP1 cell ploidy status. Biology Open 9(11). https://doi.org/10.1242/bio.057174
13. Yaguchi K, Yamamoto T, Shimada M et al (2018) Ploidy-dependent change in cyclin D2 expression and sensitization to cdk4/6 inhibition in human somatic haploid cells. Biochem Biophys Res Commun 504(1):231–237. https://doi.org/10.1016/j.bbrc.2018.08.160

# Chapter 15

# A Method for Analyzing Acentrosomal Mitotic Spindles in Human Cells

**Shotaro Okuda, Takumi Chinen, and Daiju Kitagawa**

## Abstract

Centrosomes, the major microtubule organizing centers, facilitate mitotic spindle formation. However, recent studies have revealed that some cancer cells lack centrosomes. These findings suggest that certain types of cancer cells drive centrosome-independent mechanisms for the assembly of mitotic spindles. Therefore, analyzing the systems of acentrosomal spindle formation is beneficial for understanding the divergence of mitotic processes in cancer cells. Here, we focus on a method for generating acentrosomal cells using the PLK4 inhibitor centrinone and describe the proper conditions, spindle pole staining markers, and microscopy settings for observing acentrosomal spindle formation in human cells.

**Key words** Centrosome, Acentrosomal spindle, Spindle pole, NuMA, Centrinone

## 1 Introduction

The cell divides into two daughter cells with identical genetic information through the equal segregation of chromosomes. The proper chromosome segregation is ensured by the mitotic spindle, a microtubule-based cytoskeleton structure [1]. During mitosis, the centrosomes act as a major microtubule organizing center and facilitate the spindle formation in animal somatic cells. Therefore, centrosome abnormalities can induce chromosome mis-segregation, leading to various diseases such as cancer and microcephaly [2–5]. Centrosome amplification is a well-established hallmark of several types of cancers [6, 7]. Moreover, recent evidence has revealed that centrosome reduction can also occur in prostate and ovarian cancers [2, 8]. The loss of centrosomes has been shown to promote tumorigenesis in murine models [2]. Therefore, studying the cells with a reduced number of centrosomes could provide valuable insights into the mechanisms driving aberrant cell proliferation associated with malignancies.

Tomomi Kiyomitsu (ed.), *The Mitotic Spindle: Methods and Protocols*, Methods in Molecular Biology, vol. 2872,
https://doi.org/10.1007/978-1-0716-4224-5_15,

Cancer cells have compensatory mechanisms for spindle assembly in the absence of centrosomes. The existence of the compensatory mechanism was first indicated in 2000 by using laser ablation of centrosomes in animal somatic cells [9]. Subsequently, as the understanding of the centrosome duplication mechanism advanced [10–14], it became possible to observe cells lacking centrosomes through genetic manipulation of the genes responsible for the duplication process. However, these methods have some problems such as non-specificity or the potential to induce cellular damage. In such a situation, a recent study developed a more convenient approach to generate non-centrosomal cells using a small molecule inhibitor [15].

Centrinone is a potent inhibitor of polo-like kinase 4 (PLK4) [15], the master regulator of centriole duplication [13, 14]. The centriole is the core structure of the centrosome and undergoes self-replication only once per cell cycle, with equal distribution during cell division, thus maintaining a constant number throughout the cell cycle [16]. Therefore, disruption of centriole duplication results in a gradual reduction in the number of centrioles per cell with each subsequent division. Thus, the treatment of centrinone allows the observation of cells lacking centrioles after several cell cycles (Fig. 1) [15]. This small molecule-based centriole removal has revealed compensatory mechanisms in mitotic spindle formation in human cultured cells, including cancer cells [15, 17–21]. Therefore, observing the acentrosomal mitotic cells induced by centrinone treatment can be useful for understanding the regulation system of mitotic processes in cancer cells.

In this chapter, we describe the optimal conditions for observing spindle formation in cells without centrosomes when treated with centrinone, using RPE1 and HeLa cell lines as examples. We also provide detailed information on suitable markers for identifying spindle poles in these acentrosomal cells based on current knowledge.

## 2 Materials

### 2.1 Cell Lines

1. hTERT RPE-1 USP28 knockout cells (RPE-1 USP28 KO) [21] (*see* **Note 1**).
2. hTERT RPE-1 53BP1 knockout cells (RPE-1 53BP1 KO) [21] (*see* **Note 1**).
3. HeLa wild-type cells (ECACC, 93021013).
4. HeLa cells stably expressing EGFP–centrin1 [22].
5. HeLa cells expressing NuMA endogenously tagged with mCherry and stably expressing EGFP–centrin1 [20].

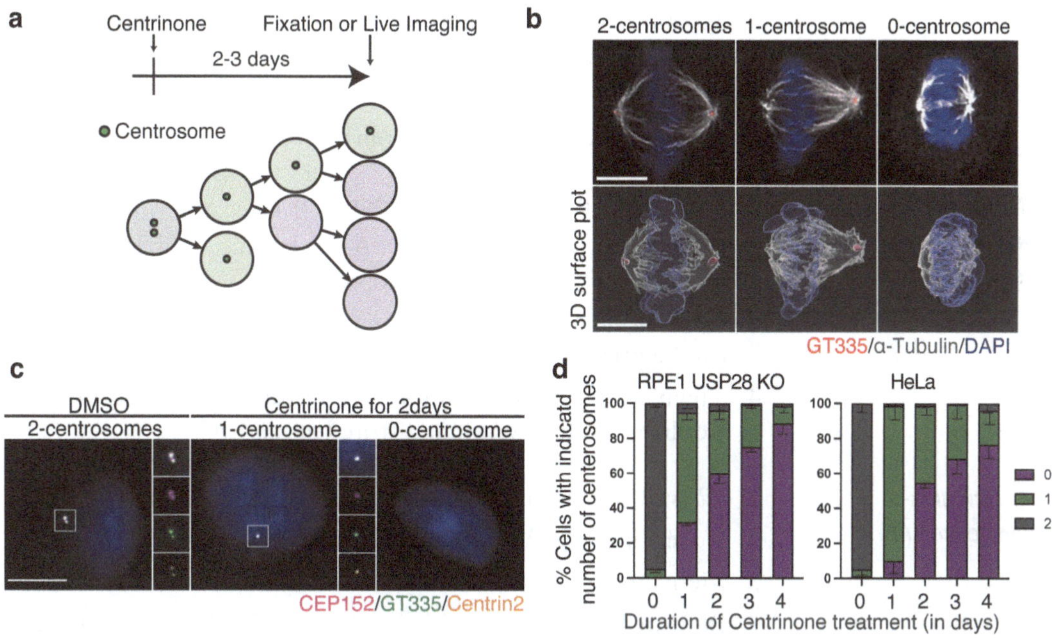


**Fig. 1** Centrinone treatment leads to a reduction in the number of centrosomes with each cell cycle. (**a**) A schematic illustration of centrinone-induced removal of centrosomes. (**b**) Microscopic images of mitotic HeLa cells with the indicated number of centrosomes (upper images, single z-section) and reconstructed images of those in 3D, surface-rendered with Imaris software (lower images). DMSO-treated control spindle (two centrosomes) and centrinone B–treated centrosome-depleted spindles (one or zero centrosomes). Red, gray, and blue represent GT335 (polyglutamylated centriole microtubules), α-tubulin, and DNA, respectively. Scale bar, 5 μm. Images are reproduced from Ref. [17] with permission from The Rockefeller University Press. (**c**) Microscopic images of interphase RPE-1 USP28 KO cells with the indicated number of centrosomes. DMSO-treated control cell (two centrosomes) and centrinone-treated centrosome-depleted cell (one or zero centrosomes). Magenta, green, yellow, and blue represent CEP152 (centrosome marker), GT335 (polyglutamylated centriole microtubules), Centrin2 (centriole marker), and DNA, respectively. For clarity, single-slice images are provided, but the numbers of centrosomes were counted by moving through the Z-stack and observing the entire cells. Scale bar, 10 μm. (**d**) Frequency of cells with the indicated number of centrosomes 1–4 days after the addition of centrinone or DMSO control (cultured with DMSO for 4 days) in (**c**). Triple co-localization foci of CEP152, GT335, and Centrin2 were counted as centrosomes. Values are mean percentages ± SD. n = 3 biological replicates, 100 cells each

6. HeLa cells expressing NuMA endogenously tagged with mClover [20].
7. HeLa cells expressing PCNT endogenously tagged with mCherry and stably expressing EGFP–centrin1 [17].
8. HeLa cells expressing CDK5RAP2 endogenously tagged with mCherry and stably expressing EGFP–centrin1 [17].

### 2.2 Cell Culture

1. Dulbecco's modified Eagle's medium (DMEM) with high glucose.
2. DMEM/F-12 medium.
3. Fetal bovine serum (FBS).

4. 10,000 U/mL penicillin, and 10,000 μg/mL streptomycin.
5. 0.5 g/L-Trypsin/0.53 mmol/L-EDTA Solution, with Phenol Red.
6. Cell Culture Dish (φ93*20 mm).

### 2.3 Chemicals for Analyzing Acentrosomal Cells

1. Dimethyl sulfoxide (DMSO) for control.
2. Centrinone (PLK4 inhibitor). A stock solution at 100 μM can be prepared in DMSO and stored at −30 °C.
3. Centrinone B (PLK4 inhibitor). A stock solution at 500 μM can be prepared in DMSO and stored at −30 °C.
4. Thymidine. A stock solution at 2 M can be prepared in filter-sterilized Milli-Q water and stored at −30 °C.

### 2.4 Immuno-fluorescence (IF)

1. 12-well cell culture plate.
2. Coverslips (e.g. Matsunami, No.1 or No.1S).
3. Methanol. Stored at −30 °C.
4. Blocking Buffer: PBS with 1% BSA, 0.05% Triton X-100, and 0.03% NaN3.
5. Primary antibodies, listed in Table 1.
6. Secondary antibodies, listed in Table 2.
7. Prolong$^{TM}$ Gold Antifade Mountant with DNA stain DAPI (Invitrogen, P36931).
8. Slide glass.
9. An Axioplan2 upright fluorescence microscope (Carl Zeiss) with 63× plan apochromatic lens objective.
10. A Leica TCS SP8 system equipped with a Leica HCX PL APO 63×/1.4 oil CS2 objective.
11. LSM980 laser-scanning microscope with AiryScan2 (Carl Zeiss) with ×63 plan apochromatic lens objective.

### 2.5 Live Cell Imaging

1. 35-mm glass-bottom dishes.
2. 24-well SENSOPLATE (Greiner-bio-one).
3. 2 M thymidine
4. SiR-tubulin (SPIROCHROME, CY-SC002). A stock solution at 100 μM can be prepared in DMSO and stored at −30 °C.
5. SiR-DNA (SPIROCHROME, CY-SC007). A stock solution at 100 μM can be prepared in DMSO and stored at −30 °C.
6. A confocal scanner box, the Cell Voyager CV1000 (Yokogawa Electric Corp) equipped with a 60× oil-immersion objective and a back-illuminated EMCCD camera.
7. CQ1 Benchtop High-Content Analysis System equipped with a 40× objective and a sCMOS camera (Yokogawa Electric Corp).

**Table 1**
**Primary antibodies list for immunofluorescence**

| Marker | Antigen | Manufacture | Host | Product ID | Dilution for IF |
|---|---|---|---|---|---|
| Centriole | Centrin2 | BioLegent | Rat | 698601 | 1/500 |
| | Centrin1 | Merck Millipore | Mouse | 04-1624 | 1/1000 |
| | Polyglutamilation modification | Adipogen | Mouse | AG-20B-0020-C100 | 1/500 |
| | CCP110 | Proteintech | Rabbit | 12780-1-AP | 1/500 |
| Centrosome | CEP152 | Bethyl Laboratories | Rabbit | A302-479A | 1/500 |
| Centrosome (and spindle pole depending on cell line) | PCNT | Abcam | Rabbit | ab4448 | 1/2000 |
| | PCNT | Santa Cruz Biotechnology | Goat | sc-28142 | 1/500 |
| | CDK5RAP2 | Bethyl Laboratories | Rabbit | IHC-00063 | 1/500 |
| | CEP192 | Bethyl Laboratories | Rabbit | A302-324A | 1/500 |
| | γ-Tubulin | Sigma-Aldrich | Rabbit | T5192 | 1/1000 |
| | γ-Tubulin | Sigma-Aldrich | Mouse | T5326 | 1/500 |
| | NEDD1 | Abnova | Mouse | H00121441-M05 | 1/500 |
| Spindle pole | NuMA | Abcam | Rabbit | ab109262 | 1/500 |
| | NuMA | Cell Signaling Technology | Rabbit | 3888S | 1/1000 |
| Spindle apparatus | α-Tubulin | Sigma-Aldrich | Mouse | T5168 | 1/500 |
| | α-Tubulin | MBL | Rabbit | PM054 | 1/500 |

**Table 2**
**Secondary antibodies list for immunofluorescence**

| Excitation/emission max | Antigen | Manufacture | Product ID |
|---|---|---|---|
| 493/518 nm | Mouse IgG(H+L) | ThermoFisher Scientific | A32766 |
| | Rabbit IgG(H+L) | ThermoFisher Scientific | A32790 |
| | Rat IgG(H+L) | ThermoFisher Scientific | A48269 |
| 558/572 nm | Mouse IgG(H+L) | ThermoFisher Scientific | A32773 |
| | Rabbit IgG(H+L) | ThermoFisher Scientific | A32794 |
| | Rat IgG(H+L) | ThermoFisher Scientific | A48270 |
| 658/675 nm | Mouse IgG(H+L) | ThermoFisher Scientific | A32787 |
| | Rabbit IgG(H+L) | ThermoFisher Scientific | A32795 |
| | Rat IgG(H+L) | ThermoFisher Scientific | A48272 |

### 2.6 Data Analysis

1. ImageJ software (version: 2.1.0/1.53c).
2. Imaris 10.1 (Oxford Instruments).

## 3 Methods

### 3.1 Routine Cell Culture

1. HeLa cell lines are cultured in DMEM with high glucose supplemented with 10% FBS, 100 U/mL penicillin, and 100 μg/mL streptomycin at 37 °C in a 5% $CO_2$ atmosphere.
2. RPE-1 cell lines are cultured in DMEM/F-12 supplemented with 10% FBS, 100 U/mL penicillin, and 100 μg/mL streptomycin at 37 °C in a 5% $CO_2$ atmosphere.

### 3.2 Immunofluorescence

#### 3.2.1 Preparation of Acentrosomal Cells for IF

1. HeLa cell lines or RPE-1 cell lines are passaged when they reached 70–90% confluency. The culture medium is discarded from the cell culture dish and washed with 10 mL of PBS; then cells are dissociated from the dish with 1.5–2 mL of Trypsin/EDTA. Cells are cultured at 37 °C with Trypsin/EDTA for 5–10 min. After dissociation, cells are suspended with 8 mL of medium and seeded on coverslips in 12 well culture dishes. Cells should be seeded onto coverslips at the dense, which will reach 80–90% confluency after 4 days, because to observe enough cells with acentrosomal spindles, cells should be cultured with centrinone for 3 days (Fig. 1d).
2. To eliminate centrosomes, 1 day after seeding cells, the culture medium is changed to the medium with 100 nM centrinone or 500 nM centrinone B (*see* **Note 2**).
3. Cells are incubated for 3 days at 37 °C with a 5% $CO_2$ atmosphere.

#### 3.2.2 Fixation, Immunofluorescence Staining, and Mounting of Samples

1. Medium is aspirated from culture wells.
2. Cold methanol is poured into culture wells, and cells seeded on coverslips are fixed for 7 min at −30 °C (*see* **Note 3**).
3. Methanol is aspirated, and cells are gently washed with room-temperature PBS three times (*see* **Note 4**).
4. After washing, cells are incubated in a blocking buffer for 30 min at room temperature.
5. Cells are then incubated with primary antibodies (primary antibodies list: Table 1) for 2 h at room temperature or overnight at 4 °C (*see* **Note 5** and Fig. 2a). Cells are gently washed three times with PBS (*see* **Note 4**) and incubated with secondary antibodies (secondary antibodies list: Table 2) for 30 min at room temperature. Primary antibody solutions are prepared by diluting in blocking buffer as a solvent to the concentrations shown in Table 1. Secondary antibody solutions are prepared

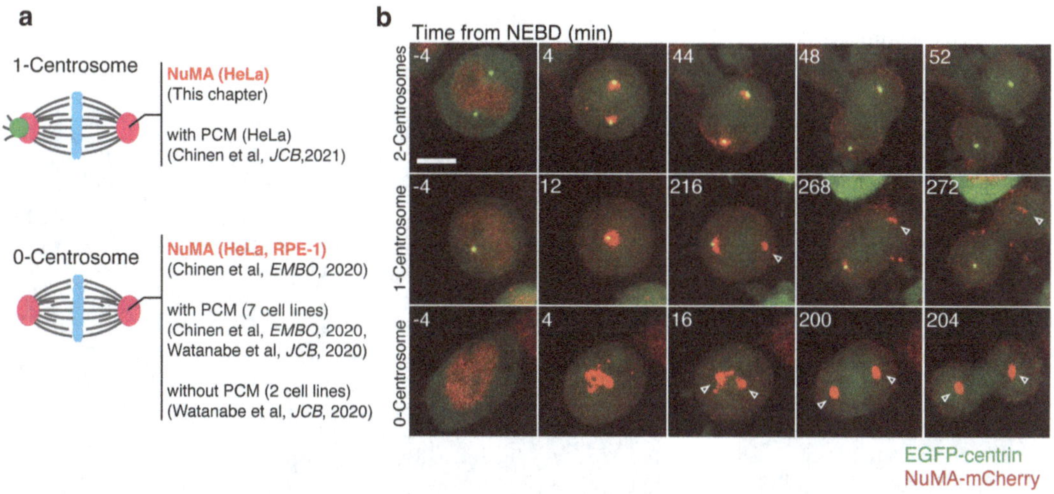


**Fig. 2** NuMA is a suitable marker to observe acentrosomal spindle poles in human cell lines. (**a**) Localization patterns of the spindle pole matrix NuMA and PCM components (CDK5RAP2, CEP192, and PCNT) at the acentrosomal spindle poles in HeLa and RPE-1 cells with 1- or 0-centrosomes. Previous studies have reported that PCM localization at the acentrosomal spindle pole in 0-centrosome cells is cell type-specific (without PCM: HeLa, A549, DU145, HCT116, PANC-1, DLD-1, U2OS [18, 20]; with PCM: RPE-1, HeLa [18]). On the other hand, NuMA localizes at the acentrosomal spindle pole in 0-centrosome cells induced in various cell lines (HeLa, RPE-1, A549, DU145, GI-1, HCT116, MCF7, PANC-1, SKOV3, and U2OS [20]. Therefore, we recommend using NuMA as a marker to observe the acentrosomal pole in human cell lines. (**b**) NuMA organizes acentrosomal spindle poles in 1- and 0-centrosome cells. The mitotic progression in acentrosomal cells. DMSO-treated two-centrosome, centrinone B-treated one- and zero-centrosome. Cell cycle synchronization using thymidine was performed as described in the methods section. HeLa cells expressing EGFP-centrin1 and mCherry-NuMA were observed with a 60× objective. Red and green represent mCherry-NuMA and EGFP-centrin1, respectively. Arrowheads indicate the acentrosomal spindle poles. Z-projections of 20 sections, every 1.2 μm. Scale bar, 10 μm. Time zero corresponds to the nuclear envelope breakdown (NEBD)

by diluting in blocking buffer as a solvent to 1:500. NuMA is a preferable marker of spindle pole regardless of the number of centrosomes (Fig. 2b).

6. Cells are washed three times with PBS (*see* **Note 4**) and mounted with Prolong™ Gold Antifade Mountant with DNA Stain DAPI onto glass slides. If 12 mm coverslips are used, 8 μL of Prolong Gold is suitable for each coverslip.

*3.2.3 Observation of IF Samples with Microscopy*

An Axioplan2 upright fluorescence microscope (Carl Zeiss) with 63× plan apochromatic lens objective is used. Since mitosis is a three-dimensional event, it is preferable that Z is relatively large; for example, Z interval is set to 250 nm and 41 sections are sequentially obtained.

Cells lacking centrosomes exhibit a disorganized microtubule structure during prometaphase, as illustrated in Fig. 3. To examine the detailed structure of microtubules, confocal microscopy would be more appropriate. We use a Leica TCS SP8 system or LSM980 laser-scanning microscope with AiryScan2.

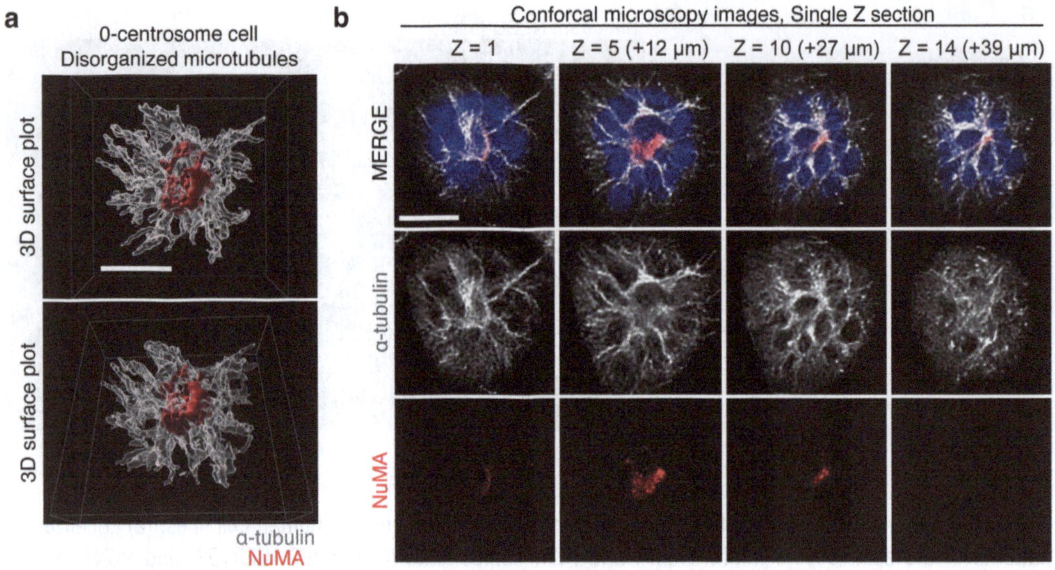


**Fig. 3** Confocal microscopy is suitable for observing disorganized microtubules in acentrosomal spindles. (**a**) Confocal microscopy images of NuMA and microtubules in centrinone-treated acentrosomal spindles of HeLa cells. Images were reconstructed in 3D and surface-rendered with Imaris software. Gray and red represent α-tubulin and NuMA, respectively. Scale bar: 5 μm. (**b**) Single Z section of (**a**). Scale bars: 5 μm. To examine the detailed structure of disorganized microtubules in 0-centrosome cells, confocal microscopy would be more appropriate because such spindle precursor-like microtubules are three-dimensionally complex

### 3.3 Live Cell Imaging

#### 3.3.1 Preparation of Acentrosomal Cells for Live Cell Imaging

1. Cells are seeded as described above into 35-mm glass-bottom dishes or 24-well SENSOPLATE.
2. To eliminate centrosomes, 1 day after seeding cells, the culture medium is changed to the medium with 100 nM centrinone or 500 nM centrinone B.
3. After 2–3 days of treatment with centrinone, cells are observed under microscopes.

Cell cycle synchronization by thymidine treatment is one option to enrich the population of cells entering mitosis. For this purpose, after 2 days of centrinone treatment, 2 M thymidine is added to the medium at a 1:1000 dilution to arrest cells in the S phase. 16 h after thymidine addition, cells are washed three times with a medium containing centrinone to remove thymidine. Following an 8 h incubation period, time-lapse analysis can be initiated.

To visualize DNA or microtubules, before imaging, the cells are incubated with SiR-DNA (20 nM for RPE-1 cell lines, 200 nM for HeLa cell lines) or SiR-tubulin (50 nM for RPE-1 cell lines, 100 nM for HeLa cell lines) for 2–3 h, respectively (*see* **Note 6**).

#### 3.3.2 Observation of Live Cell Imaging Samples with Microscopy

For long-term live-cell imaging, preventing phototoxicity is crucial. To track the cell cycle and cell division process, a suitable microscope system and settings for image acquisition are necessary to minimize phototoxicity while enabling high temporal and spatial resolution imaging. We use a spinning-disk confocal microscope such as CV1000 or CQ1. In addition, minimizing laser power, optimizing exposure time, and using a suitable binning size (we use 2 × 2 binning) are crucial for reducing phototoxicity.

During cell division, cell adhesion decreases and cells become rounded, gaining thickness in the Z-direction. Therefore, it is important to properly image in the Z-axis direction. In the images shown in Fig. 2b, the Z-interval was set to a relatively large value of 1.2 μm, and 20 consecutive optical sections were acquired for each field. By acquiring images in this way along the Z-direction, it becomes possible to observe the overall shape of the spindle apparatus and track the cell division process three-dimensionally.

## 4 Notes

1. When using hTERT RPE-1 wild-type cells, it is necessary to consider cell cycle arrest at the G1 phase due to prolonged mitosis upon centrosome removal. Since this cell cycle arrest is caused by the USP28–53BP1–p53–p21 signaling axis [21, 23], USP28 knockout or 53BP1 knockout cell lines are more suitable for the analysis of acentrosomal spindle assembly in RPE-1 cells. These cell lines are gifts from Dr. Karen Oegema, University of California, San Diego, San Diego, CA.
2. Paradoxically, low concentrations of centrinone are prone to induce centrosome amplification [24]. Therefore, it is necessary to carefully determine the appropriate centrinone concentration to achieve complete centrosome depletion.
3. Certain antibodies against centrosomal proteins, such as anti-centrin antibody, are incompatible with paraformaldehyde fixation. Thus, for immunostaining of centrosomes in cells, methanol fixation is recommended instead.
4. Mitotic cells can easily detach from culture dishes if washed too vigorously with PBS, potentially resulting in an insufficient number of mitotic cells available for observation.
5. Certain centrosomal proteins such as PCNT, CDK5RAP2, CEP192, and γ-tubulin are localized on acentrosomal spindle poles in certain cell types [17, 18]. Therefore, to judge the existence of centrosomes in mitosis, staining of appropriate markers that are not localized on acentrosomal poles such as CEP152, CP110, Centrin or polyglutamylation modification is suitable.

6. Note that a high dose of SiR-DNA or SiR-tubulin affects proper cell division processes [25, 26]. To prevent the effects, it is necessary to carefully determine the appropriate concentration of SiR-dyes.

## Acknowledgments

We thank all members of Kitagawa Laboratory for the discussion. This work was supported by the Japan Society for the Promotion of Science (19H05651, 23H02627), Japan Science and Technology Agency (JPMJCR22E1). The authors used GPT4 and Claude3 to improve language and readability.

### References

1. Prosser SL, Pelletier L (2017) Mitotic spindle assembly in animal cells: a fine balancing act. Nat Rev Mol Cell Biol 18:187–201
2. Wang M, Nagle RB, Knudsen BS et al (2020) Centrosome loss results in an unstable genome and malignant prostate tumors. Oncogene 39: 399–413
3. Marthiens V, Rujano MA, Pennetier C et al (2013) Centrosome amplification causes microcephaly. Nat Cell Biol 15:731–740
4. Levine MS, Bakker B, Boeckx B et al (2017) Centrosome amplification is sufficient to promote spontaneous tumorigenesis in mammals. Dev Cell 40(3):313–322
5. Martin CA, Ahmad I, Klingseisen A et al (2014) Mutations in PLK4, encoding a master regulator of centriole biogenesis, cause microcephaly, growth failure and retinopathy. Nat Genet 46:1283–1292
6. Sauer CM, Hall JA, Couturier D-L et al (2023) Molecular landscape and functional characterization of centrosome amplification in ovarian cancer. Nat Commun. https://doi.org/10.1038/s41467-023-41840-3
7. Zhang Y, Tian J, Qu C et al (2020) A look into the link between centrosome amplification and breast cancer. Biomed Pharmacother. https://doi.org/10.1016/j.biopha.2020.110924
8. Morretton J, Simon A, Herbette A et al (2022) A catalog of numerical centrosome defects in epithelial ovarian cancers. EMBO Mol Med. https://doi.org/10.15252/emmm.202215670
9. Khodjakov A, Cole RW, Oakley BR et al (2000) Centrosome-independent mitotic spindle formation in vertebrates. Curr Biol 10(2): 59–67
10. Leidel S, Delattre M, Cerutti L et al (2005) SAS-6 defines a protein family required for centrosome duplication in C. elegans and in human cells. Nat Cell Biol 7:115–125
11. Tang C-JC, Lin S-Y, Hsu W-B et al (2011) The human microcephaly protein STIL interacts with CPAP and is required for procentriole formation. EMBO J 30:4790–4804
12. Kitagawa D, Kohlmaier G, Keller D et al (2011) Spindle positioning in human cells relies on proper centriole formation and on the microcephaly proteins CPAP and STIL. J Cell Sci 124(22):3884–3893
13. Habedanck R, Stierhof Y-D, Wilkinson CJ et al (2005) The Polo kinase Plk4 functions in centriole duplication. Nat Cell Biol 7:1140–1146
14. Bettencourt-Dias M, Rodrigues-Martins A, Carpenter L et al (2005) SAK/PLK4 is required for centriole duplication and flagella development. Curr Biol 15(24):2199–2207
15. Wong YL, Anzola JV, Davis RL et al (2015) Reversible centriole depletion with an inhibitor of Polo-like kinase 4. Science 348(6239): 1155–1160
16. Nigg EA, Holland AJ (2018) Once and only once: mechanisms of centriole duplication and their deregulation in disease. Nat Rev Mol Cell Biol 19:297–312
17. Chinen T, Yamazaki K, Hashimoto K et al (2021) Centriole and PCM cooperatively recruit CEP192 to spindle poles to promote bipolar spindle assembly. J Cell Biol. https://doi.org/10.1083/jcb.202006085
18. Watanabe S, Meitinger F, Shiau AK et al (2020) Centriole-independent mitotic spindle assembly relies on the PCNT–CDK5RAP2

pericentriolar matrix. J Cell Biol. https://doi.org/10.1083/jcb.202006010

19. Meitinger F, Ohta M, Lee K-Y et al (2020) TRIM37 controls cancer-specific vulnerability to PLK4 inhibition. Nature 585:440–446
20. Chinen T, Yamamoto S, Takeda Y et al (2020) NuMA assemblies organize microtubule asters to establish spindle bipolarity in acentrosomal human cells. EMBO J. https://doi.org/10.15252/embj.2019102378
21. Meitinger F, Anzola JV, Kaulich M et al (2016) 53BP1 and USP28 mediate p53 activation and G1 arrest after centrosome loss or extended mitotic duration. J Cell Biol 214(2):155–166
22. Tsuchiya Y, Yoshiba S, Gupta A et al (2016) Cep295 is a conserved scaffold protein required for generation of a bona fide mother centriole. Nat Commun. https://doi.org/10.1038/ncomms12567
23. Lambrus BG, Daggubati V, Uetake Y et al (2016) A USP28–53BP1–p53–p21 signaling axis arrests growth after centrosome loss or prolonged mitosis. J Cell Biol 214(2):143–153
24. Tkach JM, Philip R, Sharma A et al (2022) Global cellular response to chemical perturbation of PLK4 activity and abnormal centrosome number. elife. https://doi.org/10.7554/eLife.73944
25. Sen O, Saurin AT, Higgins JMG (2018) The live cell DNA stain SiR-Hoechst induces DNA damage responses and impairs cell cycle progression. Sci Rep. https://doi.org/10.1038/s41598-018-26307-6
26. Lukinavičius G, Reymond L, D'Este E et al (2014) Fluorogenic probes for live-cell imaging of the cytoskeleton. Nat Methods 11:731–733

# Chapter 16

# Isolation of Mitotic Centrosomes from Cultured Human Cells

**Momoko Miyazawa, Shohei Yamamoto, Susumu Goyama, and Daiju Kitagawa**

## Abstract

The centrosome plays a crucial role in facilitating mitotic spindle assembly through its microtubule organizing capacities. Analyzing the composition, structure, and functions of mitotic centrosomes is essential for understanding the mechanisms underlying cell division and centrosome-associated diseases. Isolating centrosomes is an effective method to gain comprehensive information about them while minimizing interference from other cellular components. In this chapter, we describe a protocol for isolating mitotic centrosomes from cultured human cells. This protocol includes cell synchronization and centrosome isolation through ultracentrifugation with a sucrose gradient. We also describe a method for conducting a microtubule nucleation assay to assess the functionality of isolated centrosomes.

**Key words** Centrosome, Centriole, PCM, Microtubule, MTOC, Mitosis, Sucrose gradient

## 1 Introduction

The centrosome is an organelle that contributes to the formation of cilia, flagella, and mitotic spindles in most animal cells. It is composed of centrioles and pericentriolar material (PCM) [1, 2]. During mitosis, the PCM expands through the accumulation of PCM proteins and sequesters γ-tubulin, which nucleates microtubules [1]. Because of the robust activity of microtubule nucleation from centrosomes, they act as a driver of efficient formation of the mitotic spindle [3, 4]. Indeed, removal of centrosomes leads to delayed spindle formation and frequent chromosomal instability in vertebrate somatic cells [5]. Centrosomes also play important roles in asymmetric cell divisions [6–8]. In *Drosophila* neuronal asymmetric cell division, for example, the presence of centrosomes ensures accurate spindle orientation and positioning [7–9]. Associated with these roles of centrosomes, abnormalities of mitotic centrosomes have been implicated in various diseases, including cancer and microcephaly [6, 10–13]. Thus, investigating the

Tomomi Kiyomitsu (ed.), *The Mitotic Spindle: Methods and Protocols*, Methods in Molecular Biology, vol. 2872, https://doi.org/10.1007/978-1-0716-4224-5_16,

compositional and structural features of mitotic centrosomes in various types of cells is crucial for understanding the mechanisms underlying cell division processes and centrosome-related diseases.

Analysis of centrosomes isolated from cells could provide pure compositional, structural, and functional information about them while minimizing interference from other cellular components [14–23]. While approaches like immunostaining of cells provide insights into specific molecules, proteomic and structural analysis of isolated centrosomes offer more comprehensive information [15, 17, 18]. Furthermore, isolating centrosomes enables various *in vitro* analyses for directly assessing centrosome properties, such as microtubule nucleation activity and interactions with other cellular components [20, 22, 24, 25].

Isolation of centrosomes or equivalent structures has been reported using various cell types, including yeast [26, 27] *Trichonympha* [16], *Chlamydomonas* [28, 29], *Drosophila* embryos [30], and Chinese hamster ovary cells [23, 31, 32]. In the case of human cells, centrosomes have mainly been isolated from specific cell lines such as KE37 cells, a lymphoblastic cell line [14, 17, 33], and methods applicable to various cell lines have been limited. Furthermore, detailed protocols for the isolation and analysis of mitotic centrosomes from human cells have also been limited [15, 34]. Since the isolation process could affect centrosome composition and structure, careful experimental procedures are required to maintain the functionality of centrosomes such as microtubule nucleation activity [35]. In this chapter, we describe a protocol for isolating mitotic centrosomes from cultured human cells (Fig. 1). We utilized HeLa and RPE-1 cells, widely used cell lines for cell biological studies. This method primarily involves cell synchronization, cell lysis, and fractionation through ultracentrifugation with a sucrose gradient. The isolation of centrosomes was confirmed by Western blot, and their functionality was assessed using a microtubule nucleation assay.

## 2 Materials

### 2.1 Cell Culture and Cell Cycle Synchronization

1. DMEM High glucose medium supplemented with 10% FBS and 1% penicillin/streptomycin for HeLa cells.
2. DMEM/F-12 medium supplemented with 10% FBS and 1% penicillin/streptomycin for RPE-1 cells.
3. Cell culture dishes, 100 mm × 20 mm.
4. Phosphate-buffered saline (PBS).
5. Nocodazole dissolved in DMSO.
6. Cryopreservation Solution (e.g. Cell Reservoir One, Nacalai, 07579-24).

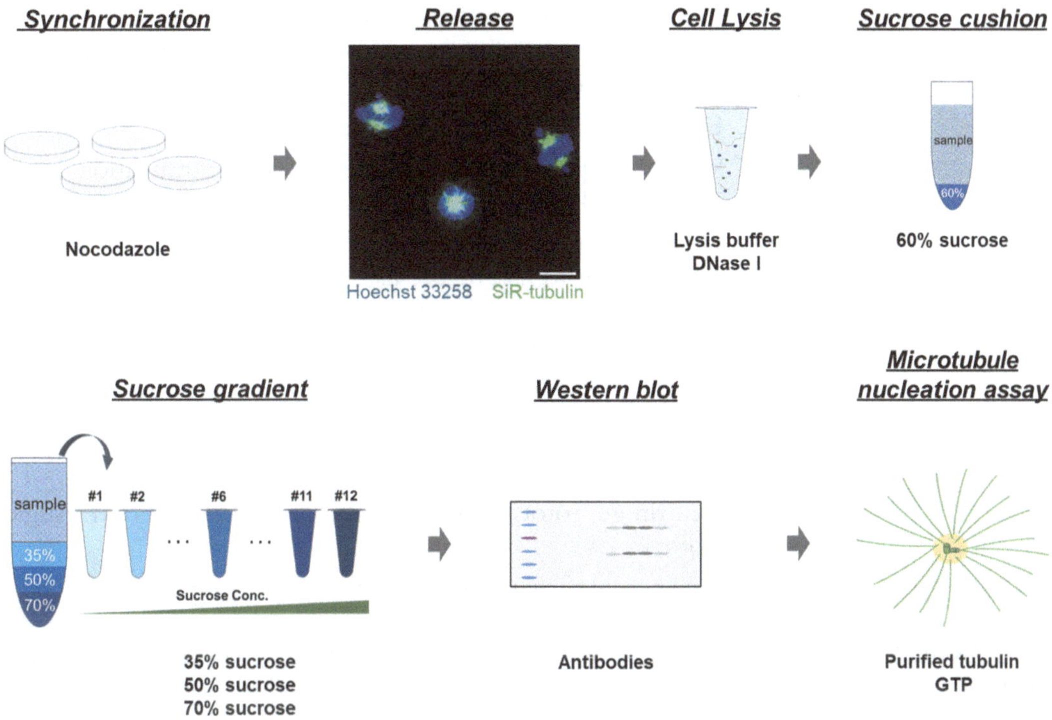


**Fig. 1** An overview of the experimental procedure for mitotic centrosome isolation. Cells are synchronized at the mitotic phase with nocodazole treatment and subsequently collected by shake-off. After being released from nocodazole and subsequently treated with Cytochalasin D, cells are lysed and then treated with DNase I. A microscopic image represents HeLa cells before cell lysis (*Green*: SiR-tubulin; *Blue*: Hoechst 33258. *Scale bar*: 10 μm). The lysates are then sedimented on a sucrose cushion and fractionated through ultracentrifugation with a sucrose gradient. Each fraction is manually retrieved from the top. The fractionated samples are analyzed using a Western blot to assess the centrosome isolation. The centrosome-enriched fraction is then utilized in a microtubule nucleation assay to confirm the functionality of the isolated centrosomes. The reagents shown below each illustration are the main materials required for the respective experimental steps

7. $CO_2$ incubator (37 °C, 5% $CO_2$).
8. Tabletop centrifuge (e.g., Eppendorf 5702).

### 2.2 Centrosome Isolation

1. Cytochalasin D dissolved in DMSO.
2. SiR-tubulin.
3. Hoechst 33258.
4. 2-Mercaptoethanol (2-ME).
5. Protease inhibitor cocktail (e.g., Nacalai, 25955-11).
6. Washing buffer: 2 mM Tris-HCl (pH 8.0), 0.2 mM $CaCl_2$, 50 mM KCl, 2 mM $MgCl_2$. Before use, add 0.1% 2-ME, Protease inhibitor cocktail (at 1:1000 dilution), and 2 μM Cytochalasin D.

7. Lysis buffer: 1 mM HEPES, pH 7.5 with KOH, 0.5% NP-40, 0.5 mM $MgCl_2$. Before use, add 0.1% 2-ME, Protease inhibitor cocktail (at 1:1000 dilution).
8. DNase I (e.g. Takara, 2270A).
9. Sucrose, applicable for density-gradient centrifugation (e.g., Nacalai, 30406-25).
10. 70%, 60%, 50%, 35% (w/w) sucrose solution in 10 mM PIPES, pH 7.2 with KOH. Before use, add 0.1% 2-ME and Protease inhibitor cocktail (at 1:1000 dilution) (*see* **Note 1**).
11. 1.5 mL microtube.
12. Tabletop micro refrigerated centrifuge (e.g., Kubota 3520).
13. Centrifuge tube (e.g., Beckman 1 mL Open-top Thickwall Polycarbonate Tube).
14. Centrifuge rotor (e.g. Beckman TLA-120.2 Fixed Angle Rotor).
15. Tabletop ultracentrifuge (e.g., Beckman Optima TLX).

### 2.3 Western Blot Analysis

Standard materials and equipment for SDS-PAGE and Western blot as well as the reagents and antibodies described below.

1. 6× SDS sample loading buffer (e.g., Nacalai, 09499-14).
2. Prestained Protein-Ladder (e.g., Nippon genetics, NE-MWP04).
3. Anti-γ-tubulin (Sigma, T5192, 1:1500).
4. Anti-ODF-2 (abcam, ab43840, 1:1000).
5. Anti-IAK1 (Aurora-A) (BD, 610938, 1:1000).
6. Anti-GM130 (CST, 12480, 1:1000).
7. Anti-β-actin (Santa Cruz Biotechnology, sc-47778, 1:500).

### 2.4 Microtubule Nucleation Assay

1. BRB80 (10×): 800 mM PIPES, 10 mM $MgCl_2$, 10 mM EGTA, pH 6.8 with KOH.
2. 0.2 M GTP in water, pH 7.0 with NaOH.
3. 1 M DTT in water, pH 7.5 with NaOH.
4. Tubulin protein, purified from porcine brain (*see* **Note 2**).
5. Tubulin protein labeled with ATTO565 (*see* **Note 3**).
6. mPEG5K-Silane solution: 1 mg/mL of mPEG5K-Silane (e.g., Sigma-Aldrich, JKA3037-1G) in 99% ethanol and 0.1% HCl.
7. Hellmanex III (Hellma).
8. Ultrasonic cleaner.
9. Air duster (e.g., Sanwa supply, 200-CD067).
10. Double-sided tape, 80 μm thickness (e.g., Nitto, No. 5608).

11. UV ozone cleaner.
12. Slide glass (e.g., Matsunami, S0318).
13. Cover glass (e.g., Matsunami, C218181).
14. Mineral oil.
15. Inverted confocal microscope (e.g., Leica TCS SP8 equipped with a 63× oil-immersion objective (NA 1.4) and an incubation system (INUBTF-WSKM, Tokai Hit)).

## 3 Methods

### 3.1 Cell Culture and Cell Cycle Synchronization

To synchronize the cell cycle at mitosis, cells are treated with nocodazole, a microtubule-depolymerizing agent. Subsequently, nocodazole is washed out from the cells to minimize contamination of nocodazole during the subsequent centrosome isolation and assays. Below, we describe the detailed procedure.

1. Seed cells in 10 cm dishes at a density of $1 \times 10^6$ cells/dish in culture media containing 10% FBS and 1% penicillin/streptomycin. Incubate at 37 °C in a 5% $CO_2$ atmosphere for 24 h (*see* **Note 4**).
2. Treat cells with nocodazole (150 nM and 300 nM for HeLa and RPE-1 cells, respectively) for 15 h.
3. Shake the culture dishes to detach mitotic cells from the bottom of the dishes.
4. Collect the culture media and centrifuge at 200× *g* for 2 min.
5. Resuspend the cells with PBS (1.25 mL for cells from one dish) and centrifuge at 200× *g* for 2 min. Repeat this step once more.
6. Incubate the cells at 37 °C for 30 min in fresh culture media (1.25 mL for cells from one dish) to release cells from nocodazole-induced arrest (*see* **Notes 5** and **6**).

### 3.2 Centrosome Isolation

After synchronization, mitotic cells are lysed using a lysis buffer containing a detergent. Centrosomes are then sedimented on a sucrose cushion and fractionated through sucrose gradient centrifugation. In this section, we detail the procedure for cell lysis, sedimentation, and fractionation.

#### 3.2.1 Preparation of Cell Lysates

1. Add 2 μM of Cytochalasin D to the medium and incubate the cells at 37 °C for 30 min.
2. Collect cells by centrifugation at 200× *g* for 2 min (*see* **Note 7**).
3. Resuspend the cells with 4 mL of PBS supplemented with 2 μM of Cytochalasin D, and centrifuge at 200× *g* for 2 min. Repeat this step once more.

4. Resuspend the cells with 1 mL of the washing buffer and transfer the cell suspension into a 1.5 mL tube. From this step, the sample should be kept on ice or at 4 °C (*see* **Note 8**).
5. Centrifuge the cells at 500× *g* for 2 min at 4 °C and remove the supernatant.
6. Add 500 μL of the Lysis buffer to the cell pellet and mix by pipetting up and down 10 times. Then, rotate the tube at 4 °C for 45 min (*see* **Note 9**).
7. Precipitate debris by centrifugation at 5000× *g* for 2 min at 4 °C and collect the supernatant. An aliquot of the supernatant is collected as the whole cell lysate for Western blot analysis.
8. Add 1 unit of DNase I (2 units/mL final concentration) and rotate the tube at 4 °C for 30 min.

#### 3.2.2 Sedimentation on a Sucrose Cushion

1. Put 80 μL of 60% sucrose solution at the bottom of the centrifuge tube.
2. Gently place the lysates (approx. 480 μL) onto the sucrose cushion.
3. Centrifuge at 30,000 rpm (equal to approx. 25,000× *g*) at 4 °C for 30 min using an ultracentrifuge with a fixed rotor (*see* **Note 10**).
4. Discard the supernatant (approx. 320 μL) until the one-third volume of the sample solution remains on the cushion.
5. Mix the remaining sample and the sucrose cushion in the tube.

#### 3.2.3 Density Gradient Ultracentrifugation

1. Prepare the sucrose density gradient by layering 200 μL of each solution into the new centrifuge tube slowly in order 70%, 50%, and 35%.
2. Gently place the mixture (obtained at Subheading 3.2.2, **step 5**) onto the sucrose gradient.
3. Centrifuge at 69,000 rpm (approx. 130,000× *g*) at 4 °C for 90 min using an ultracentrifuge with a fixed rotor (*see* **Note 10**).
4. Carefully fractionate and collect samples from the top 75 μL/fraction using a micropipette (*see* **Note 11**).
5. Freeze the fractions with liquid nitrogen and store them at −80 °C.

### 3.3 Western Blot Analysis

Western blot analysis is conducted to determine which fractions contain centrosomes using antibodies against centrosome proteins (*see* **Note 12**). Fractionated samples (4.2 μL per fraction) are subjected to SDS-PAGE with 12% polyacrylamide gel, followed by Western blot (Fig. 2).

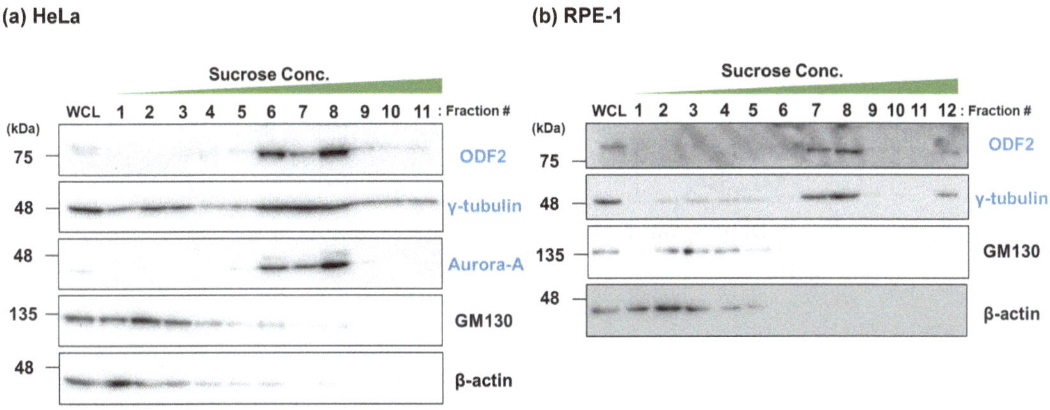


**Fig. 2** Evaluation of centrosome isolation by Western blot analysis. Fractionated samples and the whole cell lysate (WCL) were subjected to Western blot analysis. Antibodies against centrosome proteins (ODF2, γ-tubulin, and Aurora-A) and non-centrosome proteins (Golgi matrix protein GM130, and β-actin) were used. Blue letters indicate centrosome proteins. The isolation profiles from HeLa cells (**a**) and RPE-1 cells (**b**) were shown, respectively. Fraction numbers were assigned from top to bottom of the sucrose gradient sample. These profiles indicate that centrosomes were concentrated in fractions #6–8 for HeLa cells and fractions #7–8 for RPE-1 cells, suggesting their separation from fractions enriched with non-centrosome proteins

### *3.4 Microtubule Nucleation Assay*

Microtubule nucleation assay is conducted to assess the functionality of the isolated centrosomes. In this section, we describe how to observe microtubule formation from isolated centrosomes.

#### *3.4.1 Preparation of the Flow Chamber*

To minimize non-specific protein binding to the slides and cover glasses, we wash the glasses and coat the surfaces with Silane-PEG.

1. Immerse the glasses in 2% Hellamex III and sonicate for 10 min using an ultrasonic cleaner.
2. Rinse the glasses twice with 1 L of water.
3. Dry the glasses with airflow using an air duster.
4. Treat the glasses with UV-ozone using a UV-ozone cleaner for 2 min.
5. Incubate the glasses in 1 mg/mL of mPEG5K-Silane solution overnight at room temperature.
6. Rinse the glasses with 99% EtOH and then with 1 L of water, twice.
7. Dry the glasses with airflow using an air duster. The glasses are stored at 4 °C.
8. Assemble the flow chamber with the Silane-PEG coated slide and cover glass using two double-sided tapes as shown in Fig. 3 (*see* **Note 13**).

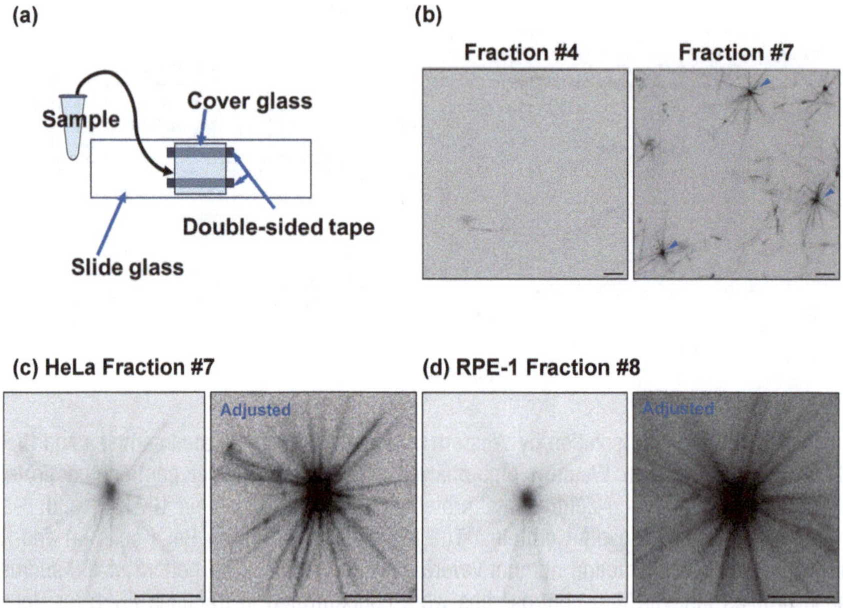


**Fig. 3** Assessing microtubule nucleation activity of the isolated centrosomes. (**a**) A scheme for preparing the flow chamber. Firstly, the flow chamber was assembled using double-sided tapes to join a Silane-PEG-coated slide and a cover glass. An aliquot of the centrosome fraction was mixed with purified tubulin, GTP, DTT, and a buffer solution. The reaction mixture was then loaded into the flow chamber and incubated at 37 °C. Microtubules were visualized by adding ATTO-565 labeled tubulin. (**b**) Microtubule aster formation in the reaction mixture containing centrosome-enriched fraction. Fraction #4 (centrosome-less) and #7 (centrosome-enriched) from HeLa cells were subjected to the assay (37 °C, 20 min). The centrosome-enriched fraction #7 exhibited the formation of multiple microtubule asters, whereas few microtubule asters were observed in fraction #4. Blue arrowheads indicate microtubule asters. (**c**, **d**) Magnified images of the microtubule asters observed in the reaction mixture containing centrosome-enriched fraction #7 from HeLa cells and fraction #8 from RPE-1 cells, respectively. *Scale bar*: 5 μm. The fluorescence signals were visualized as black. Right images in (**c**) and (**d**): The intensities of the left images in (**c**) and (**d**) were adjusted to clearly visualize the microtubules of the asters. Fraction numbers are consistent with those in Fig. 2

*3.4.2 Observation of Centrosomes Nucleating Microtubules*

1. Thaw aliquots of tubulin, GTP, DTT, and the centrosome--containing fractions on ice (*see* **Note 14**).
2. Prepare 10 μL of the following reaction mixture on ice: 1 mM GTP, 10 mM DTT, 9 μM unlabeled tubulin, 2.2 μM labeled tubulin, 4.6 μL of centrosome fraction, and 1 μL 10× BRB80.
3. Mix gently and load the reaction mixture into the flow chamber.
4. Seal the flow chamber with a drop of mineral oil to prevent the reaction mixture from drying.
5. Incubate the chamber at 37 °C (*see* **Note 15**).
6. Observe the samples using a confocal microscope with an incubation system (Fig. 3).

## 4 Notes

1. The sucrose solutions need to be prepared in advance as they take time to dissolve. It should be heated but not boiled to dissolve sucrose and prepare a 70% sucrose solution. The concentrations of the sucrose are indicated as w/w (weight/weight). To obtain 60%, 50%, and 35% sucrose solutions, the 70% solution is diluted with 10 mM PIPES and supplemented with 0.1% 2-ME and a protease inhibitor cocktail (at 1:1000 dilution). These solutions can be preserved at 4 °C. Other than the sucrose solutions, the concentrations are indicated as w/v (weight/volume).
2. Tubulin proteins were purified from porcine brains as previously described [36]. Briefly, two cycles of polymerization-depolymerization in high-molarity PIPES buffer were performed. The purified tubulin proteins were finally suspended in 1× BRB80 and frozen with liquid nitrogen for storage.
3. Tubulin proteins were fluorescently labeled with ATTO565 NHS-ester (ATTO-TEC, AD565-31) as previously described [37].
4. Approximately $1 \times 10^7$ cells can be obtained when cells are seeded at the described density in eight dishes. We usually use all of them for the subsequent cell lysis.
5. At this step, cells can be cryopreserved. At the very least, we were able to isolate the centrosomes from cells cryopreserved for two weeks. We usually collect cells by centrifugation at 200× *g* for 2 min. The cell pellet is then resuspended in a cryopreservation solution (e.g., Cell Reservoir One) and stored at −80 °C. To use the stocked cells for centrosome isolation, thaw the stock at 37 °C and wash the cells with PBS.
6. The methods for cell cycle synchronization should be adjusted to each cell line. Depending on the cell lines, alternative approaches such as the Thymidine-Nocodazole block may be more effective for efficiently collecting mitotic cells [34, 38].
7. It should be confirmed whether the collected cells are in the mitotic phase. We usually take aliquots of the cell suspension and stain the cells with SiR-tubulin and Hoechst 33258. The presence of mitotic spindles and chromosomes in these cells can be recognized, as shown in Fig. 1.
8. From this step, all centrifuge machines should be cooled at 4 °C in advance to avoid the inactivation of proteins by degradation and heat denaturation.
9. To successfully isolate centrosomes through this protocol, it can be crucial to carefully determine the lysis conditions,

including the volume of the lysis buffer and the duration of the lysis step.

10. Conversions between revolutions per minute (rpm) and gravitational force (*g*) were calculated using the minimum radius value of the Beckman TLA-120.2 Fixed Angle Rotor.
11. Samples should be collected slowly to avoid mixing the fractions. Allocate samples for subsequent Western blot analysis and the microtubule nucleation assay.
12. Proteins localized at the centrosomes during mitosis should be selected as the centrosome marker because the composition of the centrosome changes throughout the cell cycle.
13. For a reaction volume of 10 μL, the suitable spacing between two double-sided tapes is approx. 6 mm.
14. The tubulin proteins and the centrosome fractions should not be reused since the freezing-thawing cycle would affect their stability.
15. We could clearly observe microtubule asters within 20 min after loading the reaction mixture into the flow chamber. However, depending on the tubulin concentrations and incubation time, the spontaneous formation of microtubules could interfere with the observation of asters originating from centrosomes. Therefore, it is necessary to optimize the conditions such as tubulin concentration and incubation time in order to accurately analyze the microtubule structures and dynamics of the asters.

## Acknowledgments

We thank all members of the Kitagawa laboratory for their advice and discussions. This work was supported by JSPS KAKENHI grants from the Ministry of Education, Science, Sports and Culture of Japan (19H05651, 22K20624, 23K14176, 24H02284), the CREST program (JPMJCR22E1) of the Japan Science and Technology Agency, Takeda Science Foundation, Princess Takamatsu Cancer Research Fund, The Uehara Memorial Foundation, The Inamori Foundation, Astellas Foundation for research on metabolic disorders, The Kishimoto Fund Research Grant from the Senri Life Science Foundation and The Noguchi Shitagau Research Grant from the Noguchi Institute.

### References

1. Conduit PT, Wainman A, Raff JW (2015) Centrosome function and assembly in animal cells. Nat Rev Mol Cell Biol 16:611–624
2. Bettencourt-Dias M, Glover DM (2007) Centrosome biogenesis and function: centrosomics

brings new understanding. Nat Rev Mol Cell Biol 8:451–463

3. Valdez VA, Neahring L, Petry S et al (2023) Mechanisms underlying spindle assembly and robustness. Nat Rev Mol Cell Biol 24:523–542
4. Prosser SL, Pelletier L (2017) Mitotic spindle assembly in animal cells: a fine balancing act. Nat Rev Mol Cell Biol 18:187–201
5. Sir JH, Pütz M, Daly O et al (2013) Loss of centrioles causes chromosomal instability in vertebrate somatic cells. J Cell Biol 203:747–756
6. Nigg EA, Raff JW (2009) Centrioles, centrosomes, and cilia in health and disease. Cell 139: 663–678
7. Yamashita YM (2009) The centrosome and asymmetric cell division. Prion 3:84–88
8. Gómez-López S, Lerner RG, Petritsch C (2014) Asymmetric cell division of stem and progenitor cells during homeostasis and cancer. Cell Mol Life Sci 71:575–597
9. Cheng J, Türkel N, Hemati N et al (2008) Centrosome misorientation reduces stem cell division during ageing. Nature 456:599–604
10. Holland AJ, Cleveland DW (2012) Losing balance: the origin and impact of aneuploidy in cancer. EMBO Rep 13:501–514
11. Gönczy P (2015) Centrosomes and cancer: revisiting a long-standing relationship. Nat Rev Cancer 15:639–652
12. Godinho SA, Picone R, Burute M et al (2014) Oncogene-like induction of cellular invasion from centrosome amplification. Nature 510: 167–171
13. Jaiswal S, Singh P (2021) Centrosome dysfunction in human diseases. Semin Cell Dev Biol 110:113–122
14. Guichard P, Chrétien D, Marco S et al (2010) Procentriole assembly revealed by cryo-electron tomography. EMBO J 29:1565–1572
15. Carden S, Vitiello E, Rosa e Silva I et al (2023) Proteomic profiling of centrosomes across multiple mammalian cell and tissue types by an affinity capture method. Dev Cell 58: 2393–2410.e9
16. Guichard P, Hamel V, Neves A et al (2015) Isolation, cryotomography, and three-dimensional reconstruction of centrioles. Methods Cell Biol 129:191–209
17. Jakobsen L, Schrøder JM, Larsen KM et al (2013) Centrosome isolation and analysis by mass spectrometry-based proteomics. Methods Enzymol 525:371–393
18. Bauer M, Cubizolles F, Schmidt A et al (2016) Quantitative analysis of human centrosome architecture by targeted proteomics and fluorescence imaging. EMBO J 35:2152–2166
19. Firat-Karalar EN, Sante J, Elliott S et al (2014) Proteomic analysis of mammalian sperm cells identifies new components of the centrosome. J Cell Sci 127:4128–4133
20. Kinoshita K, Arnal I, Desai A et al (2001) Reconstitution of physiological microtubule dynamics using purified components. Science 294:1340–1343
21. Klena N, Le Guennec M, Tassin A et al (2020) Architecture of the centriole cartwheel-containing region revealed by cryo-electron tomography. EMBO J 39:e106246
22. Farina F, Gaillard J, Guérin C et al (2016) The centrosome is an actin-organizing centre. Nat Cell Biol 18:65–75
23. Mitchison T, Kirschner M (1984) Microtubule assembly nucleated by isolated centrosomes. Nature 312:232–237
24. Laan L, Pavin N, Husson J et al (2012) Cortical dynein controls microtubule dynamics to generate pulling forces that position microtubule asters. Cell 148:502–514
25. Gavriljuk K, Scocozza B, Ghasemalizadeh F et al (2021) A self-organized synthetic morphogenic liposome responds with shape changes to local light cues. Nat Commun 12: 1548
26. Wigge PA, Jensen ON, Holmes S et al (1998) Analysis of the *Saccharomyces* spindle pole by matrix-assisted laser desorption/ionization (MALDI) mass spectrometry. J Cell Biol 141: 967–977
27. Rout MP, Kilmartin JV (1990) Components of the yeast spindle and spindle pole body. J Cell Biol 111:1913–1927
28. Li S, Fernandez J-J, Marshall WF et al (2019) Electron cryo-tomography provides insight into procentriole architecture and assembly mechanism. elife. https://doi.org/10.7554/eLife.43434
29. Keller LC, Romijn EP, Zamora I et al (2005) Proteomic analysis of isolated chlamydomonas centrioles reveals orthologs of ciliary-disease genes. Curr Biol 15:1090–1098
30. Müller H, Schmidt D, Steinbrink S et al (2010) Proteomic and functional analysis of the mitotic Drosophila centrosome. EMBO J 29: 3344–3357
31. Mitchison TJ, Kirschner MW (1986) Isolation of mammalian centrosomes. Methods Enzymol 134:261–268
32. Reber S (2011) Isolation of centrosomes from cultured cells. Methods Mol Biol 777:107–116

33. Gogendeau D, Guichard P, Tassin A-M (2015) Purification of centrosomes from mammalian cell lines. Methods Cell Biol 129:171–189
34. Xie G, Zhou Y, Du M et al (2021) Protocol to identify centrosome-associated transcription factors during mitosis in mammalian cell lines. STAR Protoc 2(3):100495
35. Gräf R, Euteneuer U, Ueda M et al (1998) Isolation of nucleation-competent centrosomes from Dictyostelium discoideum. Eur J Cell Biol 76:167–175
36. Castoldi M, Popov AV (2003) Purification of brain tubulin through two cycles of polymerization–depolymerization in a high-molarity buffer. Protein Expr Purif 32:83–88
37. Hyman A, Drechsel D, Kellogg D et al (1991) Preparation of modified tubulins. Methods Enzymol 196:478–485
38. Habu T, Kim J (2023) Protocol for combining immunological procedures with cell cycle stage synchronization to detect cell-cycle-specific complexes. STAR Protoc 4(1):102063

# Part VI

## Spindle Positioning in Synthesized or Complicated Contexts

# Chapter 17

# Synthetic Engineering of Cortical Polarity During Mitosis Using Designed Proteins

**Lara K. Krüger, Joseph L. Watson, and Emmanuel Derivery**

## Abstract

During asymmetric cell division, cortical polarity cues drive the polarization of the microtubule cytoskeleton to ensure the generation of two daughter cells with different fates. While this a critical process for development and tissue homeostasis, the underlying molecular mechanisms orchestrating those processes are not completely understood, especially in mammals. Here, we present an assay that allows the study of the molecular mechanisms driving mammalian asymmetric cell division in a high-throughput manner by capitalizing on protein design to engineer cortical polarity of virtually any protein of interest in otherwise unpolarized mammalian culture cells.

**Key words** Asymmetric cell division, Cell polarity, Microtubule, Mitotic spindle, Synthetic biology

## 1 Introduction

Cell polarity critically depends on a polarized cell cortex, where specific polarity proteins are present only on one side, forming so-called "caps." From these cortical caps, asymmetric signaling is induced, and the intracellular space becomes polarized, with the polarization of the cell cytoskeleton and the polarized localization of cell organelles. This is critical for diverse cellular functions, in particular, for epithelial cells to maintain their function as a barrier during immune responses or asymmetric cell division.

Asymmetric cell division is a hallmark of stem cells and is the process by which one cell gives rise to two daughter cells that inherit different fate determinants and, thus, acquire distinct fates. Polar caps of conserved polarity cues induce key processes that eventually allow the generation of two daughter cells having different fates [1], in particular the profound polarization of the mitotic spindle, which rotates to align along the established polarity axis [2], and the

The original version of the chapter has been revised. A correction to this chapter can be found at https://doi.org/10.1007/978-1-0716-4224-5_20

Tomomi Kiyomitsu (ed.), *The Mitotic Spindle: Methods and Protocols*, Methods in Molecular Biology, vol. 2872, https://doi.org/10.1007/978-1-0716-4224-5_17, © The Author(s) 2025, Corrected Publication 2025

formation of an asymmetric central spindle, where one side contains a higher microtubule density as opposed to the other [3, 4].

To examine which polarity cues are *sufficient* to drive key processes of asymmetric cell division and to delineate the downstream pathways that regulate intracellular polarization in a high-throughput manner, we developed an assay that induces cortical polarity of virtually any protein of interest in a population of otherwise unpolarized cultured mammalian cells [3]. Technically, we capitalize on a de novo-designed 2D protein polymer [5] to cluster a stably expressed transmembrane segment from the outside. This allows the formation of endocytosis-resistant clusters of the transmembrane segment, to which we target intracellular proteins of interest. Remarkably, those clusters, which are uniformly spread all over the cell surface in adhered interphase cells, coalesce into an asymmetric cortical cap on round mitotic cells, thereby polarizing specifically the protein of interest in an otherwise unpolarized cultured cell [3].

In this chapter, we describe in detail how to purify the de novo-designed polymer components from bacteria, form clusters on mouse fibroblasts, and examine the effect of asymmetric caps during metaphase using live-cell imaging or during later stages of mitosis using fixed cells.

## 2 Materials

### 2.1 Reagents Polymer Component (A(d) and B(c)-GFP) Purification

1. pET plasmids containing the designed polymer component 1 (A(d)) and component 2 (B(c)-GFP), respectively.
2. BL21 *E. coli* bacteria, stored at −80 °C.
3. TYE plates with Kanamycin.
4. 4 L of 2x TY medium with Kanamycin in four 2 L baffled flasks; 1 L per flask.
5. IPTG at 1 M stock, stored at −20 °C.
6. 200 mL Lysis buffer: 25 mM Tris, 300 mM NaCl, 1 mM DTT (or 5 mM beta-mercapto ethanol), 1 mM PMSF, 0.1 mg/mL lysozyme (or 1x anti-protease tablet) in $H_2O$.
7. NiNTA purification columns.
8. 500 mL Wash buffer: 25 mM Tris, 300 mM NaCl, 1 mM DTT (or 5 mM beta-mercapto ethanol), 20 mM imidazole in $H_2O$.
9. 500 mL Elution buffer: 25 mM Tris, 300 mM NaCl, 1 mM DTT (or 5 mM beta-mercapto ethanol), 500 mM imidazole in $H_2O$.
10. 4–10% SDS gels.
11. Concentrators (AMICON ULTRA or equivalent; Molecular weight cut-off (MWCO) = 4kD)
12. Superdex 200 16/100 Gel filtration column.
13. 1 L Gel filtration buffer: 35 mM Tris, 150 mM NaCl, 5% Glycerol in $H_2O$.

#### 2.2 Equipment for A(d) and B(c)-GFP Purification

1. Ultracentrifuge with JLA 8100 & JA25.50 rotor (Beckmann; or equivalent).
2. Sonicator or homogenizer.
3. Peristaltic pump.
4. Gel filtration system.

#### 2.3 Cell Lines

1. 3T3 FlpIn mouse fibroblasts stably co-expressing GBP-TM-GBP, GFP-fused to the protein of interest and iRFP670-Jupiter (microtubule marker, *see* **Note 1**).
2. 3T3 FlpIn mouse fibroblasts stably co-expressing GBP-TM-GBP, GFP only as control and iRFP670-Jupiter (microtubule marker) (*see* **Notes 2–4**).

#### 2.4 Reagents and Equipment for Imaging

1. 4-well, glass-bottom imaging dishes.
2. Fibronectin at 1 mg/mL in PBS.
3. Imaging medium: Leibovitz 15 medium enriched with 10% Donor Calf Serum (DCS) and 20 mM HEPES (pH 7.5)
4. Spinning Disk confocal microscope (or equivalent).

## 3 Methods

#### 3.1 Purification of A(d) and B(c)-GFP (Fig. 1)

1. Transform BL21 (DE3) E. coli with pET A(d) or pET B(c)-GFP according to manufacturer's protocol.
2. Add one clone each to two flasks containing 25 mL 2xYE with kanamycin and grow over night at 37 °C.
3. Add each 25 mL pre-culture into 1 L of 2xYE with kanamycin and grow at 37 °C.
4. When the main cultures are at exponential growth ($OD_{600nm}$ = ~0.6) add 0.5 mM IPTG and incubate for 3 h at 37 °C.
5. Transfer cultures into 1 L centrifuge bottles and centrifuge for 20 min at 4000 × *g* at 4 °C using a JLA8100 rotor (or equivalent).
6. Discard the supernatant and resuspend the cell pellets in lysis buffer. Use 50 mL lysis buffer for 1 L of culture (100 mL in total for 2 L of culture). *From this step on, perform all steps on ice.*
7. Sonicate the resuspended pellets, kept on ice, with 20 Watt for 5 min total 'ON' time, using cycles of 10 s on, 10 s off, amplitude 40%. Alternatively, a homogenizer can be used.
8. Transfer the lysed pellets into 25 mL ultracentrifuge tubes and centrifuge at 20,000 × *g* for 30 min at 4 °C.

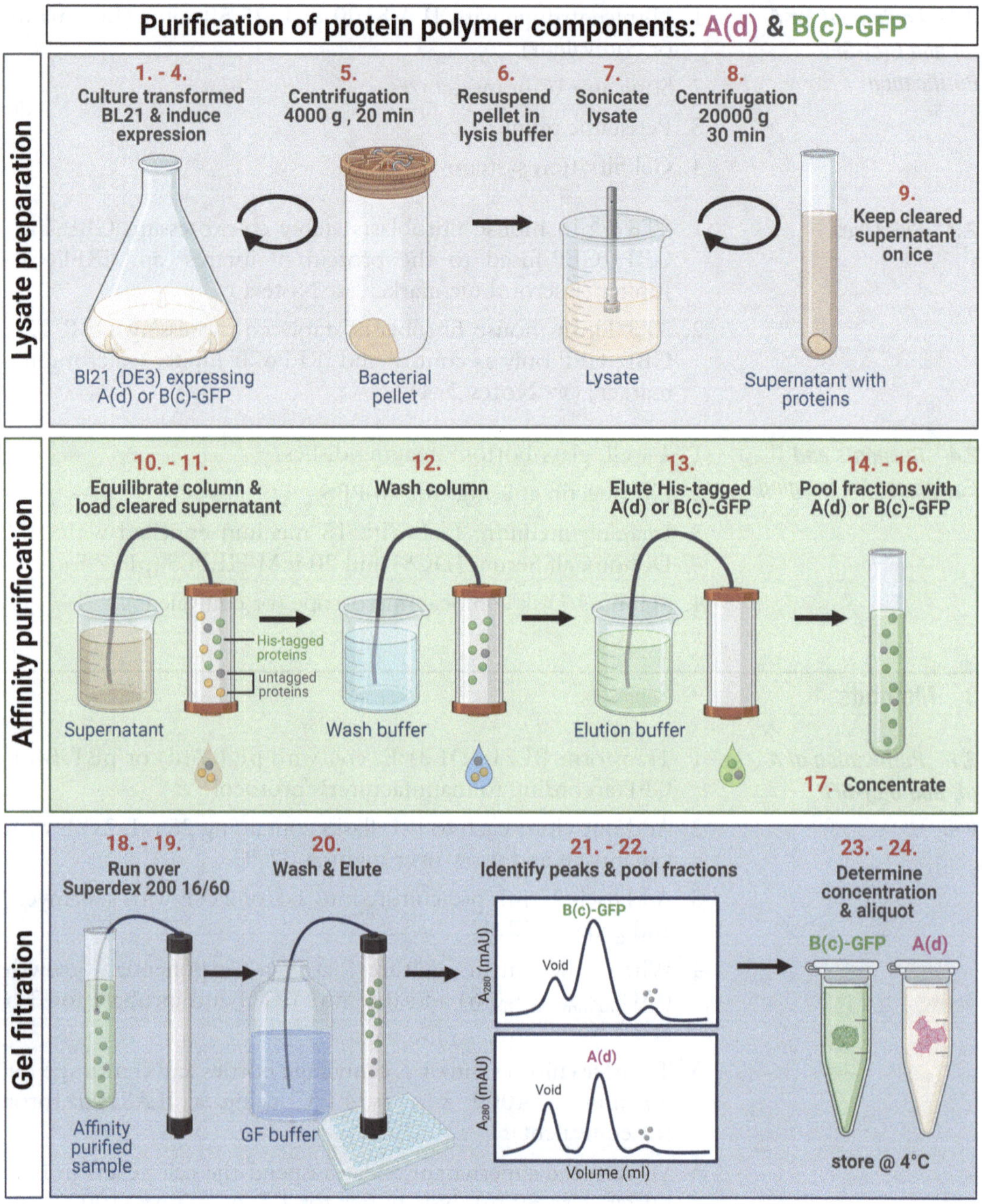


**Fig. 1** Schematic of the purification protocol of the protein polymer components A(d) and B(c)GFP

9. Pool the supernatants and keep on ice. *In the case of B(c)-GFP the supernatant will be green.*
10. Attach the NiNTA purification column to a peristaltic pump and equilibrate according to manufacturer's instructions with wash buffer.

11. Load the cleared supernatant.
12. Wash the column with approximately 100–200 mL of wash buffer.
13. Elute the His-tagged A(d) or B(c)-GFP by flowing elution buffer and collecting 8 fractions of 5 mL in 15 mL tubes.
14. Run 12 μL of each fraction on an SDS gel and stain with colloidal Coomassie blue.
15. Identify 2–3 of the 5 mL fractions that contain the majority of either A(d), which runs at 30 kDa, or B(c)-GFP, which runs at 70 kDa.
16. Pool those fractions.
17. Concentrate the pooled fractions to approximately 5 mL using a centrifugal concentrator with a molecular weight cut-off (MWCO) of 10 kDa.
18. Equilibrate Superdex 200 16/60 column with gel filtration (GF) buffer following the manufacturer's instructions.
19. Run the concentrated fractions over the equilibrated Superdex 200 16/60 column.
20. Wash and elute into 96 deep well plates.
21. Run 12 μL of fractions that give significant peaks in the $A_{280}$ absorption profile on a SDS gel to identify in which fractions A (d) or B(c)-GFP is present.
22. Pool the fractions that contain most of the protein.
23. Determine the concentration of the protein solution by measuring the absorbance at 280 nM and calculate the concentration using the extinction coefficient (A(d): $\varepsilon_{280} = 21620\ M^{-1}\ cm^{-1}$ ; B(c)-GFP: $\varepsilon_{280} = 42800\ M^{-1}\ cm^{-1}$).
24. Aliquot in 1 mL tubes and store at 4 °C. Azide can be added to prevent the growth of contaminants. The proteins are stable for at least one year.

### 3.2 Cluster Formation on Cells for Live Cell Imaging (Fig. 2)

1. Coat the imaging dish with 50 μg/mL fibronectin in PBS for 30–60 min at room temperature.
2. Add trypsinized cells stably co-expressing the transmembrane construct (GBP-TM-GBP) and the GFP-tagged protein of interest or GFP only as a control (*see* **Notes 2–4**).
3. Let the cells adhere for at least 30 min in the incubator at 37 °C and 5% $CO_2$.
4. Add 250 μL of 0.5 μM B(c)-GFP in medium per 4-well and incubate for 1 min.
5. Wash with pre-warmed medium.

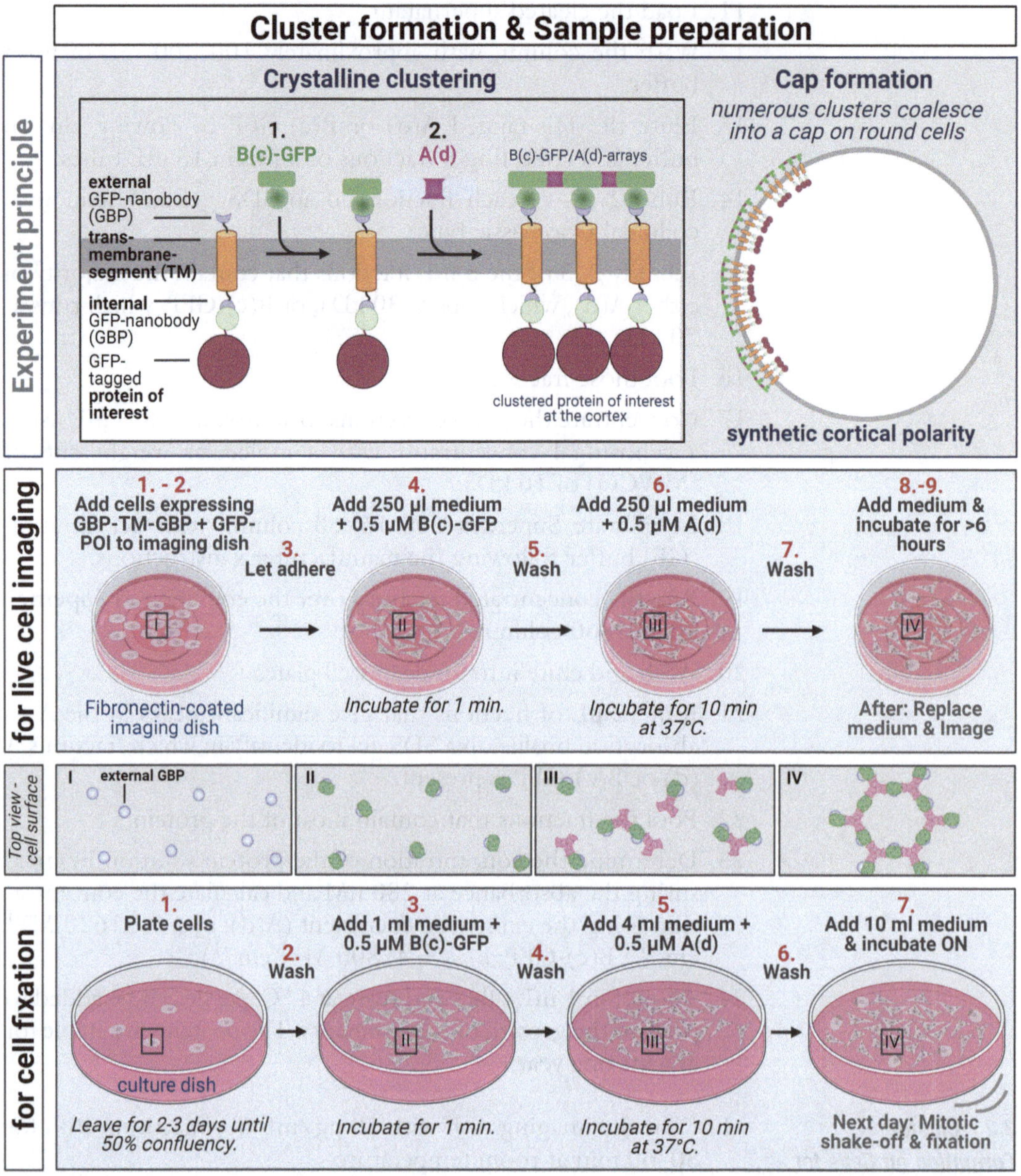


**Fig. 2** Schematic of the cluster formation on NiH3T3 cells for live cell imaging or for cell fixation after mitotic shake-off

6. Add 250 µL 0.5 µM A(d) in medium per 4-well and incubate for 10 min at 37 °C (*see* **Note 5**).
7. Wash with pre-warmed medium.

8. Add fresh, warm medium. Optional: To increase the number of mitotic cells, synchronization can be achieved by adding 30 nM nocodazole in the medium.
9. Incubate for at least 6 h.

### 3.3 Cluster Formation on Cells for Analysis of Fixed Mitotic Cells (Fig. 2)

1. Plate the cells in a round 10 cm dish 2–3 days before the experiment so that they reach ~50% confluency on the day of the experiment (*see* **Note 6**).
2. The evening before the experiment, wash the cells with pre-warmed PBS.
3. Add 1 mL of 0.5 μM B(c)-GFP in a medium per dish and incubate for 1 min.
4. Wash with pre-warmed PBS.
5. Add 4 mL of 0.5 μM A(d) in medium per dish and incubate for 10 min at 37 °C (*see* **Note 5**).
6. Wash with pre-warmed PBS.
7. Add 10 mL of medium and incubate overnight at 37 °C.

### 3.4 Measuring the Effect of Cortical Polarity on Spindle Orientation Using Live Cell Imaging

1. Pre-warm the microscope stage incubator to 37 °C.
2. Gently remove the medium.
3. Wash gently with imaging medium.
4. Add 250 μL imaging medium to each 4-well.
5. Take the dish to the microscope, find round mitotic cells with GFP caps and start imaging (*see* **Note 7**).
6. Caps will be visualized in the GFP channel, and the mitotic spindle can be imaged in the far-red channel using the iRFP670-jupiter microtubule marker (Fig. 3a).

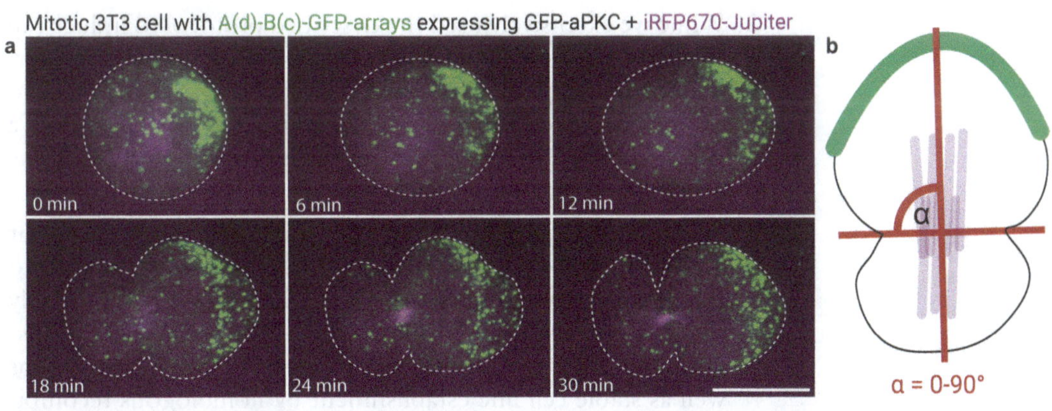


**Fig. 3** Effect of cortical aPKC caps on mitotic spindle orientation. (**a**) GFP caps assembled on 3T3 cells stably co-expressing iRFP670-Jupiter, GBP-TM-GBP, and GFP-aPKC. Scale bar, 10 μm. (**b**) The orientation angle α is the angle between the division plane and the mitotic spindle or the GFP cap

7. Spindle orientation angle can subsequently be determined by computing maximum intensity projections in the preferred image analysis program (e.g., ImageJ/Fiji) and by measuring the angle between the division plane and the mitotic spindle. Alternatively, if no microtubule markers are used, the angle between the division plane and the GFP cap can be used to identify the spindle orientation angle (Fig. 3b).

### 3.5 Measuring the Effect of Cortical Polarity in Late Mitosis in Fixed Cells

1. The morning after cluster formation, coat imaging dishes with 50 μg/mL fibronectin in PBS for 30–60 min at room temperature.
2. Wash the imaging dish with PBS and keep it at 37 °C until mitotic cells are added.
3. Wash the cells with pre-warmed PBS.
4. Add 0.5 mL of warm medium to each 10 cm dish.
5. Tap the sides of the dish hard, 5–10 taps per side.
6. Gently move the 0.5 mL of mitotic cells to the imaging dish.
7. Let the mitotic cells adhere at 37 °C for 10–20 min.
8. Gently remove the medium and wash with PBS. Then proceed with the preferred cell fixation protocol.
9. Stain with tubulin antibodies to examine the central spindle architecture upon induction of cortical polarity. In addition, the localization of any protein of interest during mitosis upon induction of cortical polarity can be examined.

## 4 Notes

1. In NIH-3T3 cells, iRFP670-Jupiter labels microtubules quite faintly. Microtubule dyes such as SiR-tubulin or other fluorescent mitotic spindle markers like microtubule-associated proteins could also be used.
2. While we extensively used the described protocol to induce cap formation in murine 3T3 fibroblasts, we have tested cap formation on human U2OS, mouse embryonic stem cells, and *Drosophila* S2 cells [3]. The assay can likely be adapted to a variety of other cell types, albeit slight modifications of the protocol may be required (*see* **Note 5**).
3. Since the transmembrane segment is linked to an intracellular anti-GFP-nanobody, any protein of interest can be targeted to the cap by simply co-expressing it as a GFP fusion. We routinely use bicistronic plasmids, allowing co-expression of the transmembrane segment and the GFP-tagged protein of interest, as well as stable cell line establishment by homologous recombination (FlipIn system). By experience, direct fusion of the protein of interest to the transmembrane domain is not recommended as it can lead to expression defects [3].

4. Cells with caps of GFP should always be included in every experiment as a control to rule out that the cap itself (instead of the clustered protein) influences the examined process.
5. Importantly, the concentrations of A(d) and B(c)-GFP described above have been precisely adjusted to lead to efficient cluster formation given the density of transmembrane segments achieved by stable expression of our plasmid in 3T3 cells. Different expression levels of the transmembrane segment may require adjustment of either the expression level of the transmembrane segment itself or the concentration of the designed proteins, particularly the amounts of A(d) used to induce clustering. We also recommend to always use the stable expression for the transmembrane anchors as it limits variability within the population.
6. As cluster formation is induced on the cell surface, the assay is restricted to individual cells in culture and cannot be adapted to tissues.
7. Artificial polarization with this assay can only be achieved using round cells. Indeed, cap formation is not dependent on mitosis but rather seems to depend on cell shape: cluster coalescence was only observed on round cells, irrespective of whether they are in mitosis, or not [3]. In contrast, during interphase, when 3T3 cells are not round, but rather spread, clusters are homogeneously distributed along the cell surface. Hence, the system is not adapted to reconstitute polarity in interphase.

## Acknowledgments

This work has been supported by the Medical Research Council (MC_UP_1201/13 to E.D), the Human Frontier Science Program (Career Development Award CDA00034/2017-C to E.D), and EMBO (ALTF 876-2021 Postdoctoral fellowship to L.K.K.).

### References

1. Suzuki A, Ohno S (2006) The PAR-aPKC system: lessons in polarity. J Cell Sci 119:979–987
2. Sunchu B, Cabernard C (2020) Principles and mechanisms of asymmetric cell division. Development 147(13)
3. Watson JL, Krüger LK, Ben-Sasson AJ et al (2023) Synthetic Par polarity induces cytoskeleton asymmetry in unpolarized mammalian cells. Cell 186:4710–4727
4. Derivery E, Seum C, Daeden A et al (2015) Polarized endosome dynamics by spindle asymmetry during asymmetric cell division. Nature 628:280–285
5. Ben-Sasson AJ, Watson JL, Sheffler W et al (2021) Design of biologically active binary protein 2D materials. Nature 589:468–473

# Chapter 18

# High-Resolution Imaging of Spindle Orientation Dynamics in 3D Intestinal Organoids

**Amlan Barai and Delphine Delacour**

## Abstract

Live imaging of microtubules (MTs) allows obtaining mechanistic insights into cell division. In literature, mitotic spindle dynamics have been investigated in mammalian systems largely focusing on established cell lines. Here, we describe a detailed protocol that investigates MT dynamics during cell division in a 3D mouse intestinal organoid model that more accurately captures the in vivo system.

**Key words** Cell division, Spindle dynamics, Spindle orientation, Microtubules, Organoid model, Live imaging, Intestinal organoid, Intestinal epithelium

## 1 Introduction

During epithelial morphogenesis, the regulation of cell divisions plays a critical role in determining both the pace and direction of tissue expansion and its proper arrangement. Mitotic defects frequently underlie various pathologies, notably cancer [1–3]. Despite decades of research on cell division and tissue architecture, our understanding of how mitotic spatial-temporal properties are orchestrated at a tissue level remains in its infancy. Model invertebrate epithelia have proven invaluable in unraveling the mechanisms governing cell division and its repercussions on proliferative tissues. These studies have elucidated how tissue architecture is intricately shaped by factors such as the orientation of the mitotic spindle and its regulation by polarity effectors, tissue forces, and cell shapes [4, 5]. In stark contrast, our knowledge regarding the regulation of epithelial cell division in mammalian tissues is considerably limited, limiting our comprehension of how aberrant cell divisions contribute to human-relevant disorders. Furthermore, most investigations into cell division in mammals have thus far relied on studies conducted on isolated cells or two-dimensional (2D) epithelial cell lines. However, such approaches, which are

Tomomi Kiyomitsu (ed.), *The Mitotic Spindle: Methods and Protocols*, Methods in Molecular Biology, vol. 2872, https://doi.org/10.1007/978-1-0716-4224-5_18,

largely dependent on transformed or cancer cell lines, fail to adequately capture the complexities of cell division within multicellular tissues under physiological conditions.

Primary organoid cultures emerged about 15 years ago as reference models for studying cellular processes in a 3D tissue context [6–8]. Prepared from isolated organ compartments or induced pluripotent stem (iPS) cells, these specific cultures develop mini-organs in culture that retain many architectural and physiological properties of the original organ [9–11]. Thus, primary organoid culture protocols have been developed for various organs such as the kidney, brain or gut [6, 12, 13]. The mouse intestinal model, a mammalian columnar epithelium, is one of the most rapidly regenerating tissues in adults and therefore serves as a suitable model for studying division mechanisms [14–17]. Cell divisions are crucial for constantly renewing the intestinal epithelium and maintaining proper expansion and functional integrity of proliferative crypts. However, they must be tightly controlled at the tissue level to preserve epithelial architecture and homeostasis. Typically, murine intestinal organoids are routinely generated from isolated adult mouse crypts and develop regular crypt-like structures with self-renewing capacity within 3 days of culture, thereby reproducing the organization and dynamics of the in vivo crypt [12, 18]. Working with organoids bypasses the complexity of in situ mouse models and allows access to studying tissue dynamics from subcellular to multicellular scales [6, 19]. However, it is difficult to introduce fluorescent constructs into these primary cultures that remain hardly transfected with conventional methods. Moreover, the 3D nature of these primary organoid cultures traditionally grown in hydrogel makes real-time acquisition more challenging.

In our recent study with mouse intestinal organoids, the use of such a tissue-mimicking intestinal organoid model allowed us to understand how cell division is regulated in a complex multicellular tissue environment. In this context, we developed a method for visualizing microtubules (MTs) live from 3D organoids. Using this method, we identified an original mechanism for the regulation of planar mitotic spindle orientation in the intestinal epithelium, based on the limitation of astral MT length [17]. Here, we describe the detailed protocol we used to investigate mitotic spindle dynamics in 3D intestinal organoids [17]. Specifically, we describe the process of MT probing in the organoid line and the microscopic modality used to study spindle orientation dynamics.

## 2 Materials

### 2.1 Experimental Models: Mouse Line/Organoid Line

1. Mouse intestinal organoid derived from Wild-type C57/Bl6 mice. The preparation of organoids from mice is described in detail elsewhere [17].

### 2.2 Cell and Organoid Culture Reagents

1. Advanced DMEM/F12.
2. Cell culture medium: DMEM supplemented with 10% FBS, 1% penicillin/streptomycin and glutamine.
3. IntestiCult™ Organoid Growth medium.
4. Matrigel® Matrix.
5. TrypLE Express.
6. Y-27632.
7. Puromycin.
8. 4-hydroxytamoxifen.
9. CHIR99021.
10. Nicotinamide.
11. Polybrene.

### 2.3 Live Imaging Markers

1. LentiBrite™ GFP-Tubulin Lentiviral Biosensor for stable labeling.
2. Spy650-Tubulin for transient labeling.
3. Hoechst 33342 for nucleus labeling.

### 2.4 Organoid Culture Wares

1. Regular organoid culture wares and access to cell culture facility for culture, maintenance and expansion of organoid lines.
2. 24-well plate.
3. Glass bottom Petri dishes or chambered slides.

### 2.5 Spinning Disk Microscope with a Live Imaging Setup

High-resolution 3D organization of mitotic spindle can be captured using a confocal microscope that uses a small pinhole to block out-of-focus light. While a confocal laser-scanning microscope (CLSM) allows a higher-resolution image acquisition, the relatively slower scanning speed often prevents having enough temporal resolution to capture live dynamics. A spinning disc confocal microscope uses multiple moving pinholes, allowing fast capture of live dynamics. The lower excitation energy of the spinning disc microscope also reduces sample phototoxicity and photobleaching, which makes it ideal for live imaging. The microscope also has to be equipped with a stage-top incubator that fits the microscope XY stage. The incubator should be equipped with temperature, gas, and humidity to maintain organoid culture conditions, namely 5% $CO_2$ and 37 °C temperature with appropriate humidity. Live images can be acquired using such spinning disc microscope that has 40×, and 60× or 63× immersion objectives.

For the protocol demonstrated here, we have used an inverted Zeiss microscope equipped with a CSU-X1 spinning disk head (Yokogawa—Andor) equipped with a live cell stage-top incubator.

## 3 Methods

Live high-resolution imaging of cell division and spindle dynamics involves two essential steps. The first step is to fluorescently label organoids that can then be seeded on a glass bottom dish or chamber slide suitable for the imaging platform. The second step is to acquire the live dynamics in the microscope, which includes appropriate imaging modality. The acquired images may need further processing, which includes re-constitution before further analysis.

Using a stable organoid line expressing fluorescently labeled alpha-tubulin allows long-term imaging. This can be done using lentiviral vector systems, adding extra steps of organoid line preparation to the experimental framework (namely steps in Subheading 3.1) that extend the experiment duration. The major advantage is that once generated, the organoid line can be expanded for multiple experiments, passaged several times and even cryopreserved for future use. Hence, for short-term experiments, transient live labeling using probes like SPY-tubulin is recommended instead of lentivirus-mediated stable cell line generation (Subheading 3.2, **step 3**).

### *3.1 Preparation of Fluorescently Labeled mouse Intestinal Organoid Line*

Lentiviral vector systems are widely used to introduce gene products into cells. Lentiviral transfection offers higher transfection efficiency and long-term stable expression of the transgene. Several such lentiviral systems are commercially available that can be used to generate organoid lines expressing fluorescent-tagged proteins. Here we have used the LentiBrite™ GFP-Tubulin system, a lentiviral vector system that can be used to generate an organoid line that expresses GFP-tubulin. Generating such organoid lines enables recording of long-term live dynamics of mitotic spindle without significant loss of fluorescent signal.

1. Preparation of transduction medium: Transduction media can be prepared by adding 10 μM Y-27632, 10 μM CHIR-99021, 10 mM nicotinamide, and 8 μg polybrene to organoid culture medium.
2. Preparation of Lentivirus particle: LentiBrite™ GFP-Tubulin Lentiviral Biosensor is prepared by diluting one vial in 550 μL transduction medium. The diluted lentivirus can be stored as

50–100 μL aliquots at −80 °C for up to 4 months without significant loss of efficacy.

3. Preparation of stem cell enrichment media: Stem cell enrichment media increases the population of stem cell-enriched cystic organoids. This can be prepared by enriching regular organoid culture media by adding CHIR-99021 and nicotinamide. To prepare the media, add 10 μM CHIR-99021 and 10 mM nicotinamide in IntestiCult™ Organoid Growth medium. Alternatively, the L-WRN medium prepared using the protocol described elsewhere [20] can also be used as stem cell enrichment media.
4. Two days prior to transduction, split organoids into 4 wells in a 24-well plate and add stem cell enrichment media to increase the stem cell-enriched cystic organoids population.
5. On the day of transduction, harvest organoids by disrupting the Matrigel domes with ice-cold DMEM/F12 media and collect them in a 15 mL tube. Centrifuge for 5 min at 100× *g* for 3–5 min to pellet the organoids and remove the excess media. Resuspend the organoids in 500 μL DMEM/F12 and break the organoids using P200 pipette by pipetting 80 times.
6. Centrifuge for 4 min at 100× *g* (800 rpm on an Eppendorf A-4-38 rotor), discard supernatant, then resuspend pellet in 500 μL of pre-warmed TrypLE and incubate in a 37 °C water bath for 5 min.
7. After incubation, inactivate TrypLE by adding 3.5 mL of cell culture medium and then centrifuge for 5 min at 130× *g* (~900 rpm on an Eppendorf A-4-38 rotor).
8. Discard the supernatant, add lentivirus (resuspended in transduction medium), and resuspend the organoid pellet. The lentivirus amount added depends on transfection efficiency. Optimized high titer volume allows efficient transfection without toxicity. Here we use 200 μL of lentivirus from Subheading 3.1, **step 1** per 4 Matrigel domes, i.e., 50 μL lentivirus per organoid dome. Then, seed them on 2 wells in a 48-well plate previously coated with a layer of Matrigel (60–80 μL Matrigel added in each well of a 48-well plate and incubated at 37 °C for 15 min). Add up to 250 μL transduction medium in each well and incubate at 37 °C overnight.
9. Next day carefully remove the medium without disrupting the Matrigel layer and overlay with another layer of Matrigel (60–80 μL) on top, and let it solidify at 37 °C for 20 min. Once the top Matrigel layer solidifies, add warm (37 °C) IntestiCult™ Intestinal Organoid Growth Medium and grow them for 3 days.

10. Small disrupted organoid fragments will start to form small cystic organoids within 24 h. In 3 days, they will start to form larger organoids. At this stage, refresh the medium. Depending on the lentivirus construct, a suitable selection antibiotic can be added to the culture media to remove non-transduced organoids.
11. Once the organoids mature, the immunofluorescent labeling can be checked using an epifluorescence microscope before passaging them to expand and cryopreserve as needed. As they are fluorescently labeled, they can be prepared for live immunofluorescent imaging described in the next steps.

### 3.2 Organoids Culture on Glass Bottom Disc for Live Imaging

Organoids have to be placed on a glass-bottom Petri dish or glass chamber slides that fit the live cell imaging setup equipped to the microscope. This can be done in two ways. While passaging, they can directly be seeded on those glass bottom dish/chamber slides (Subheading 3.2, **step 1**), or organoids cultured in regular culture ware (24-well plate) can be carefully transferred to the glass bottom dish/chamber slides without disrupting them (*see* Subheading 3.2, **step 2**).

1. Organoids are dissociated and passed as small 45 μL Matrigel® domes at the center of a glass bottom disc until they have grown to the desired stage with distinctive crypts (typically 3–4 days).
2. High-magnification immersion objectives that allow high-resolution imaging often have a very small working distance. As a result, only organoids close to the glass bottom base in the Matrigel dome fall within the objective working distance. Many organoids from the upper layer are usually out of the microscope's objective reach and cannot be imaged. If this becomes a major challenge with imaging, an alternative method can be used during sample preparation. In this method,
   (a) Overlay a thin layer of Matrigel (50–80 μm thickness) on the glass bottom Petri plate and incubate for 15 min in a 37 °C incubator. Such a thin layer can be made by putting a small amount of Matrigel on the ice-cold plate and then spread with an ice-cold pipette. The plate can be optionally surface treated for 5 min with a vacuum plasma cleaner that eases the Matrigel spreading by reducing the contact angle of the gel droplet.
   (b) Harvested organoids from Matrigel domes (from the 24-well plate) by depolymerizing Matrigel domes with ice-cold DMEM/F12 using a 1 mL pipette gently without breaking the organoids. Collect them in a 15 mL tube and mix well with ice-cold media by pipetting with 1 mL

pipette a few times and then centrifuge at 80× *g* (~700 rpm on an Eppendorf A-4-38 rotor).

(c) Carefully remove the supernatant and put the organoids palette on the Matrigel® layer using a cut pipette. Let them settle 2–3 min and then put a second layer of Matrigel on top of the organoids. Once the top layer solidifies (15–20 min incubation at 37 °C) add media and proceed to sample labelling (Subheading 3.2, **step 3**).

3. Organoids prepared on a glass-bottom dish/chamber can be taken for live imaging. Alpha-tubulin-GFP labeled organoids have GFP-tagged MTs and can be taken for time-lapse microscopy without additional sample labeling. If the sample is not labeled, SPY-Tubulin probe (like SPY650-tubulin) can be used to label MTs according to the manufacturer's protocol. To label organoids with SPY650-tubulin, incubate organoids for 1 h with 1X SPY650-tubulin in the organoid culture media at 37 °C. Additionally, 5 μg/mL Hoechst 33342 can be used to stain nuclei in live cells by adding the required amount to the culture media 15 min before live imaging.

### 3.3 Time-Lapse Microscopy

Live mitotic spindle dynamics and cell division progression can be recorded from fluorescently labeled organoids using a spinning disc microscope. Here, we used an inverted Zeiss microscope equipped with a CSU-X1 spinning disk head to acquire live movies from the alpha-tubulin-GFP organoid and the SPY650-Tubulin-labeled organoid. Hoechst 33342 is used to label nuclei, and organoids are visualized using Zeiss 40× and 63× water objectives.

1. Start the microscope live imaging setup 20–30 min before live imaging. 30-min wait helps to equilibrate the sample chamber and stabilize the temperature to achieve a stable 5% $CO_2$ level and a stable 37 °C temperature.
2. Put the sample on a sample holder and locate some healthy intestinal organoids using 20× or 25× objective. Spindle dynamics can be captured better with higher magnification objectives, 40× or 60×/63× with low numerical aperture. Add the appropriate immersion medium (oil) for the objective and locate the organoid crypt (Fig. 1). The crypt can be easily identified based on the organoid morphology (*see* **Note 1**).
3. Set up the appropriate z-stack, and acquisition time interval to record cell division and MT dynamics. We have recorded 50 μm Z-depth thickness at 0.5 μm intervals, i.e., 11 Z-frames at 6 s time intervals for 60 min for each crypt. To avoid phototoxicity, it is recommended to use a lower laser power.
4. After recording 60 min, another crypt (ideally from a different organoid) can be imaged using the same acquisition parameters (*see* **Note 2**).

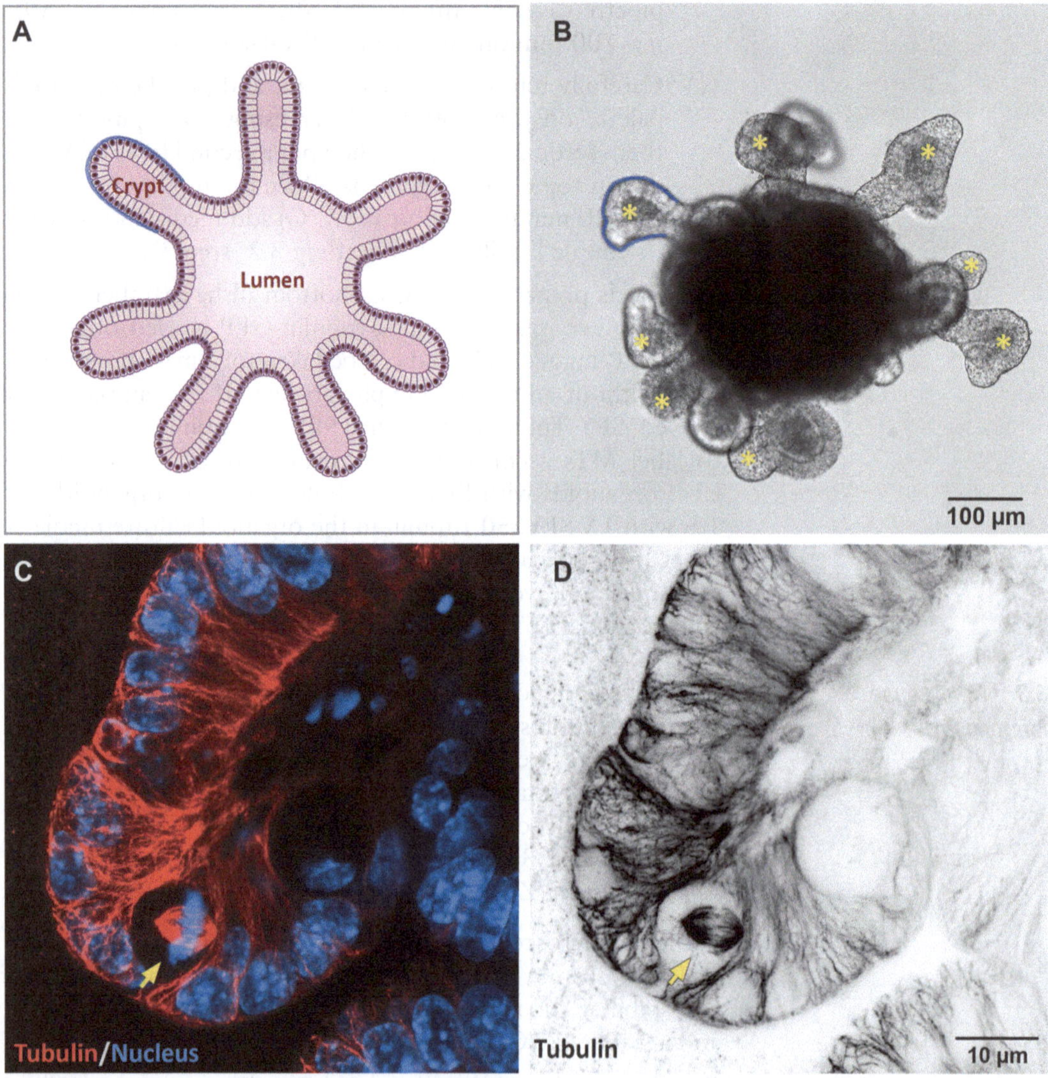


**Fig. 1** Representative image showing intestinal organoids and crypts. (**a**) Schematic and (**b**) phase contrast image showing a healthy organoid morphology. Yellow asterisk symbols (*) indicate crypt compartments and blue curved line indicates one crypt that can be imaged at high magnification. (**c**) and (**d**) shows confocal microscopy images of one crypt with a dividing cell indicated with a yellow arrow. Alpha-tubulin is shown in red with Hoechst 33342 stained nuclei in blue in (**c**). The mitotic spindle is better visualized using the inverted greyscale image of alpha-tubulin, as shown in (**d**)

### *3.4 Data Analysis*

The data analysis framework depends on the experiment objectives. A 3D image can be rendered in image processing programs like Fiji ImageJ (3D projection, Volume viewer plugin, etc.) or Bitplane Imaris. The obtained dynamics can be represented as time-lapse movies using those image-processing programs (Fig. 2). Spindle properties can be analyzed using various analysis tools and plugins

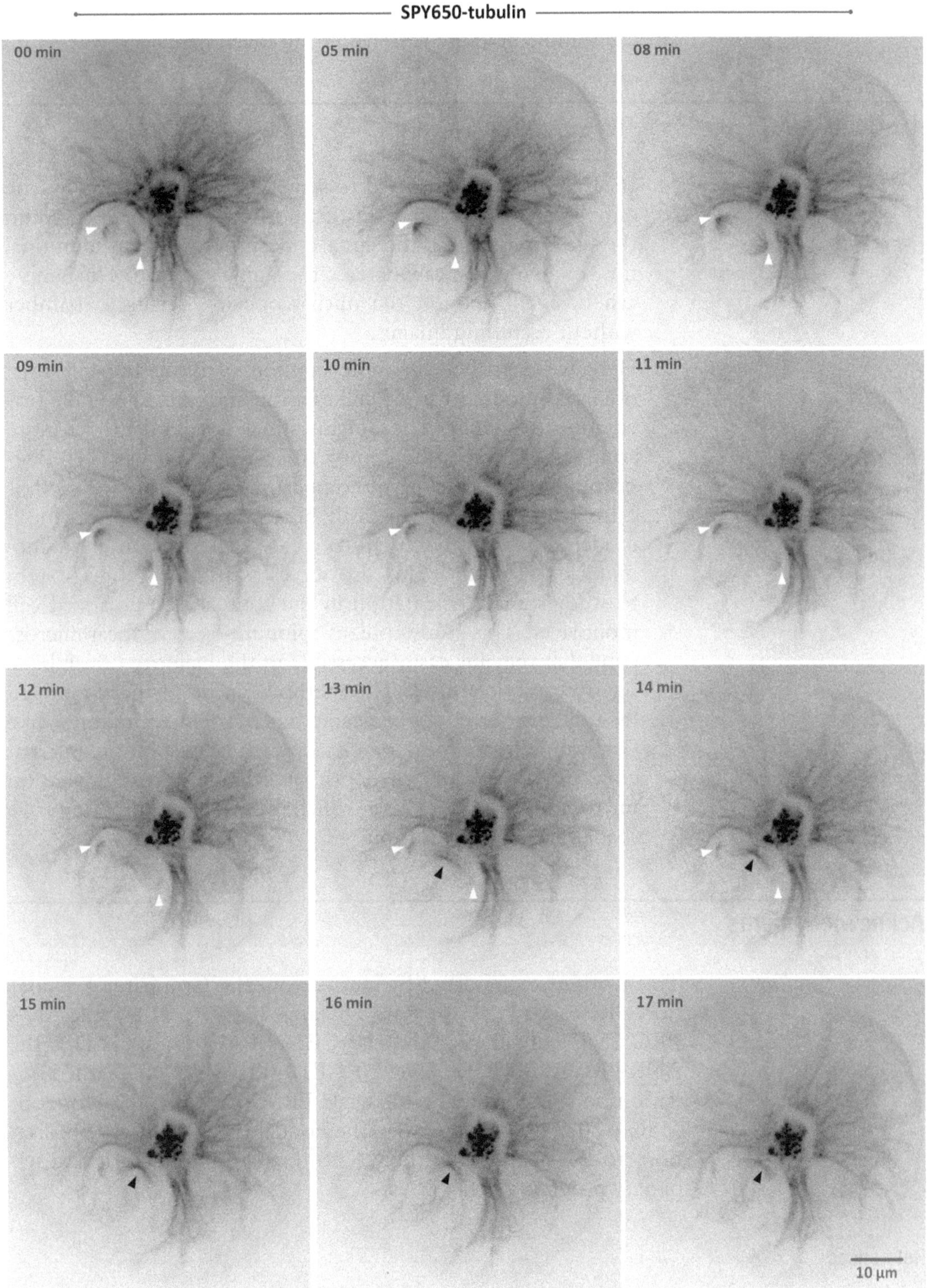


**Fig. 2** Representative images from time-lapse spinning disc microscopy showing spindle dynamics in mouse intestinal organoid. Organoid labeled with SPY650-tubulin showing spindle orientation dynamics with spindle poles indicated by the white arrowheads, and progression of telophase central microtubule bundle/intercellular bridge indicated by black arrowhead

available in Fiji. For example, MTrack J plugin can be used to analyze spindle pole oscillations during metaphase.

## 4 Notes

1. If SPY650-Tubulin is used to stain MTs in live organoids, the sample can be checked in this step for staining before beginning the live acquisition. If the signal appears too weak, an additional time of incubation can increase the signal strength. The sample can be incubated on the microscope live imaging chamber without recording images.
2. Here we described a simple method of using the intestinal organoid model to study cell division and mitotic spindle orientation dynamics. The 3D intestinal organoid model more accurately captures the complexities associated with cell division in the intestinal crypt compartment. With minor modification, this protocol can easily be adapted for other organoid models and can be used to study cell division in tissue-mimicking systems. This also serves as an ideal model system to study spindle orientation in the context of a polarized cell monolayer. This study utilizes spinning-disc confocal microscopy that remains easily accessible in the majority of cell biology research institutes. The method can be further improved by utilizing more sophisticated microscopy techniques, like fast-scanning confocal microscopy, super-resolution microscopy [21], or extended depth of field random illumination microscopy (RIM) [22], which can significantly improve spatial-temporal resolution.

## Acknowledgments

This work was supported by grants from the Groupama Foundation—Research Prize for Rare Diseases 2017 (to D.D), the ANR PRC LUCELL grant (ANR-19-CE13-0014-01) (to D.D.), the ANR PRC ACTRICE (ANR-20-CE13-0015) (to D.D.), the Fondation pour la Recherche Médicale FRM (to A.B.), the Horizon-MSCA-2022-PF-01 (to A.B.), the Fondation ARC PGA Labellisation (to D.D), and from the CNRS through the MITI Interdisciplinary programs (to D.D.).

### References

1. Levine MS, Holland AJ (2018) The impact of mitotic errors on cell proliferation and tumorigenesis. Genes Dev 32(9–10):620. https://doi.org/10.1101/GAD.314351.118
2. Morin X, Bellaïche Y (2011) Mitotic spindle orientation in asymmetric and symmetric cell divisions during animal development. Dev Cell

21(1):102–119. https://doi.org/10.1016/J.DEVCEL.2011.06.012

3. Bergstralh DT, Johnston DS (2014) Spindle orientation: what if it goes wrong? Semin Cell Dev Biol 34:140. https://doi.org/10.1016/J.SEMCDB.2014.06.014
4. Lechler T, Mapelli M (2021) Spindle positioning and its impact on vertebrate tissue architecture and cell fate. Nat Rev Mol Cell Biol 22(10):691–708. https://doi.org/10.1038/s41580-021-00384-4
5. di Pietro F, Echard A, Morin X (2016) Regulation of mitotic spindle orientation: an integrated view. EMBO Rep 17(8):1106–1130. https://doi.org/10.15252/EMBR.201642292
6. Clevers H (2016) Modeling development and disease with organoids. Cell 165(7):1586–1597. https://doi.org/10.1016/J.CELL.2016.05.082
7. Bock C et al (2020) The organoid cell atlas. Nat Biotechnol 39(1):13–17. https://doi.org/10.1038/s41587-020-00762-x
8. Mei Tian C et al (2023) Stem cell-derived intestinal organoids: a novel modality for IBD. Cell Death Dis 9(1):1–16. https://doi.org/10.1038/s41420-023-01556-1
9. Garreta E et al (2020) Rethinking organoid technology through bioengineering. Nat Mater 20(2):145–155. https://doi.org/10.1038/s41563-020-00804-4
10. Zhao Z et al (2022) Organoids. Nat Rev Methods Prim 2(1):1–21. https://doi.org/10.1038/s43586-022-00174-y
11. Hofer M, Lutolf MP (2021) Engineering organoids. Nat Rev Mater 6(5):402–420. https://doi.org/10.1038/s41578-021-00279-y
12. Sato T et al (2009) Single Lgr5 stem cells build crypt-villus structures in vitro without a mesenchymal niche. Nature 459(7244):262–265. https://doi.org/10.1038/NATURE07935
13. Eiraku M, Sasai Y (2012) Self-formation of layered neural structures in three-dimensional culture of ES cells. Curr Opin Neurobiol 22(5):768–777. https://doi.org/10.1016/J.CONB.2012.02.005
14. Guevara-Garcia A, Soleilhac M, Minc N, Delacour D (2023) Regulation and functions of cell division in the intestinal tissue. Semin Cell Dev Biol 150–151:3–14. https://doi.org/10.1016/J.SEMCDB.2023.01.004
15. Barker N (2013) Adult intestinal stem cells: critical drivers of epithelial homeostasis and regeneration. Nat Rev Mol Cell Biol 15(1):19–33. https://doi.org/10.1038/nrm3721
16. McKinley KL et al (2018) Cellular aspect ratio and cell division mechanics underlie the patterning of cell progeny in diverse mammalian epithelia. elife 7. https://doi.org/10.7554/ELIFE.36739
17. Saleh J et al (2023) Length limitation of astral microtubules orients cell divisions in murine intestinal crypts. Dev Cell 58(17):1519–1533.e6. https://doi.org/10.1016/J.DEVCEL.2023.06.004
18. Kretzschmar K, Clevers H (2016) Organoids: modeling development and the stem cell niche in a dish. Dev Cell 38(6):590–600. https://doi.org/10.1016/J.DEVCEL.2016.08.014
19. Clevers H (2013) The intestinal crypt, a prototype stem cell compartment. Cell 154(2):274–284. https://doi.org/10.1016/J.CELL.2013.07.004
20. Miyoshi H, Stappenbeck TS (2013) In vitro expansion and genetic modification of gastrointestinal stem cells in spheroid culture. Nat Protoc 8(12):2471–2482. https://doi.org/10.1038/nprot.2013.153
21. Valli J, Garcia-Burgos A, Rooney LM, de Melo e Oliveira BV, Duncan RR, Rickman C (2021) Seeing beyond the limit: a guide to choosing the right super-resolution microscopy technique. J Biol Chem 297(1):100791. https://doi.org/10.1016/J.JBC.2021.100791
22. Mazzella L et al (2023) Extended-depth of field random illumination microscopy, EDF-RIM, provides super-resolved projective imaging. bioRxiv:2023.10.30.564754. https://doi.org/10.1101/2023.10.30.564754

# Chapter 19

# Time-Lapse Imaging of Asymmetric Spindle Positioning During Endothelial Tip Cell Division in Angiogenesis In Vivo

**Holly E. Lovegrove and Shane P. Herbert**

## Abstract

The branching of new blood vessels by angiogenesis is critical to the development, growth, and repair of most vertebrate tissues and is frequently dysregulated in disease. At its core, angiogenesis is driven by the collective migration of leading "tip" and follower "stalk" endothelial cells. Recent work reveals that this collective movement is coordinated by asymmetric tip cell divisions that generate daughters of distinct size, signaling capacity and tip-stalk behaviors. Polarized mitotic spindle positioning is critical to generating such asymmetries in daughter cell size. However, the spatiotemporal dynamics of vertebrate spindle movement are often difficult to explore using in vivo systems. Here we describe a method for the sample preparation, live-imaging and data analysis of endothelial cell mitotic spindle positioning in developing zebrafish embryos. This method enables single-cell and population-level spindle dynamics to be monitored and quantified, both in wild-type or genetically/pharmacologically perturbed embryos. Moreover, this approach can be easily adapted for live imaging of spindle dynamics in other zebrafish embryonic tissues that experience similar asymmetric divisions, such as the trunk neural crest.

**Key words** Zebrafish, Live imaging, Transgenic, Endothelial, Angiogenesis, Spindle, Tubulin

## 1 Introduction

Cell division is fundamental to the generation and growth of tissues and organs [1–3]. When cell divisions predominantly contribute to tissue growth, they are often carefully controlled to ensure that two identical daughter cells are generated. However, beyond these symmetric divisions employed in growth, cell division can also be highly asymmetric [4, 5]. At their most extreme, these asymmetric cell divisions (ACDs) create daughter cells of very distinct fate and act to generate the broad heterogeneity of cell types needed to build complex biological systems. Such asymmetries in daughter cell fate are often established by active mechanisms that ensure key determinants are primarily inherited by just one daughter cell over the other, for example, as observed during *Drosophila* neuroblast

Tomomi Kiyomitsu (ed.), *The Mitotic Spindle: Methods and Protocols*, Methods in Molecular Biology, vol. 2872, https://doi.org/10.1007/978-1-0716-4224-5_19, 

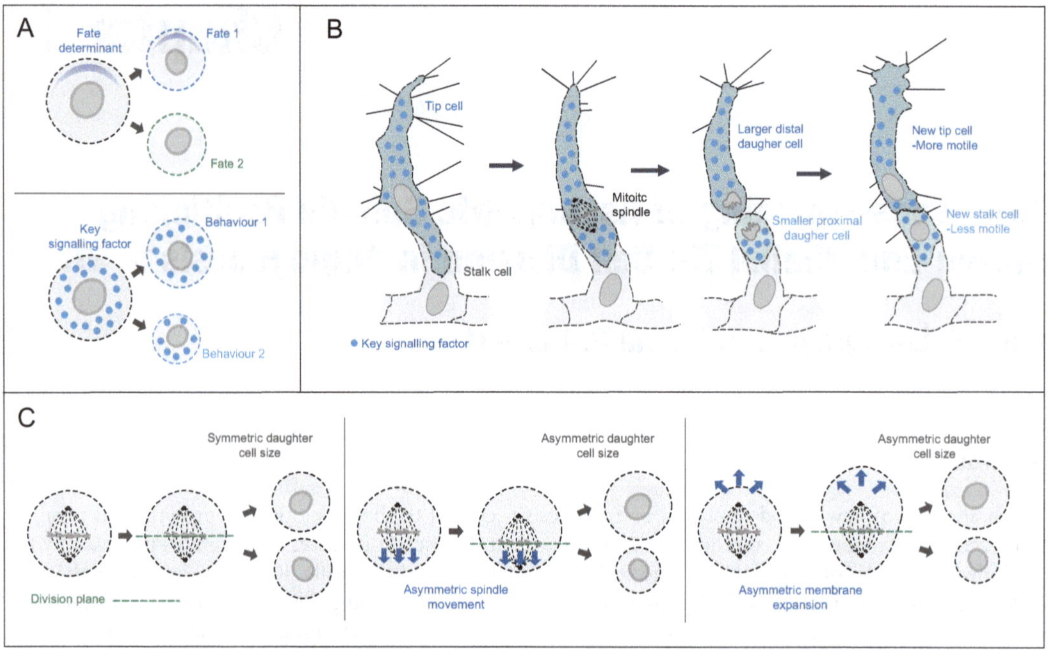


**Fig. 1** Modes and mechanisms of asymmetric cell division. (**a**) Illustration of the differences between fate-determining asymmetric divisions driven by inheritance of fate determinant predominantly by one daughter cell (top) and cell-size-dependent asymmetric divisions that induce differential daughter cell behavior via more subtle differences in the inheritance of key signaling components (bottom). (**b**) Diagram of the asymmetric division of endothelial tip cells (blue) during angiogenesis in a developing zebrafish intersegmental vessel. Asymmetry in daughter size and signaling capacity generates daughters of differential tip and stalk cell behavior. (**c**) Illustration of distinct methods for generating symmetrically (left)/asymmetrically (middle and right) sized daughter cells following mitosis

divisions (Fig. 1a, top) [6, 7]. However, in addition to these fate-determining divisions, recent work demonstrates that ACD can also generate more subtle differences in signaling factors that elegantly tune post-mitotic cell identity and behavior [8] (Fig. 1a, bottom). In doing so, these ACDs act to self-generate or maintain more nuanced differences between cells in tissues consisting of a single cell type, ultimately functioning to coordinate the collective behavior of groups of cells. Tissues that form via collective cell migration represent a great example of when such ACDs are important to generating transient heterogeneity in cell identity. In particular, during collective movement, cells must maintain the robust hierarchy of "leader" and "follower" cellular behaviors that maintains tissue organization, even during concomitant proliferative tissue growth [9, 10]. For example, in these systems the strictly symmetric division of leader cells could result in excessive leader cell formation and/or a breakdown of the important leader-follower cellular differences needed to organize collective tissue movement. In this case, ACD and the subtle differential partitioning of core signaling components have been shown to play a key role in generating/maintaining the distinct leader-follower cell identities that coordinate these processes [8].

Recent work demonstrates that post-mitotic asymmetries in daughter cell size are a key trigger for ACDs coordinating leader-follower identity decisions in collective migration [8]. Using the branching of intersegmental vessels (ISVs) in Zebrafish larvae as a model for collective movement, it was demonstrated that during angiogenesis, “tip” (leader) cells undergo ACDs that generate daughters of differential size (Fig. 1b). Following these divisions, the new distal tip cell daughter is approximately 1.5 times larger than the proximal following “stalk” cell daughter. This, in turn, results in the new tip cell inheriting larger amounts of key signaling factors compared to the stalk cell, thereby generating distinct behaviors in the two daughter cells. Consequently, these ACDs elegantly restore the tip-stalk cell hierarchy necessary to coordinate collective cell migration and efficiently generate a new blood vessel. Additional studies of neural crest collective cell migration further indicate that this form of ACD, which generates asymmetries in daughter size, plays a highly conserved role in defining the leader-follower hierarchy during the collective movement of other tissues [11].

During cell division, daughter size is chiefly determined by the position of the division plane (green dashed line Fig. 1c) [12]. Therefore, to generate two equally sized daughter cells, the division plane must form across the center of the dividing cell (Fig. 1c, left) [13]. This plane of division is itself determined by the position of the spindle midzone during metaphase/anaphase, which results in a localized activity of the Centralspindlin complex and the formation of the contractile ring and cleavage furrow at the cell center [14, 15]. To generate asymmetrically sized daughter cells, previous investigations have demonstrated that the resultant position of the mitotic spindle can be altered via two broad mechanisms: [1] displacement of the mitotic spindle to one side of the cell and/or [2] alterations to the boundary of the cell around the spindle upon asymmetric expansion of the cell membrane [4, 13, 16, 17] (Fig. 1c, middle and right). Both of these mechanisms are hard to identify using fixed imaging, due to their highly dynamic nature, and are difficult to explore using vertebrate in vivo systems.

To explore ACD during collective endothelial cell migration, the live imaging and image analysis method described below was developed in the zebrafish model system. This method enables the positioning of the mitotic spindle to be monitored live in dividing endothelial cells during vertebrate angiogenesis. Hence, this chapter describes a robust method for (1) the generation of zebrafish embryos containing endothelial cells that mosaically express GFP-tagged α-tubulin, (2) the high spatiotemporal resolution live imaging of mitotic spindle positioning during endothelial cell division in vivo, and (3) the quantitative image analysis of spindle dynamics both at single cell and population level. Not only is this method useful to researchers interested in the control of endothelial cell spindle positioning in vivo, but the described approach can also be readily adapted to monitor many key aspects of cell behavior and function, both during angiogenesis and in other vertebrate tissues that arise during zebrafish development.

## 2 Materials

### 2.1 Stock Solutions

1. *Embryo buffer*: 5 L Milli-Q $H_2O$ plus 0.35 g Tropic Marin salt; leave on magnetic stirrer overnight to dissolve. Decant 1 L and add 6 drops of Methyl blue (1% Methylene blue hydrate in $ddH_2O$ w/v).
2. *Embedding agarose*: To make 40 tubes dissolve 200 mg of 1% low melt agarose in 20 mL of embryo buffer (without Methyl blue, *see* Subheading 2.1, **item 1**) in a microwave, make 500 μL aliquots and keep in a heating block at 42 °C to keep liquid.
3. *0.4% Tricaine/MS222*: Dissolve 0.4% (w/v) Tricaine (a.k.a. MS222) (Ethyl 3-aminobenzoate methanesulfonate) in embryo buffer (without Methyl blue, *see* Subheading 2.1, **item 1**) with 20.1% (v/v) 1M Tris (15 mM, pH 8). Adjust to pH7 with NaOH. Store in 5 mL aliquots at −20 °C.
4. *0.045% N-Phenylthiourea (PTU)*: Dissolve 0.045% (w/v) N-Phenylthiourea (PTU) in embryo buffer (without Methyl blue, *see* Subheading 2.1, **item 1**) overnight on a magnetic stirrer, store at 4 °C.

### 2.2 Injecting One-Cell-Stage Embryos with Expression Vectors

1. *Injection dish*: Melt 0.5 g of 1% low melt agarose in 50 mL embryo buffer (minus Methyl blue, *see* Subheading 2.1, **item 1**) in a microwave, pour into petri dish until almost full, leave to cool for 2 min, float injection mold (Fig. 2a, b) (parallel line microinjection Z mold; World Precision Instruments) on top and leave overnight to set at room temperature. Carefully remove mold. Keep at 4 °C covered with embryo buffer.
2. *Injection mix*: 1 μL RNAse free $H_2O$, 1 μL Phenol red solution, 1 μL Tol2 RNA (12.5 ng/μL) and 1 μL *fli1*:α-tubulin-GFP plasmid (125 ng/μL) to enable vascular-specific expression of GFP labeled α-tubulin [8]. Keep on ice until loaded into the injection needle.
3. One-cell-stage embryos of an appropriate Zebrafish transgenic line, i.e., with the vasculature labeled with a complimentary fluorophore containing a membrane localization sequence to allow identification of the cell outline (e.g., *Tg(kdrl:rasCherry)*$^{s896}$) [18].

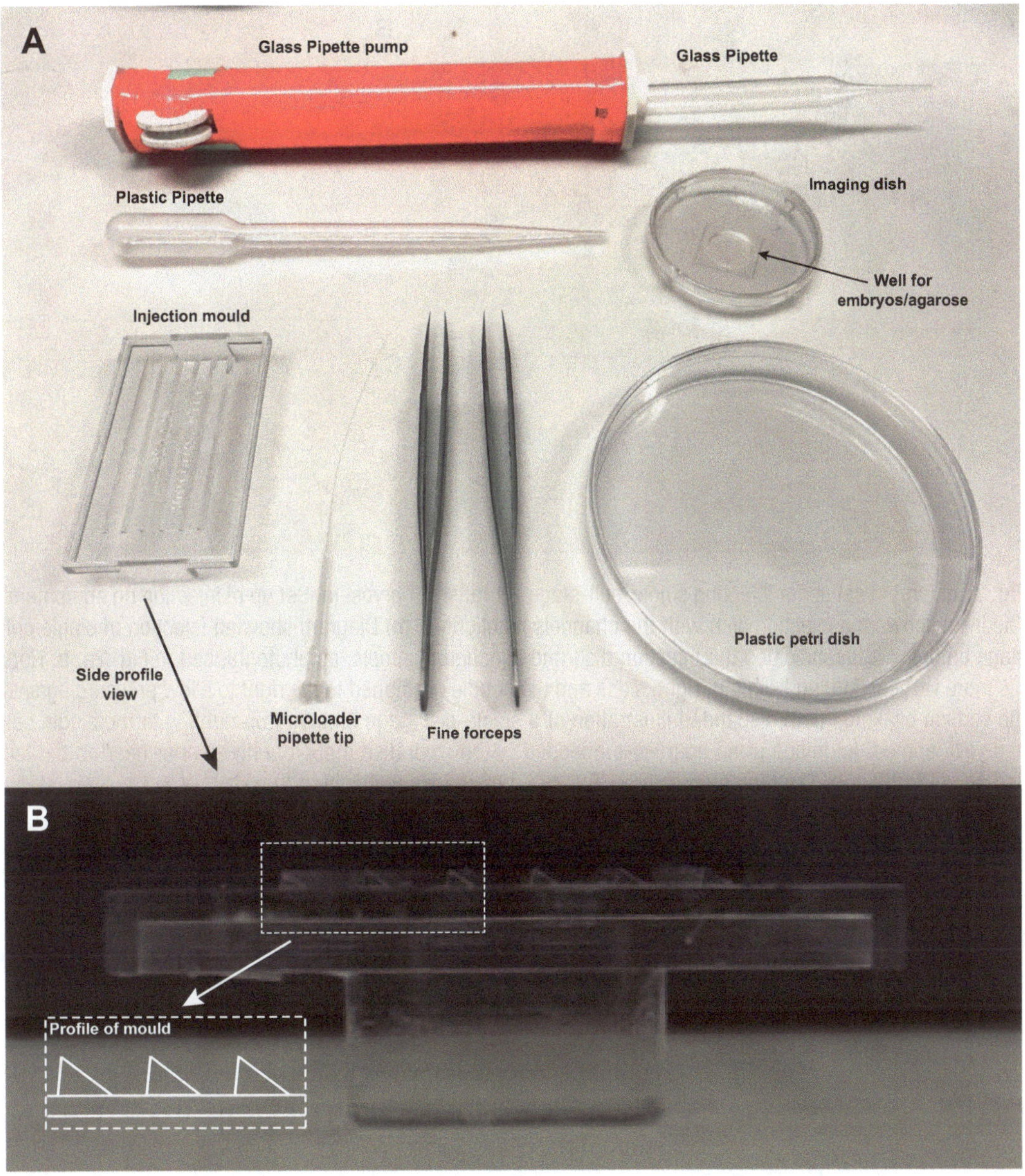


**Fig. 2** Key materials. (**a**) Photograph of the key items required. (**b**) Side view of the injection mold in **a**, with an inset diagram of the profile of the channels. Note the more vertical (left) and sloping (right) sides

4. Small sieve (tea strainer size)
5. Embryo Buffer (Subheading 2.1, **item 1**)
6. Glass Pasteur pipets (Kimble Chase) and pump ("The Pipette pump" from Bel-Art Products) or plastic Pasteur pipettes (Fig. 2a)
7. Sharp forceps (e.g., 5.5 dumount, Fine Scientific Tools) (Fig. 2a)

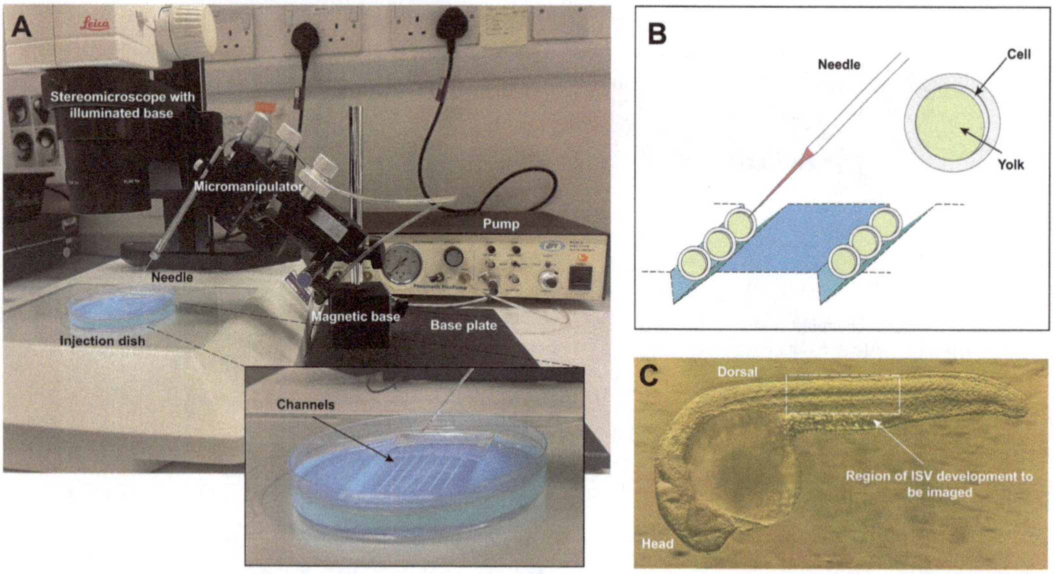


**Fig. 3** Equipment set up for injecting single-cell-stage zebrafish embryos. (**a**) Set up of injection rig equipment. The inset shows the injection dish with the channels highlighted. (**b**) Diagram showing injection of single cell stage embryos. Blue indicates the injection dish and injection channels, as photographed in Fig. 2a, b. Note the more vertical (left) and sloping (right) sides and the needle positioned to the right to allow pressure against the vertical backstop (left). Expanded illustration of a single-cell stage embryo (top right) with the single cell and yolk labeled. (**c**) Image of an agarose-embedded 24 hpf zebrafish embryo with anterior head to the left and dorsal to the top. The region containing ISVs to be imaged is highlighted (box)

8. Single barrel borosilicate glass capillaries (1B100F-3, World Precision Instruments) pulled into micropipettes (Flaming/Brown micropipette puller (model P-97), Sutter Instruments Co. heated to 574, pull at 50 with velocity 100).
9. Microloader tips (Eppendorf) (Fig. 2a)
10. p20 pipette
11. Stereomicroscope, illuminated from the base (e.g., Leica M80)
12. Picopump with foot pedal (SYS-PV820; World Precision Instruments)
13. Injection rig (manual micromanipulator, magnetic base and steel base plate; World Precision Instruments) set up next to stereomicroscope (Fig. 3a).
14. S21 micrometer scale
15. Mineral oil
16. Petri dishes (plastic) (Fig. 2a)
17. Incubator set at 28 °C

### 2.3 Embedding Embryos in Agarose for Live Imaging

1. Embryo buffer (*see* Subheading 2.1, **item 1**)
2. Stereomicroscope with lamp and fluorescence filters (e.g., Leica M80) for embryo screening
3. 2 pairs of sharp forceps (e.g., 5.5 Dumount) (Fig. 2a)
4. Microloader tips (Eppendorf) (Fig. 2a)
5. 50 mm glass bottom culture dishes for imaging (e.g., MatTek Corporation) (Fig. 2a)
6. Embedding agarose (*see* Subheading 2.1, **item 2**)
7. 0.4% Tricane/MS222 (*see* Subheading 2.1, **item 3**)
8. P1000, P20 pipettes and corresponding tips
9. Glass Pasteur pipets (Kimble Chase) and pump ("The Pipette pump" from Bel-Art Products) or plastic Pasteur pipettes (Fig. 2a)

### 2.4 Live Imaging of Developing Embryos

1. *Imaging media*: 100 mL of 0.0045% PTU (v/v) (*see* Subheading 2.3, **item 2**) and 0.1% Tricaine (v/v) (*see* Subheading 2.2, **item 2**) in Embryo buffer minus Methyl blue (*see* Subheading 2.1, **item 1**), in a 100 mL borosilicate glass laboratory bottle (e.g. Duran).
2. Glass Pasteur pipets (Kimble Chase) and ("The Pipette pump" from Bel-Art Products) or plastic Pasteur pipettes (Fig. 2a)
3. Upright confocal microscope capable of high spatiotemporal resolution imaging (e.g., Zeiss LSM 980 with Airyscan II)
4. Corresponding imaging software (e.g., Zeiss Zen Blue)
5. 20× dipping objectives with long working distance (W Plan-Apochromat 20×/1.0 DIC Water; Zeiss)
6. 50/60 mm Quick exchange heated microscope imaging platform and heater controller cable assembly (QE-2 and CC-28; World Precision Instruments)
7. Microscope stage adaptor for QE-2 platform (SA-30KZ; World Precision Instruments)
8. SloFlo In-line solution heater (SF-28; World Precision Instruments)
9. Dual channel heater controller (TC-344C; World Precision Instruments)
10. Mid-flow peristaltic pump (120S from Watson Marlow)

### 2.5 Image Analysis of Live Imaging Movies

1. Laptop or desktop computer
2. FIJI/ImageJ—free online image analysis software
3. Spreadsheet/graphical/statistical analysis software (e.g., Excel/Prism)

## 3 Methods

### *3.1 Injection of Embryos with Expression Vectors*

1. Load the injection needle with half of the injection mix (the second half can be used if the needle blocks/breaks) using a microloader pipette tip and p20 pipette. May need to gently flick the needle to get the liquid to the end of the needle.
2. Fix the needle into the pipette holder of the Picopump injection rig tight enough to hold the needle in place when the injection gas is dispensed but being careful not to crush the needle.
3. Add a drop of mineral oil to the surface of the micrometer.
4. Snip the pulled end of the needle with forceps and adjust the injection time until a 1 nL drop can be generated. If the caliber of the snipped needle is too small, re-snip to generate a larger caliber.
5. Collect embryos within 5–10 min of them being laid in a small sieve and wash them into a plastic petri dish with embryo buffer (*see* **Note 1**).
6. Aspirate approximately 20 embryos with a glass/plastic pipette, tilt the injecting dish to about 45° and slowly eject the embryos to line the embryos along one of the agarose channels. Re-aspirate any embryos that are accidentally flushed out of the channels. Repeat as required to line up as many embryos for injection as needed.
7. On the stereomicroscope (see Fig. 3a for setup) use a microloader tip with the end pinched off to gently organize the embryos along the channel and orient the channel such that the sloped side faces the injecting needle (the vertical side provides a backstop to push the embryo against to help the needle break through the chorion) (Fig. 3b).
8. Inject 1 nL of injection mix into the single cell of each embryo (*see* **Note 2**) (Fig. 3b).
9. Use embryo buffer to rinse the injected embryos into a new petri dish and half-fill the dish with embryo buffer.
10. Once the single cell starts to divide at about 30–45 min post-fertilization, discard the uninjected embryos, collect freshly fertilized embryos and repeat the above steps as required.
11. Keep embryos at 28 °C for the rest of the day in dishes of no more than ~50 embryos.
12. After approximately 4–5 h post fertilization (hpf) (once the embryos have reached ~50% epiboly), remove any unfertilized/damaged embryos. This will improve the overnight survival rate.
13. Incubate embryos at 28 °C overnight.

### 3.2 Embedding Embryos in Agarose for Live Imaging

1. At 20–22 hpf remove any additional dead/malformed embryos (*see* **Notes 3** and **4**).
2. Defrost a tube of tricaine (*see* Subheading 2.1, **item 3**).
3. Use two pairs of forceps to remove the chorions from the injected embryos, being careful not to cause damage to the embryo. Pinch one side of the chorion, use the other pair of forceps to pinch close to the first pair and gently tear the chorion open. The yolk sack is particularly delicate and easy to damage, and this is not always apparent until a few minutes later. At any stage of the process, discard any embryos leaking viscous yolk from the yolk sac.
4. On a stereomicroscope with a lamp and fluorescence filter set, select embryos that are positive for α-tubulin-GFP expression in the vasculature, indicating delivery of the injected plasmid (*see* **Note 5** and Fig. 3c for best region of interest).
5. Gather 6–10 positive embryos together and transfer them using a glass/plastic Pasteur pipette to one side of the central well in a 50 mm glass bottom dish imaging dish in as little liquid as possible. Removal of the excess liquid is required to prevent the agarose from becoming too dilute to set, but work quickly to avoid embryos drying out.
6. Add 18.75 μL of 0.4% tricaine (*see* Subheading 2.1, **item 3**) to a 500 μL tube of agarose (*see* Subheading 2.1, **item 2**) and invert several times to mix.
7. Carefully fill the central well with liquid agarose to suspend the embryos. Remove some agarose if overfilled such that the agarose domes are above the top of the central well (*see* **Note 6**).
8. Using a stereomicroscope, quickly arrange embryos (using a microloader tip with the end pinched off) into two rows of embryos in the center of the glass-bottomed dish well (this enables embryos to be easily found on the confocal microscope). Orient them laterally on their sides, with their heads to the left and dorsal side upward (*see* Fig. 3c).
9. Continue to monitor the embryos and as the agarose sets keep carefully adjusting the embryos such that they remain horizontal until the agarose begins to set and the embryo remains stationary.
10. Leave the dish to one side to set for 20 min (*see* **Note 7**).

### 3.3 Set-Up of Perfusion and Heating Rig (See Fig. 4a, b) for Live Imaging Microscopy

1. This subheading can be started whilst the agarose is setting.
2. Fill an empty 50 mm glass bottom imaging dish with imaging media (Subheading 2.4, **item 1**).
3. Place the dish in the heated microscope imaging stage and secure it with the retainer ring.

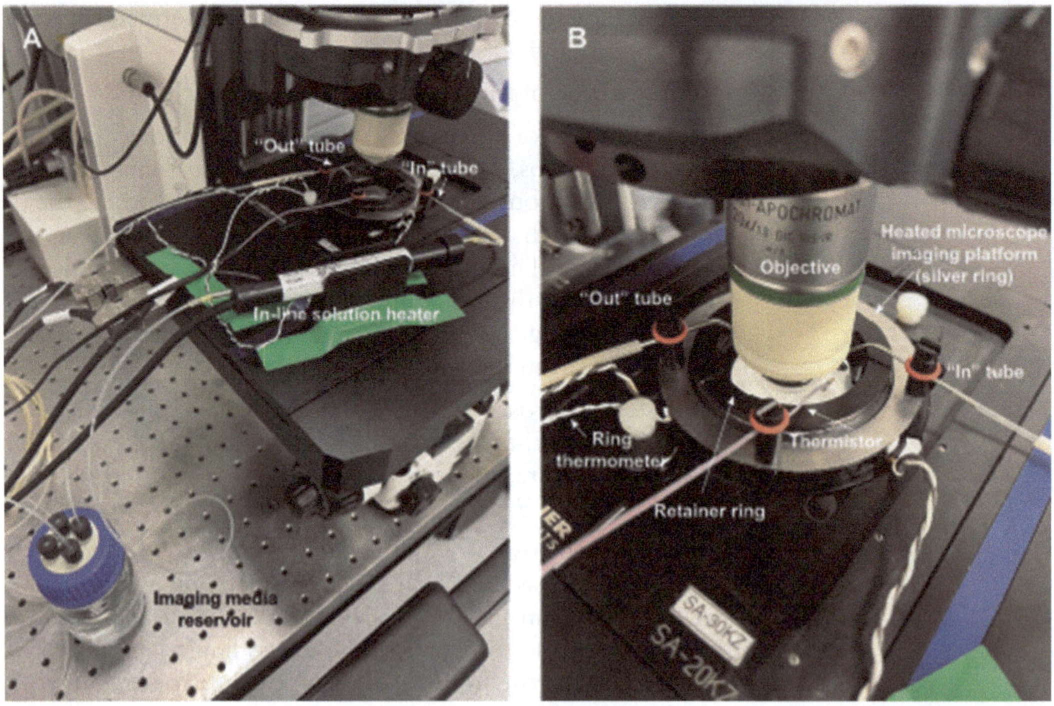


**Fig. 4** Equipment set up for live imaging of agarose-embedded zebrafish embryos. (**a**) Photograph showing perfusion rig, heated microscope imaging platform and In-line solution heater in situ on microscope. (**b**) Close-up photograph of live imaging set up showing positions of the heated microscope imaging platform, perfusion tubes and temperature monitoring thermistor

4. Use a mid-flow peristatic pump to establish a steady flow of imaging media from the 100 mL media reservoir, through the in-line solution heater, to the glass bottom dish imaging dish and then back to the 100 mL media reservoir.
5. Turn on the heating rig, put the monitoring thermistor into imaging media, and calibrate to an in-dish temperature of 28 °C.
6. Leave for 15 min to warm up and ensure a stable temperature of 28 °C.

### 3.4 Live Imaging of Developing Embryos

1. Stop the peristaltic pump to cease perfusion, remove the glass bottom dish from the microscope and then carefully transfer the warmed imaging media to the embedded embryos by very gently tipping it into the glass bottom imaging dish containing agarose embedded embryos from Subheading 3.2.
2. Position the glass bottom dish containing embryos onto the heated microscope imaging stage and secure it with the retainer ring.
3. Repeat **steps 4–5** of Subheading 3.3.

4. Screen for *fli1*:α-tubulin-GFP expression in cells of interest (for example, migrating tip or stalk cells) and *see* **Note 8** for how to optimize cell selection.
5. Set up confocal microscope image acquisition using Z slices 1–2 μm apart (*see* **Note 9**).
6. Live image with a time resolution of at least 1 min per frame, preferably 20–30 s, to capture key aspects of mitotic spindle dynamics (*see* **Note 10**).

### 3.5 Analysis of Spindle Movements (See Figs. 5 and 6)

1. Apply any additional image post-acquisition software processing, as required (e.g. airyscan or deconvolution processing).
2. Use ImageJ to open the movie as a Z stack.
3. Maximum project the Z stack (Image > Stacks > Z project... > Projection type > Max Intensity).
4. Correct for any X/Y drift over time (e.g., Plugins > Registration > Correct 3D Drift).
5. See Fig. 5a for examples of the different stages of cell division and spindle dynamics that can be visualized by playing through the movie, for example, nuclear envelope breakdown (NEB), spindle assemble, movement of the spindle and anaphase.
6. To start quantifying spindle movements, rotate the image so that the spindle is vertical (Image > Transform > Rotate...) (Fig. 5b, top).
7. Measure spindle movements relative to a fixed point in time and space. This will vary depending on the time frame of interest, for example:
   - For prometa/metaphase spindle movements compare positioning relative to the middle of the spindle at the last point of metaphase (i.e., the frame before anaphase spindle movements start (See lower left panel Fig. 5a))
   - To analyze relative to the final division plane (including prometa/metaphase and anaphase spindle movements), use the contact site between the two daughter cells in the first time frame after cytokinesis when two daughter cells are apparent (Fig. 5b, bottom left)
8. Mark this reference point with a line on all time frames (line tool > draw line > command D > Yes, process all images) (white line, Fig. 5b, bottom left).

*To investigate the whole spindle movement:*

9. Measure the position of the center point between the 2 spindle poles (spindle midzone) from the reference line (at 90°) for each time point to look at full spindle movement (grey line, Fig. 5b, bottom right).

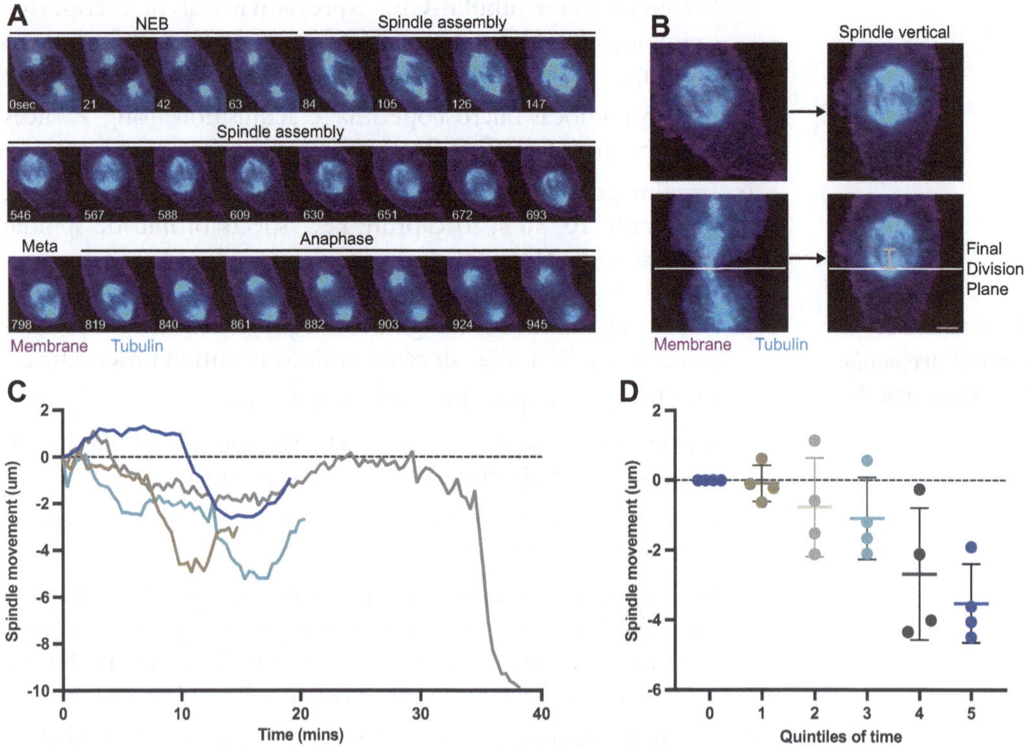


**Fig. 5** Mitotic spindle dynamics in an endothelial tip cell during angiogenesis in a developing zebrafish embryo. (**a**) Features of mitotic spindles that can be visualized using this method. Images were taken every 21 s. Membrane in magenta and tubulin in cyan. Abbreviations: *NEB* nuclear envelop breakdown, *Meta* metaphase. (**b**) Method for quantifying mitotic spindle movements. The image should be rotated such that the spindle and cell body are vertical (top). Draw a line through the reference point, which is, in this case, the final division plane (white line, bottom left & right). Measure from the center of the spindle to the reference line (grey line, bottom right) at each time point. Membrane in magenta and tubulin in cyan. (**c**) Graph of spindle movements quantified using four independent cells (as measured in (**b**)). Time 0 is the frame after nuclear envelope breakdown begins and final time point is when two distinct daughter cells are apparent (as in (**b**) bottom left). (**d**) Spindle movements from (**c**) binned into quintiles to represent general trends in spindle dynamics. All scale bars 3 μm

10. Generate an X/Y scatter plot of these distances versus time to visualize spindle dynamics over time. If plotting multiple data from spindles, setting all the 0-time points to 0 μm can help with visualization as all spindles will start in the same position on the y-axis (Fig. 5c).
11. While it is important to look at the dynamics of individual spindles, it is also necessary to look at general population trends. However, as divisions will take different lengths of time in individual cells/embryos, this can be difficult. Therefore, to compare population-level spindle/centrosome dynamics, it is useful to bin each mitotic event into quintiles (or more if time resolution is sufficiently high) and average the spindle movements within each quintile. X/Y scatter plots of

the average movement of multiple spindles against the same 5-time points can then be generated to identify general trends (Fig. 5d).

*To investigate individual spindle pole movements:*

12. Following individual spindle poles movement can be particularly informative when wanting to analyze the dynamics of anaphase, as it can determine whether the spindle poles are displaced equally or not.
13. To best visualize this process kymographs can be very informative. Select a thin vertical box/ROI (using the ImageJ rectangle tool) (white box Fig. 6a, left) that will include the spindle poles at every time of interest and duplicate (Image > Duplicate). Use the magic montage feature to line all the timeframes next to each other (Fig. 6a, right).
14. In order to quantify spindle pole movements, as above, a reference line is needed. In the example in Fig. 6b, the center of the spindle (midpoint between the 2 spindle poles) at metaphase is used (white line).
15. Measure the position of each spindle pole from the reference line (at 90°) for each time point (blue and gold lines Fig. 6b). In Fig. 6c measurements start at metaphase, in particular, the last time frame before the first obvious anaphase spindle movements carry the spindle poles away from each other (see kymograph in Fig. 6a).
16. Generate an X/Y scatter plot of these distances versus time to visualize individual spindle pole movements over time. In the example in Fig. 6c, the proximal measurements were given negative values to better visualize the spindle poles moving away from each other. Additionally, the 0-time point value has been subtracted from each value to better visualize spindle movement over time.

## 4 Notes

1. Plastic petri dishes can be rinsed with water, left to dry and reused after use. This prevents the wasting of plastic, but dechorionated embryos are also less likely to stick to pre-used plasticware. Plastic pipettes are also best filled with liquid a few times before they are used to transfer embryos, as embryos can stick to the inside of new plastic pipettes.
2. If the needle is not easily passing through the chorion, place the tip of the needle against the chorion so that it dents it and lightly tap the manual micromanipulator to penetrate the chorion. If this fails, the injection needle may need to be replaced.

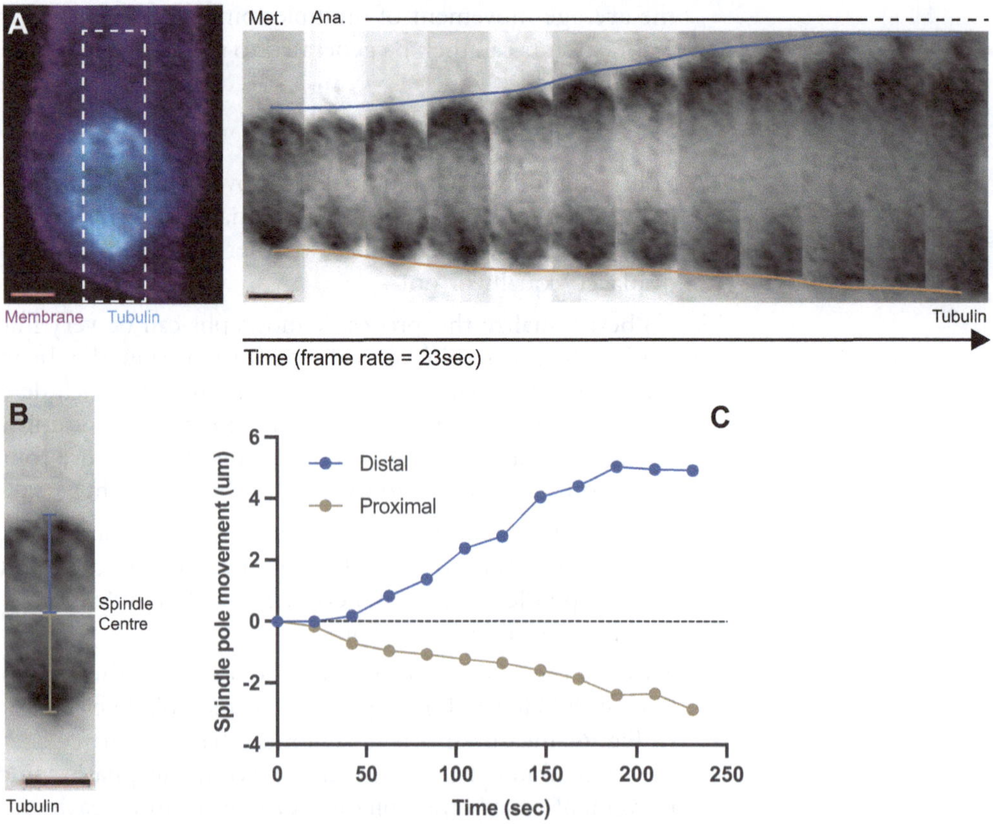


**Fig. 6** Measuring relative mitotic spindle pole movements in an endothelial tip cell during angiogenesis in a developing zebrafish embryo. (**a**) (left) Mitotic spindle rotated to vertical with the region of interest highlighted (white box). This region of interest was used to generate a kymograph (right). Blue and gold lines in the kymograph track positions of distal and proximal spindle poles, respectively. The membrane in magenta and tubulin in cyan (left) and tubulin in grey (right). Abbreviations: *Met.* metaphase, *Ana.* anaphase. (**b**) Measuring mitotic spindle pole movements. Draw a line drawn through the reference point, in this case the spindle center at metaphase (white line), then measure from the reference line to each spindle pole (blue line distal, gold line proximal), at each time point. Tubulin in grey. (**c**) Graph of mitotic spindle pole movements from kymograph in (**a**), measured as described in (**b**). Time points were normalized to the 0 time point to better visualise spindle pole movements. All scale bars are 3 μm

3. If there is high embryo death:
   - Inject less plasmid.
   - Use different fish pairs—they might not be generating healthy eggs. If possible, keep a dish of uninjected eggs from the same clutch to compare survival rates.
   - Only attempt to inject each egg once. Inserting the needle multiple times may cause excess damage.
4. If a different developmental stage is of interest or imaging at a different time of day is required, then development can be slowed down or slightly speed up by placing the embryos at

26 or 30 °C respectively [19]. However, embryos should be left at 28 °C until ~50% epiboly as early stages of development are more sensitive to temperature. Injections are best done first thing in the morning as the fish are more likely to breed closer to "sunrise."

5. If having trouble finding labeled cells under the stereomicroscope:
   - Inject the plasmid as early as possible after the eggs are laid, preferably before the cell looks "domed." At this early stage the cell is a thin clear region to one side of yolk (*see* Fig. 3b). The egg might need to be rotated to find it.
   - Inject a higher concentration of plasmid.
   - Look for positive cells elsewhere in the vasculature, particularly in the head region and dorsal aorta. If positive cells are present in these regions, there may also be lower expressing cells in the ISVs, which can only be seen when using confocal microscopy.
6. Make sure the agarose is not domed above the level of the well in the imaging dish or the objective can hit the agarose while trying to find the correct focal plane. If this is still a problem, try to embed embryos higher in the agarose and/or switch to an objective with a longer working distance.
7. If the agarose is not setting properly/keeps lifting or dispersing:
   - Leave it to set for longer; it will look glassy when set.
   - Transfer embryos into the glass-bottomed imaging dish with as little liquid as possible—allow them to sink to the bottom of the pipette tip and dispense them in a single drop. If there is excess liquid carefully remove it with a P200 Pipette but be careful to add the agarose quickly as the embryos will be damaged if allowed to dry out.
   - Make sure there are no gaps around the edges of the agarose in the central well, as imaging media may penetrate the side/underneath the disk of agarose and cause it to lift.
   - When adding the warmed imaging media to the imaging dish, to avoid dislodging the agarose and embedded embryos, do not allow the media to hit the agarose directly. Add it to the side of the dish and allow it to run over the agarose gradually.
8. If finding that cells selected for imaging are not undergoing divisions in the timeframe of imaging, cell selection can be optimized by:
   - Avoiding cells that appear to have recently divided. Tip cells very rarely divide twice during ISV formation, and so if it is

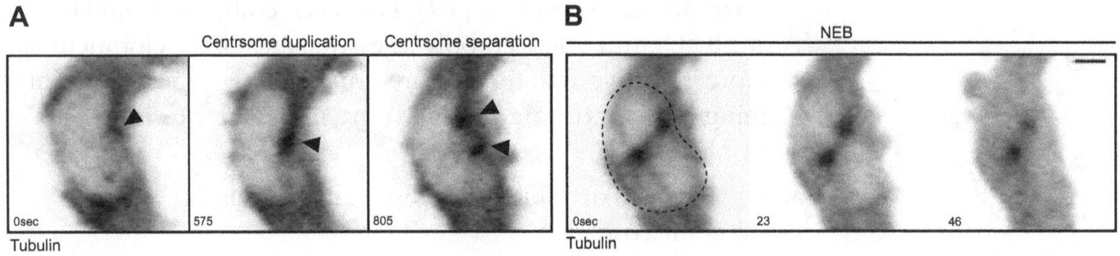


**Fig. 7** Examples of features that aid the selection of cells for live imaging (as mentioned in **Note 8**). (**a**) The centrosome will appear as a bright region of tubulin near the nucleus (arrowhead left). During duplication, the centrosome(s) will increase in intensity (arrowhead middle) and will separate into two bright spots during centrosome separation (two arrowheads right). Tubulin in grey. (**b**) Example of nuclear envelope breakdown (NEB). The microtubule-free nuclear space (black dashed line), which indicates the nucleus, is rapidly lost as the nuclear envelope breaks down. Images were taken every 23 s. Tubulin in grey. Scale bar is 3 μm

apparent a division has just occurred (indicated by a two-cell clone labeled with tubulin and/or the observation of a microtubule-rich midbody, which can often be seen between newly formed daughter cells) do not choose that tip cell for further imaging.

- Imaging cells in ISVs adjacent to ISVs in which a division recently occurred. Neighboring ISVs will often divide within a similar timeframe. If a tip cell is dividing/recently divided (see above), then its neighboring tip cells are more likely to divide soon afterward. This phenomenon more often occurs anterior to posterior, though not exclusively.
- Select cells with duplicated centrosomes. The centrosome position should be apparent as the emanation point of the microtubules (Fig. 7a left). The centrosome will become much brighter as it goes through duplication prior to mitotic spindle formation (Fig. 7a middle). Alternatively, look for cells with two centrosomes (two bright foci of α-tubulin-GFP, Fig. 7a right) within a single cell. These cells will form a mitotic spindle within minutes.
- Identify cells entering prophase. In prophase and immediately before mitotic spindle formation, nuclear envelope breakdown (NEB) will occur. This can be seen via the rapid loss of the microtubule-free nuclear space where the nucleus usually sits (Fig. 7b). Mitotic spindle formation will occur immediately after NEB.

9. Cell/part of the cell is lost in the Z plane during imaging:
   - If this is only slight, it is most likely due to subtle developmental displacement of the embryo and a slightly larger Z stack is needed.
   - If this movement is large (i.e., most/all of the cell is lost), it suggests that the focal plane of the microscope is shifting.

This could be due to fluctuations in the temperature of the imaging media or room (heating rig/room temperature control should be checked) or changes in the level of the imaging media (check the "in" and "out" rates of the perfusion rig are even).

   - Movement of the agarose can also cause this problem, *see* **Note 7**.

10. To maximize temporal resolution without compromising X/Y resolution:
    - Reduce the number of Z planes imaged to the minimum required to capture a single cell. Mitosis is a relatively quick process, so Z movement of the cell should be limited. However, if imaged cells move out of the imaging field, refer to **Note 9**.
    - If Z spatial resolution is not critical, the distance between Z slices can be increased to 2 μm, which is sufficient to observe key aspects of spindle dynamics.

## Acknowledgments

This work was supported by a Wellcome Trust Senior Research Fellowship grant (219500/Z/19/Z) (SPH) and funds from a British Heart Foundation Accelerator Award (AA/18/4/34221) (HEL).

### References

1. Godard BG, Heisenberg CP (2019) Cell division and tissue mechanics. Curr Opin Cell Biol 60:114–120
2. Guevara-Garcia A, Soleilhac M, Minc N et al (2023) Regulation and functions of cell division in the intestinal tissue. Semin Cell Dev Biol 150-151:3–14
3. Lechler T, Mapelli M (2021) Spindle positioning and its impact on vertebrate tissue architecture and cell fate. Nat Rev Mol Cell Biol 22(10):691–708
4. Sunchu B, Cabernard C (2020) Principles and mechanisms of asymmetric cell division. Development 147(13)
5. Knoblich JA (2008) Mechanisms of asymmetric stem cell division. Cell 132(4):583–597
6. Gallaud E, Pham T, Cabernard C (2017) Drosophila melanogaster neuroblasts: a model for asymmetric stem cell divisions. Results Probl Cell Differ 61:183–210
7. Homem CC, Knoblich JA (2012) Drosophila neuroblasts: a model for stem cell biology. Development 139(23):4297–4310
8. Costa G, Harrington KI, Lovegrove HE et al (2016) Asymmetric division coordinates collective cell migration in angiogenesis. Nat Cell Biol 18(12):1292–1301
9. Mayor R, Etienne-Manneville S (2016) The front and rear of collective cell migration. Nat Rev Mol Cell Biol 17(2):97–109
10. Haeger A, Wolf K, Zegers MM et al (2015) Collective cell migration: guidance principles and hierarchies. Trends Cell Biol 25(9): 556–566
11. Alhashem Z, Camargo-Sosa K, Kelsh RN et al (2022) Trunk neural crest migratory position and asymmetric division predict terminal differentiation. Front Cell Dev Biol 10:887393
12. Rappaport R (1985) Repeated furrow formation from a single mitotic apparatus in cylindrical sand dollar eggs. J Exp Zool 234(1): 167–171
13. Kiyomitsu T (2015) Mechanisms of daughter cell-size control during cell division. Trends Cell Biol 25(5):286–295

14. White EA, Glotzer M (2012) Centralspindlin: at the heart of cytokinesis. Cytoskeleton (Hoboken) 69(11):882–892
15. Pollard TD, O'Shaughnessy B (2019) Molecular mechanism of cytokinesis. Annu Rev Biochem 88:661–689
16. Kiyomitsu T, Cheeseman IM (2013) Cortical dynein and asymmetric membrane elongation coordinately position the spindle in anaphase. Cell 154(2):391–402
17. Kotak S (2019) Mechanisms of spindle positioning: lessons from worms and mammalian cells. Biomolecules 9(2)
18. Chi NC, Shaw NC, De Val S et al (2008) Foxn4 directly regulates tbx2b expression and atrioventricular canal formation. Genes Dev 22(6):734–739
19. Urushibata H, Sasaki K, Takahashi E et al (2021) Control of developmental speed in zebrafish embryos using different incubation temperatures. Zebrafish 18(5):316–325

# Correction to: Synthetic Engineering of Cortical Polarity During Mitosis Using Designed Proteins

Lara K. Krüger, Joseph L. Watson, and Emmanuel Derivery

Correction to:
Chapter 17 in: Tomomi Kiyomitsu (ed.), *The Mitotic Spindle: Methods and Protocols*, Methods in Molecular Biology, vol. 2872,
https://doi.org/10.1007/978-1-0716-4224-5_17

The acknowledgement section in the chapter has been published with an error. The initially published version has now been corrected.

The updated version of this chapter can be found at
https://doi.org/10.1007/978-1-0716-4224-5_17

Tomomi Kiyomitsu (ed.), *The Mitotic Spindle: Methods and Protocols*, Methods in Molecular Biology, vol. 2872,
https://doi.org/10.1007/978-1-0716-4224-5_20,

# Index




Tomomi Kiyomitsu (ed.), *The Mitotic Spindle: Methods and Protocols*, Methods in Molecular Biology, vol. 2872, https://doi.org/10.1007/978-1-0716-4224-5,